Advanced Information and Knowledge Processing

Series Editors
Professor Lakhmi Jain
Lakhmi.jain@unisa.edu.au
Professor Xindong Wu
xwu@cs.uvm.edu

T0137852

For other titles published in this series, go to
http://www.springer.com/4738

Michel Chein   Marie-Laure Mugnier

Graph-based Knowledge Representation

Computational Foundations of Conceptual Graphs

 Springer

Michel Chein
Laboratory of Informatics, Robotics, and
  Micro-electronics (LIRMM)
FRANCE

Marie-Laure Mugnier
Laboratory of Informatics, Robotics, and
  Micro-electronics (LIRMM)
FRANCE

AI&KP ISSN 1610-3947
ISBN: 978-1-84996-769-3   e-ISBN: 978-1-84800-286-9
DOI 10.1007/978-1-84800-286-9

British Library Cataloguing in Publication Data
A catalogue record for this book is available from the British Library

Printed on acid-free paper

9 8 7 6 5 4 3 2 1

Springer Science+Business Media
springer.com

Preface

This book provides a definition and study of a knowledge representation and reasoning formalism stemming from conceptual graphs, while focusing on the computational properties of this formalism.

Knowledge can be symbolically represented in many ways. The knowledge representation and reasoning formalism presented here is a graph formalism – knowledge is represented by *labeled graphs*, in the graph theory sense, and reasoning mechanisms are based on *graph operations*, with graph homomorphism at the core.

This formalism can thus be considered as related to semantic networks. Since their conception, semantic networks have faded out several times, but have always returned to the limelight. They faded mainly due to a lack of formal semantics and the limited reasoning tools proposed. They have, however, always rebounded because labeled graphs, schemas and drawings provide an intuitive and easily understandable support to represent knowledge.

This formalism has the visual qualities of any graphic model, and it is *logically founded*. This is a key feature because logics has been the foundation for knowledge representation and reasoning for millennia. The authors also focus substantially on *computational facets* of the presented formalism as they are interested in knowledge representation and reasoning formalisms upon which knowledge-based systems can be built to solve real problems. Since object structures are graphs, naturally graph homomorphism is the key underlying notion and, from a computational viewpoint, this moors calculus to combinatorics and to computer science domains in which the algorithmic qualities of graphs have long been studied, as in databases and constraint networks.

The main notions are intuitively introduced from a knowledge representation viewpoint, but the authors have also tried to give precise definitions of these notions along with complete proofs of the theorems (readers only require a few simple mathematical notions, which are summarized in the appendix).

This book does not present a methodology for knowledge representation nor describe actual applications (except for one chapter) or software tools. It presents theoretical bases that merge numerous previous studies carried out in the conceptual graph domain since Sowa's seminal book and links basic reasoning issues to other fundamental problems in computer science.

In a nutshell, the authors have attempted to answer the following question: *"How far is it possible to go in knowledge representation and reasoning by representing knowledge with graphs (in the graph theory sense) and reasoning with graph operations?"*

Organization

The book is divided into 3 parts, 13 chapters and 1 appendix.

The first part is devoted to the kernel of the formalism. In Chap. 2, Basic conceptual Graphs (BGs) are defined. BGs are structured by a subsumption relation defined by the homomorphism notion between BGs, as well as by elementary specialization and generalization operations. In Chap. 3, Simple conceptual Graphs (SGs) are introduced. SGs can be sketchily defined as BGs augmented with equality. Chapter 4 is devoted to set and FOL semantics for SGs. The soundness and completeness of homomorphism with respect to entailment relation and FOL deduction is proven. Chapter 5 relates BG homomorphism with homomorphisms of other structures (e.g., hypergraphs, relational structures, conjunctive database queries) and solving a constraint network.

The second part develops computational aspects of basic conceptual graphs. Chapters 6, and 7 are devoted to algorithms for BG homomorphism. Chapter 8 presents other specialization and generalization operations that are not covered in the first part, e.g., least generalization, maximal join and other extended joins.

The third part pools the kernel extensions. All of these extensions are provided with logically sound and complete reasoning mechanisms based on graph homomorphism. Chapter 9 focuses on nested conceptual graphs, which are hierachically structured graphs. In Chap. 10, the important rule notion, which allows representation of implicit knowledge, for instance, is studied. The introduction of rules gives the powerfulness of a computability model to the formalism. Positive and negative constraints are considered in Chap. 11, and the BG family of models, combining facts, rules and constraints, is presented. Chapter 12 is devoted to negation in conceptual graphs. The last chapter presents semantic annotations—a currently favored and potentially fruitful application.

Finally, the Appendix summarizes the basic mathematical notions used in the book.

Each chapter begins with an overview of its content. It ends with bibliographical notes devoted mainly to studies in the conceptual graph domain. Concerning the references, the authors chose to cite studies from other domains throughout the text and provide the conceptual graph references in the bibliographical notes.

Implementation

This book is about a KR formalism that can serve as a theoretical foundation for knowledge-based systems. A distinction is made between a KR formalism (e.g., a fragment of first order logic) and a KR programming language (e.g., PROLOG). Even when a KR programming language is based on a KR formalism, the KR language presents variations to the KR formalism (limitations, e.g., Horn clauses or extensions, second order features, etc.), and it has a concrete syntax, contains specific programming constructs (e.g., the cut in PROLOG) and is equipped with software tools. Between a KR formalism and a programming language implementing this formalism, there may be KR tools and platforms. This is the current situation of the graph-based KR formalism presented in the book. CoGITaNT [cog01a] contains a library of C++ classes that implement most of the notions and reasoning mechanisms presented here. CoGITaNT also contains a graphical interface and a client–server architecture. CoGITaNT thus allows a programmer to build knowledge-based systems grounded on the formalism presented in this book. Let us also mention COGUI [cog01b], a graphical user interface dedicated to the construction of a knowledge base and which integrates CoGITaNT.

Audience

The book is intended for computer science students, engineers and researchers. Some of the materials presented in this book have been used for several years at different academic levels, ranging from AI courses for graduate students to professional and research master's level courses. The mathematical prerequisites are minimal, and the Appendix outlines the mathematical notions used in the book.

Here are some suggestions for reading paths of this book depending on readers' specific interests.

Chapter 1 (introduction), Chap. 2 (basic conceptual graphs), Chap. 3 (simple conceptual graphs), the two first sections of Chap. 4 (set and first order logic semantics) are the core chapters and should be considered as the basis.

For programming and algorithmic purposes the following materials can be added to the base: Chap. 6 (basic algorithms for BG homomorphism), the last section of Chap. 5 (relationships with constraint programming techniques), Chap. 7 (techniques for trees and other tractable cases), Chap. 8 (algorithms for other specialization/generalization operations, especially maximal joins), Chap. 10 (rule processing), and Chap. 12 (algorithms for processing BGs with atomic negation).

For modeling purposes the following parts can be added to the base: Chap. 9 (nested conceptual graphs), Chap. 10 (definition and use of rules), Chap. 11 (definition and use of constraints and their combination with rules), Chap. 12 (definition and use of atomic negation) and Chap. 13 (semantic annotations).

For more theory-oriented readers, expressivity, decidability and complexity results, as well as stating equivalence with problems of other domains, are presented throughout the book, except in the last chapter.

Acknowledgments

We are indebted to John Sowa, who is the founding father of conceptual graphs, and the present book would not have existed without his pioneering work [Sow84].

We began working on conceptual graphs in 1992, and over the years many of our colleagues and students have contributed, in one way or another, to this work.

We would like to thank:

Friends and colleagues who have supported our approach over the years: Franz Baader, Marie-Christine Rousset, Pierre Marquis, Fritz Lehmann, Daniel Kayser, Michel Habib, Guy Chaty;

Colleagues or students with whom we have directly collaborated on topics covered in this book: Jean-François Baget, Boris Carbonneill, Olivier Cogis, David Genest, Olivier Guinaldo, Ollivier Haemmerlé, Gwen Kerdiles, Michel Leclère, Anne Preller, Eric Salvat, Geneviève Simonet, Rallou Thomopoulos;

Colleagues from the conceptual graph community with whom we have had fruitful discussions: Tru Cao, Fritjof Dau, Gerard Ellis, Jean Fargues, Bikash Gosh, Fritz Lehmann, Robert Levinson, Dickson Lukose, Guy Mineau, John Sowa, Mark Willems, Vilas Wuwongse.

Our approach has benefited from numerous collaborations on applied research projects, and we would like to thank: Christian Baylon, Alain Bonnet, Bernard Botella, Olivier Corby, Patrick Courounet, Rose Dieng, Steffen Lalande, Philippe Martin, Patrick Taillibert.

We acknowledge Jean-François Baget, David Genest, Alain Gutierrez, Michel Leclère, Khalil Ben Mohamed, Nicolas Moreau, Fatiha Saïs, Eric Salvat, who reviewed parts of this book, with special thanks to Geneviève Simonet (some of her comments could be the seeds for a new book ...).

We would also like to thank David Manley for rectifying our Franglais and Beverley Ford and her team at Springer London for their support.

Obviously, all remaining errors are our fault alone. We welcome corrections and suggestions on all aspects of the book; please send these to us at: Michel.Chein@lirmm.fr and Marie-Laure.Mugnier@lirmm.fr. A web site companion to the book can also be queried at: http://www.lirmm.fr/gbkrbook

La Boissière,   *Michel Chein and Marie-Laure Mugnier*
                                                                      March 2008

Contents

Part II Computational Aspects of Basic Conceptual Graphs

Chapter 1
Introduction

In Sect. 1.1, we place the book in the "Knowledge Representation and Reasoning" (KR) Artificial Intelligence (AI) domain. We first briefly outline key concepts of the KR domain, and then review KR formalism properties that we consider to be essential. The second section is devoted to an intuitive presentation of Conceptual Graphs that were initially introduced by Sowa in 1976 [Sow76] and developed in [Sow84]. In the third section, we introduce the graph-based KR formalism that is detailed in the book. This KR formalism is based on a graph theoretical vision of conceptual graphs and complies with the main principles delineated in the first section.

1.1 Knowledge Representation and Reasoning

Knowledge Representation and Reasoning has long been recognized as a central issue in Artificial Intelligence. Very generally speaking, the problem is to symbolically encode human knowledge and reasoning in such a way that this encoded knowledge can be processed by a computer via encoded reasoning to obtain intelligent behavior. Human knowledge is taken here in a very broad sense. It can be the knowledge of a single person, of an expert in some domain, shared knowledge of ordinary people (common sense knowledge), social knowledge accumulated by generations, e.g., in a scientific domain, etc. Thus, we will not distinguish the *modeling view* of KR, which involves studies on how to computationally represent knowledge about the world, or the *cognitivist view* of KR, which assesses how to computationally represent cognitive capacities of a human being.

Moreover, we shall carefully avoid specifying the exact meanings of the notions of "human knowledge," "reasoning," "intelligence" or "representation." All of these issues have been discussed by philosophers since the Greek ancient times (at least in the western world), and a discussion of such topics would be far beyond the scope of this book.

1.1.1 Knowledge-Based Systems

KR is the scientific domain concerned with the study of computational models able to explicitly represent knowledge by symbols and to process these symbols in order to produce new ones representing other pieces of knowledge. Systems built upon such computational models are called knowledge-based systems. Their main components are a knowledge base and a reasoning engine.

Example (a photo of children). Assume we want to represent common sense knowledge about a photo depicting children playing in a room containing toys and furniture. The formalism studied in the book can be used to represent elements in this photo (e.g., there is a girl and a boy, a car is on the table), and knowledge about elements in the photo (e.g., Mary is the name of the girl, Mary is a sister of the boy). It is also used to represent general background knowledge (e.g., a building block is a toy, if a person A is a sister of a person B then A and B are relatives).

Knowledge representation and reasoning formalism can also express problems to be solved concerning the facts and general knowledge represented. For instance, one may ask with what kind of toy Mary's brother is playing. Answering such questions requires descriptive knowledge but also reasoning capabilities (e.g., *modus ponens*, which states that if *A* holds and if *B* can be deduced from *A*, then *B* holds).

Main Components of a Knowledge Base

A knowledge base (KB) gathers symbolic knowledge representation about an application domain. We use the expression "application domain" to denote the part of the world (which can be real or fantasy, a sophisticated model of a system or a model of an expert competence) about which we represent knowledge and reasoning.

A KB generally contains different kinds of knowledge, typically an *ontology*, *facts*, *rules* and *constraints*. From an epistemological viewpoint, an ontology provides an answer to the question "What kinds of things exist in the application domain?" or expressed in a more generic way, "How can we think about the world ?" A computational *ontology* provides a symbolic representation of objects, classes of objects, properties and relationships between objects used to explicitly represent knowledge about an application domain. It is the cornerstone of a knowledge representation since all other pieces of knowledge (e.g., facts, rules or constraints) are represented by computational structures built with ontology terms.

Besides a KB, a knowledge-based system contains a *reasoning engine*. The reasoning engine processes knowledge in a KB in order to answer some question or to solve some goal. A reasoning engine is composed of algorithms processing elements of the KB in order to construct "new" knowledge, i.e., new symbolic constructs that are only implicit in the KB. We should stress that in a knowledge-based system we cannot have knowledge representation without having reasoning mechanisms. A large part of KR research consists of finding a tradeoff between expressivity or generality of knowledge representation formalism and the efficiency of the reasoning mechanisms.

Knowledge Incompleteness

An essential point is that a KB is not assumed to provide a complete picture of the world. The fundamental reasons are that any real thing, e.g., a human face or a pebble, cannot be described by a finite set of symbolic structures, and also that a thing does not exist in isolation but is included in unlimited sets of encompassing contexts. Thus, incompleteness of descriptions is a central feature of knowledge-based systems, and is a main distinction with respect to databases: For some sentences, it cannot be determined whether they are true or false given the knowledge in the base. For instance, a KB representing the photo of children can be queried by an unlimited number of questions (e.g., "Is the house where the photo was taken located in a village?" "How old is the boy?" "Who are the children's parents?" "In what country were the toys built?" and so on *ad libitum*). To be answered, these questions would need an unlimited amount of knowledge.

1.1.2 Requirements for a Knowledge Representation Formalism

In a nutshell, we are interested in KR formalisms that comply, or aim at complying, with the following requirements:

1. to have a *denotational* formal semantic,
2. to be *logically* founded,
3. to allow for a *structured representation* of knowledge,
4. to have good *computational* properties,
5. to allow users to have a maximal *understanding and control* over each step of the KB building process and use.

The graph-based KR formalism presented in this book has the first three properties, parts of the formalism have the fourth property, and we think that, at least for some application domains, it has the last one too. We think that presently there is no universal KR formalism. Indeed, such a formalism should represent natural languages, and the present systems are far from being able to do that. Thus, every existing KR formalism, including the one presented here, can be efficiently used only for specific reasoning on specific knowledge (e.g., a privileged application domain, namely semantic annotation, will be briefly presented). The end of this section is devoted to a brief discussion on the previous five requirements.

1.1.2.1 Denotational Semantics: *What* Rather than *How*

A KR formalism should allow us to represent knowledge in an *explicit* and a *declarative* way: The meaning of the knowledge represented should be definable independently of the programs processing the knowledge base. Namely, it should not be necessary to precisely understand how reasoning procedures are implemented to build a knowledge representation, and one should be able to update the knowledge base content without modifying any program. Ideally, the result of inferences should depend only on the semantics of the given data and not on their syntax, i.e., semantically equivalent knowledge bases should lead to semantically equivalent results, regardless of their syntactical forms. Thus, having a denotational semantics is an essential KR formalism feature.

A set (or model) semantics is appreciable, particularly whenever the KR formalism has to be used by informatics non-specialists. Indeed, the basic notions of (naive) set theory: element and membership, subset and inclusion, application, relation, etc., are easily understood by many people. In addition, a set semantics should provide the notions of truth and entailment, so that what holds in the modeled world can be determined. This leads us to logic, since logic is the study of entailment, or in other words, reasoning.

1.1.2.2 Logical Foundations

Generally speaking, doing an inference consists of producing a new expression from existing ones. The correctness of an inference mechanism can be defined relative to a logic, and in this book we essentially consider logical entailment, or logical deduction.

What does it mean for a KR formalism to be logically founded? First, the expressions of the formalism are translatable into formulas of a logic. Such a mapping gives a logical semantics to the formalism. Secondly, the reasoning engine contains an inference mechanism, which should have two essential properties with respect to deduction in the target logic: *soundness* and *completeness*. Let $\mathcal{K}$ be a knowledge base expressed in some KR formalism, and let f be a logical semantics of $\mathcal{K}$, i.e., f is a mapping from $\mathcal{K}$ to formulas of some logic. The inference mechanism is *sound* with respect to this semantics if for each expression i inferred from $\mathcal{K}$, $f(i)$ is actually logically deduced from $f(\mathcal{K})$. It is *complete* if, for each expression i such that $f(i)$ is logically deduced from $f(\mathcal{K})$, i is actually inferred from $\mathcal{K}$. In other words, a procedure $\mathcal{P}$ (or algorithm, or system of rules, etc.) is sound with respect to a logical semantics if every inference made by $\mathcal{P}$ corresponds to a logical deduction. It is complete with respect to a logical semantics if every logical deduction in the logical language target of the formalism can be done by $\mathcal{P}$.

Soundness is usually ensured. But not all reasoning algorithms are complete. If an incomplete, nevertheless sound, system answers "yes" to the question "can i be retrieved from $\mathcal{K}$?" then the answer is correct. If it answers "no," then because of incompleteness, it should preferably have answered "I don't know." If the system also computes answers to a query then, if it is sound, all the computed answers are correct answers, and if it is incomplete some answers can be missed.

The incompleteness of an algorithm can be motivated by efficiency concerns when a complete reasoning is too time consuming. It can also be due to the undecidability of the deduction problem. For instance, deduction in First Order Logic (FOL) is undecidable, which means that it is not possible to build an algorithm deciding in finite time for any pair of formulas (f, g) whether f can be deduced from g. More precisely, FOL deduction is only semi-decidable: One can build an algorithm guaranteed to stop in finite time if the answer is "yes," but which may run indefinitely if the answer is "no." Thus, algorithms that stop in finite time in all cases are necessarily incomplete.

Different logics have been developed for KR purposes. FOL has been adopted as a reference formalism for knowledge representation in the AI community. Its model semantics is described with simple mathematical objects (sets and relations). However, its computational complexity has led to study of fragments of it with good computational properties.

KR formalisms can be compared according to different criteria, such as expressiveness, ease of use, computational efficiency, etc. Logical semantics facilitate expressiveness comparisons between KR formalisms and can avoid doing something that is already known. Indeed, KR deals with knowledge and reasoning, and logic (not only classical logic) is precisely the study of reasoning. The large corpus of results and techniques accumulated by logicians for more than two millennia cannot be ignored.

From a modeling standpoint, other important properties of a KR formalism are its *empirical soundness* and its *empirical completeness* with respect to an application domain. A formalism is empirically sound if any "true" expression in the formalism corresponds to a "true" fact of the application domain. It is empirically complete if any true fact of the application domain can be coded in a true expression of the KR formalism. Naturally, when there is no mathematical model of the application domain, these notions are informal, rather subjective and difficult to evaluate. Having different mathematical semantics of the KR formalism (e.g., a set semantics and a logical semantics) can help, since each semantics can be used to study, with different notions, the correspondence between the KR formalism and the application domain.

1.1.2.3 Knowledge Structuring

A KR formalism should provide a way of structuring knowledge. Knowledge structuring can be motivated by model adequacy (i.e., its "conformity" to the modeling of the application domain) and by efficiency concerns (algorithmic efficiency of reasoning procedures, facility for managing the knowledge base).

One aspect of knowledge structuring is that semantically related pieces of information (e.g., information relative to a specific entity) should be gathered together. This idea was underlying the first KR formalisms, frames and semantic networks, that were far from logics. Frames have been introduced in [Min75] as record-like data structures for representing prototypical situations and objects. The key idea was to group together information relevant to the same situation/object. Semantic networks were originally developed as cognitive models and for processing the semantics of natural language (cf. [Leh92] for a synthesis on semantic networks in AI). A semantic network is a diagram that represents connections (relationships) between objects (entities) or concepts (classes) in some specific knowledge domain. Basic links are the *ISA* link that relates an object and a concept of which it is an instance, the *AKO* (A-Kind-Of) link, that relates two concepts (with one being a kind of the other), and the *property* link that assigns a property to an object or concept. Inferences are done by following paths in the network. For instance, properties of an object are inherited following the ISA and AKO links. The main criticism concerning semantic networks was their *lack of a formal semantics*: What's in a link? [Woo75] What's in a concept? [Bra77]. The same network could be interpreted in different ways depending on the user's intuitive understanding of its diagrammatical representation.

Description logics (DLs), formerly called terminological logics or concept languages, are rooted in semantic networks, particularly in the KL-ONE system [BS85], and have been a successful attempt to combine well-defined logical semantics with efficient reasoning [BCM+03]. This is one of the most prominent KR formalism families. Let us point out, however, that DLs have lost the graphical aspects of their ancestors. Conceptual graphs represent another family of formalisms issued from semantic networks (at least partially, because they are also rooted in other domains), which we shall consider in more detail in the next section.

Another aspect of knowledge structuring is that different kinds of knowledge should be represented by different KR formalism constructs. Ontology, facts, rules, constraints, etc., are distinct sorts of knowledge that are worth being differently represented. An important distinction is between *ontological* and *factual* knowledge. Ontological knowledge is related to general categories, also called concepts or classes. Factual knowledge makes assertions about a specific situation (e.g., "this specific entity belongs to a certain category, and has a certain relationship with another entity, ..."). This distinction is essential in description logics and conceptual graphs.

Another kind of knowledge is *implicit knowledge* described, for instance, by rules of form "if *this* then *that*" (e.g., "if there is a relation r from x to y, then the same relation r also holds from y to x"). Different kinds of rules can be considered. Some rules can be included in an ontology (e.g., the transitivity of a binary relation), other rules, for instance representing possible transformations, are not ontological knowledge. Constraints, i.e., conditions that are to be fulfilled by some pieces of knowledge in order to be correctly processed by a reasoning engine, frequently appear in modeling and should be differentiated from other kinds of knowledge.

1.1.2.4 Good Computational Properties

We are interested in KR formalisms that can be used for building systems able to solve *real problems* and not only toy problems. It is thus essential to anchor these formalisms in a computational domain having a rich set of efficient algorithms. This is a key point. AI aims at building systems (or agents) for solving complex tasks and, from a complexity theory viewpoint, one can say that simple AI problems are computationally difficult to solve—they are often NP-complete or NP-hard. Thus, if on one hand a KR formalism must firmly moor to logics in order to avoid reinventing the wheel, on the other hand it must also moor to a rich algorithmic domain so that usable systems can be built.

1.1.2.5 Knowledge Programming and the Semantic Gap

A KR system should allow a user to have a maximal understanding and control over each step of the reasoning process. It should make it easy to enter the different pieces of knowledge (e.g., ontological knowledge as well as factual knowledge) and to understand their meaning and the results given by the system, and also (if asked by the user) how the system computed the results. Any computing system should have these qualities, i.e., should limit the semantic gap between real problems and their formulation in a programming language.

The correspondence between knowledge about an application domain and an expression in the KR formalism representing this knowledge must be as tight as possible. Due to the importance of natural language and schemas in the description of knowledge, a KR formalism should allow the user to easily represent simple phrases in natural language and simple schemas. The ability for describing such a correspondence, i.e., the *natural semantics* of a formal expression, is a good empirical criteria for delimiting the usability of the formalism.

As already said, in order to understand the results given by the system, a precise description of *what* is obtained (a denotational semantics) is mandatory. However, in some situations, especially whenever the knowledge deals with a poorly formalized application domain, it may be useful to understand *how* the results have been obtained, i.e., to have an operational semantics. This point is important because there can be a gap between the program and the knowledge represented in the system, i.e., a formal system, and the knowledge itself. There should be a tight correspondence between what is seen by the user and how objects and operations are implemented. This correspondence should be tight enough to enable faithful modeling of the actual data and problems, and to understand why and how results have been obtained. A way to limit the semantic gap is to use a homogeneous model—the same kinds of object and the same kinds of operation occur at each fundamental level (formal, user interface, implementation). Such a correspondence is sometimes called an "isomorphism," but this designation can be misleading. Indeed, this correspondence is not a mathematical function between two mathematical objects but is rather a correspondence between "reality" and a mathematical object. There is a gap between a reality and the concepts describing it, as well as between a conceptual modeling and its implementation, i.e., a representation of this modeling by computational objects. Moreover, as in this book we use "isomorphism" in its usual mathematical sense, we avoid using it metaphorically.

1.2 Conceptual Graphs

This section provides an intuitive introduction to conceptual graphs. Precise definitions are given in subsequent chapters. The conceptual graph model was introduced
by Sowa in 1976 [Sow76], developed in [Sow84], and then enriched and developed
by the conceptual graph community (cf. the proceedings of the International Conference of Conceptual Structures). It is the synthesis of many works in AI, but its
roots are mainly found in the following areas: natural language processing, semantic networks, databases and logics, especially the existential graphs of Pierce, which
form a diagrammatical system of logics.

We use the term "conceptual graphs" (CGs in short) to denote the *family* of formalisms rooted in Sowa's seminal work, rather than a precise formalism, and we
use specific terms—e.g., basic conceptual graphs, simple conceptual graphs, positive nested conceptual graphs—for notions which are mathematically defined and
studied in this book.

1.2.1 Basic Notions

For an intuitive introduction to CGs, let us consider again the photo of children example. We would like to represent some features of this photo and some knowledge
necessary to answer non-trivial questions about the content of the photo.

The basic vocabulary is composed of two partially ordered sets: a partially ordered set of concepts or classes (called concept types in the CG community) and
a partially ordered set of relation symbols (also called relation types). The partial
order is interpreted as a specialization or *AKO* relation: $t_1 \leq t_2$ means that t_1 is a
specialization of t_2 (or t_2 is a generalization of t_1, t_2 subsumes t_1, t_1 is a subtype
of t_2 , or every entity of type t_1 is of type t_2). There is also a set of names, called
individual markers, used for denoting specific entities. This basic vocabulary can be
considered as a rudimentary ontology.

A basic conceptual graph (BG) is composed of two kinds of nodes, i.e., *concept*
nodes representing entities and *relation* nodes representing relationships between
these entities. Nodes are labeled by types, and one can indicate that a concept node
refers to a specific entity by adding an individual marker to its label. Otherwise the
concept node refers to an unspecified entity.

For instance, the conceptual graph K in Fig. 1.1 asserts that:

- there are entities (represented by rectangles): Mary who is a Girl, a Boy, who is
  unspecified, and a Car, which is also unspecified.
- there are relations (represented by ovals) between the entities: a relation asserting
  that Mary is the sister of the Boy, and two relations asserting that Mary and
  the Boy play with the Car. The numbers on edges are used to totally order the
  neighbors of each relation node. There is also a unary relation asserting that Mary
  is smiling.

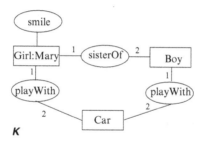

Fig. 1.1 A basic conceptual graph

This BG could be translated by the sentence "Mary (who is a girl) and her brother are playing with a car; Mary is smiling."

Important differences between the conceptual graph model and its semantic networks ancestors are to be pointed out: Firstly, there is a clear distinction between ontological knowledge (e.g., concept or relation types) and other kinds of knowledge (such as factual or implicit knowledge); secondly, relations can be of any arity, whereas the edges of semantic networks represent binary relations only; thirdly, CGs have a logical semantics in FOL.

1.2.2 Subsumption and Homomorphism

The fundamental notion for studying and using BGs is *homomorphism*. In the CG community, a BG homomorphism is traditionally called a *projection*. We prefer the term "homomorphism" for two reasons. First, it corresponds to the classical homomorphism notion in mathematics, and we will see that a BG homomorphism is a mapping between BGs preserving the BG structure. Secondly, there is an operation called "projection" in the relational database model, which is quite different from a BG homomorphism.

A *homomorphism* from a BG G to a BG H is a mapping from the nodes of G to the nodes of H, which preserves the relationships between entities of G, and may specialize the labels of entities and relationships. In graph-theoretic terms, it is a labeled graph homomorphism that we will precisely describe later. For the moment, it is only necessary to know that a *generalization/specialization* relation (or *subsumption*) over BGs can be defined with this notion: G is more general than H (or G subsumes H, or H is more specific than G) if there is a homomorphism from G to H.

Let us consider the graphs G and K in Fig. 1.2. The following mapping from the node set of G to the node set of K (pictured in dashed lines) defines a homomorphism from G to K:

the concept node [Girl] is mapped to the concept node [Girl: Mary] (the unidentified girl is specialized into the girl Mary),

the concept node [Child] is mapped to the concept node [Boy], with Boy being a subtype of Child (the child is specialized into a boy),
the concept node [Toy] is mapped to the concept node [Car], with Car being a subtype of Toy,
the relation node labeled (relativeOf) is mapped to the relation node (sisterOf) (the relation "to be a relative of someone" is specialized into the relation "to be a sister of someone"),
each relation node (actWith) is mapped to a relation node (playWith): The node (actWith) with a first neighbor [Girl] is mapped to the node (playWith) with a first neighbor [Girl:Mary], while the node (actWith) with a first neighbor [Child] is mapped to the node (playWith) with a first neighbor [Boy].

G is a generalization of K (or K is a specialization of G) since each node of G is mapped to a specialized (or identical) node of K, and because the relationships between concept nodes in G are specialized by (or identical to) relationships between the image nodes in K.

This homomorphism maps G to a subgraph of K with the same structure as G. This property is generally not fulfilled by a homomorphism, as illustrated by the graphs H and K in Fig. 1.3: there is a homomorphism from H to K, which maps both concept nodes [Car] in H to the same node [Car] in K. Graph H states that "Mary is playing with a car and a boy is playing with a car" and K specializes it by adding that they play with the same car. Moreover, K contains other relations, which are not involved in the homomorphism from H to K.

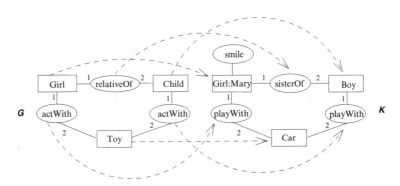

Fig. 1.2 G is a generalization of K

The fundamental problem for BGs is as follows: Given two BGs G and H, is G a generalization of H? i.e., is there a homomorphism from G to H? We call it BG-HOMOMORPHISM. We will see that this problem is NP-complete and possesses interesting polynomial cases. Furthermore, we will see that it is equivalent to other fundamental problems in AI and databases, such as the constraint satisfaction problem, which has been very well studied from an algorithmic viewpoint, or the conjunctive query inclusion problem in database theory.

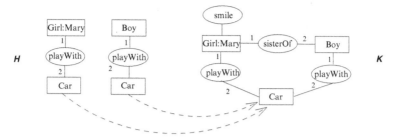

Fig. 1.3 *H* is a generalization of *K*

1.2.3 Formal Semantics

Another essential point is that BGs possess a set semantics and a logical semantics, with the former corresponding to the model theory of the latter.

Let us briefly outline how we provide BGs with a set semantics. First, one has to define a model of a vocabulary. A model of a vocabulary consists of a non-empty set D (the objects of the application domain), called the domain of the model, and the definition δ of the meaning of each element of the vocabulary. δ assigns a part of D (the set of objects of type t) to any concept type t, δ assigns a k-ary relation over D (i.e., a part of D^k composed of the tuples of objects which are related by the relation r) to any relation r of arity k and δ assigns an element of D to any individual marker. As an example let us consider a situation described by the graphs G and K in Fig. 1.2.

- The individual marker Mary is translated by an element in the domain D (i.e., $\delta(Mary)$ is the element in D representing the individual marker Mary).
- The concept types in the vocabulary, Girl, Boy, Car, etc., are translated by subsets of D (e.g., $\delta(Girl)$ is the subset of D representing the concept type Girl).
- The binary relation symbols in the vocabulary, actWith, playWith, sisterOf, relativeOf, etc., are translated by binary relations over D (e.g., $\delta(sisterOf)$ is the binary relation over D representing the binary relation symbol sisterOf), and the unary relation symbol smile is represented by a subset $\delta(smile)$ of D.

Secondly, we define BG models and the meaning of a BG that is satisfied by a model (e.g., the type of the individual marker Mary is Girl, thus $\delta(Mary)$ must be in $\delta(Girl)$; Mary is smiling, thus ($\delta(Mary)$ must be in $\delta(smile)$). Then, an entailment relation between BGs can be defined and, finally, its relationships with homomorphism is stated: Given two BGs G and H, there is a homomorphism from G to H if and only if H entails G.

Let us now outline the logical semantics, classically called Φ. The vocabulary is logically interpreted as follows. A predicate t is assigned to each type t (a unary predicate to a concept type and a k-ary predicate to a k-ary relation type) and a constant m to each individual marker m (for simplicity, here we use the same symbol for an object in the CG world and for its corresponding object in the FOL language, e.g., a unary predicate t is assigned to a concept type t).

Given two k-ary types t_1 and t_2, $t_1 \leq t_2$ is interpreted by the formula $\forall X$ ($t_1(X) \rightarrow t_2(X)$), where X is a tuple of k variables, e.g., $\forall x$ ($Girl(x) \rightarrow Child(x)$) or $\forall x \forall y$ ($sisterOf(x,y) \rightarrow relativeOf(x,y)$). An existentially closed formula is assigned to a BG, where terms (variables or constants) correspond to concept nodes.

For instance, the formula assigned to G in Fig. 1.1 is:
$\Phi(G) = \exists x \exists y (Girl(Mary) \wedge Boy(x) \wedge Car(y) \wedge smile(Mary) \wedge sisterOf(Mary,x) \wedge playWith(Mary,y) \wedge playWith(x,y))$.

Homomorphism is *sound* and *complete* with respect to logical deduction, i.e., given two BGs G and H, there is a homomorphism from G to H if and only if the formula $\Phi(G)$ can be deduced from the formula $\Phi(H)$ and the logical translation of the type hierarchies. The BG-HOMOMORPHISM problem can thus be identified with a deduction problem. We will show later that basic conceptual graphs strongly correspond to the existential, positive, conjunctive fragment of FOL (which we denote by FOL($\exists$, $\wedge$)) [1].

Basic conceptual graphs constitute the kernel of CGs. They can be used as such to represent facts and queries. They are also basic bricks for more complex constructs, such as nested graphs or rules, corresponding to more expressive CGs.

1.2.4 Full CGs

The most expressive conceptual graphs we shall consider, and call *full conceptual graphs*, were introduced by Sowa in [Sow84] and are inspired from Peirce's existential graphs (cf. [Dau02] for a mathematical study of full CGs and [Rob92] for a presentation of Peirce's existential graphs). The basic idea of the extension from BGs to full CGs is that every FOL formula can be written as a formula by solely using the existential quantifier and the conjunction and negation logical connectors. By adding to BGs *boxes* representing negation and *lines* (called co-reference links) indicating that two nodes represent the same entity, one obtains the same expressiveness as FOL. For instance, the graph of Fig. 1.4 shows that the relation r is transitive on entities of type t, i.e., for all x, y, z of type t, if $r(x, y)$ and $r(y, z)$ then $r(x, z)$. Its logical translation is more precisely:
$\Phi(G) = \neg(\exists x \, \exists y \, \exists z(t(x) \wedge t(y) \wedge t(z) \wedge r(x,y) \wedge r(y,z) \wedge \neg(r(x,z))))$,
which is equivalent to: $\forall x \forall y \forall z((t(x) \wedge t(y) \wedge t(z) \wedge r(x,y) \wedge r(y,z)) \rightarrow r(x,z))$

Peirce's existential graphs are provided with a sound and complete set of inference rules that can be adapted to CGs [Sow84] [Wer95a] [Dau02]. However, these rules do not directly lead to automated reasoning because they heavily rely on human intuition [2].

---

[1] We will see that the universally quantified formulas associated with the vocabulary can be dropped without restraining logical expressivity.

[2] For instance, the insertion rule allows one to insert *any* graph (at a place obeying specific conditions), which leads to an infinite number of choices.

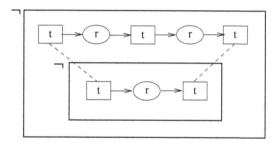

G

For binary relation nodes, directions on edges may replace numbers 1 and 2.

Fig. 1.4 A full CG

In this book, we will only briefly present full CGs, which are unsuitable for our approach, as will become clear in the next section. Instead, we will build limited extensions of BGs (e.g., with atomic negation), in an attempt to keep the essential properties of BGs.

1.3 A Graph-Based Approach to KR

Since 1991 (cf. [CM92]) our aim has been to develop and study a KR formalism respecting, as far as possible, the five requirements presented in Sect. 1.1. Our work belongs to a logical approach to KR (and to the KR scientific community main-stream as presented, for instance, by Brachman and Levesque in [BL04] or Baader in [Baa99]), but it is also *graph-based* as explained hereafter. The formalism is based on graphs and graph-theoretic notions and operations. It is logically founded, but in some way is "autonomous" from (existing) logics. Stated differently, our KR formalism is a pure graph-theoretic formalism, whose core corresponds to a FOL fragment, and most extensions of this core correspond to FOL fragments. Since it is embedded in graph theory, it is easy to define new operations, simple from a graph viewpoint, and having simple intuitive semantics, but that do not necessarily have a formal semantics expressed in a classical logic.

1.3.1 Motivations

Let us first outline our motivations for a graph-based approach to KR. They can be divided into two categories: qualities of graphs for knowledge modeling and qualities of graphs for computations.

From a *modeling viewpoint*, we see two essential properties in the basic conceptual graph model. The *objects*, i.e., basic graphs, are easily understandable by users (typically knowledge engineers or specialists in an application domain), at least if the graphs are reasonably small (note that it is always possible to split up a large conceptual graph into smaller ones while keeping its semantics). This feature partially explains the success of semantic networks and, more generally, the success of graphical models, such as the entity/relationships model, UML, Topic Maps, etc. Many people, and not only computer scientists, are now familiar with kinds of labeled graphs (mainly trees, but not exclusively). This fact is especially important in knowledge acquisition. In our approach to CGs, this quality does not only concern the descriptive facet of the formalism. *Reasoning* mechanisms are also easily understandable, for two reasons. First, homomorphism is a "global" graph-matching notion that can be easily visualized (we will also see that homomorphism is equivalent to a sequence of elementary graph operations which are very simple and easy to visualize). Secondly, the same language is used at interface and computing levels. In particular, no transformation has to be done on the components of the knowledge base before reasoning with it.

Thus, reasoning can be explained on the user's representation itself, and explanations can be given at any level from the user's level to the implementation level. At the implementation level, a graph can be represented by a structure with pointers, which is a graph too! To sum up, using a graph-based KR should reduce the semantic gap mentioned in Sect. 1.1.2.

From a *computational viewpoint*, labeled graph homomorphism firmly moors BGs to combinatorics. The graph homomorphism notion (or its variant, relational structure homomorphism) was recognized in the 90s as a central notion, unifying complexity results and algorithms obtained in several domains (e.g., cf. [Jea98] and [FV93]). On the other hand, considering graphs instead of logical formulas provides another view of knowledge constructs (e.g., some notions like path, cycle, or connected components are natural on graphs) and provides other algorithmic ideas, as we hope is illustrated throughout this book.

1.3.2 Extensions of the Basic Formalism

Full CGs *à la Peirce* are no longer graphs (in the graph theory sense); they are diagrams. Associated inference rules are not graph-based operations either. In our opinion, qualities of the BG model from a knowledge representation and reasoning perspective (as presented above) are at least partially lost: namely, the readability of objects as well as the easy understanding of the inference mechanism, and relationships with combinatorial problems.

Rather than jumping from BGs to full CGs, we prefer, depending on the kind of knowledge we would like to represent, to build extensions of the BG model, while keeping its essential properties. These properties represent our motto and can be summarized as follows:

1. objects are *labeled graphs* (mathematically defined with graph-theoretic notions),
2. reasoning mechanisms are based on graph-theoretic operations, mainly relying on graph homomorphism,
3. efficient reasoning algorithms exist for important specific cases,
4. objects and operations have graphical representations, which make them easily understandable by users (limitation of the semantic gap),
5. the BG model is logically founded, with the inference mechanism being sound and complete with respect to FOL semantics.

Let us briefly give an example of extension: BG rules. A rule represents information of the type: "if information H is found, then information C can be added." H is called the hypothesis of the rule and C its conclusion. This notion of a rule has been widely used in AI to represent implicit knowledge, which can be made explicit by applying the rules, on facts for instance.

Rules could be represented as full CGs, but in so doing they would lose their specificity and could not be processed in a particular manner. A BG rule can be defined as a bicolored BG (a more general rule definition will be given). One color (white in the figures) defines the hypothesis, and the other color defines the conclusion (gray in the figures). For instance, the rule in Fig. 1.5 has the same semantics as the full CG in Fig. 1.4, i.e., it says that the relation r is transitive on type t entities. A more complex rule is pictured in Fig. 1.6. This rule allows us to decompose the ternary relation *give* into simpler relations: If there is a relation *give* with first argument a *Human x*, with second argument a *Thing y* and with third argument an *Animate* entity z, then there is a *Gift* act, whose x is the *agent*, y is the *object* and z is the *recipient*.

The notion of a rule application is very simple. A rule R is said to be applicable to a BG G if its hypothesis H can be mapped by a homomorphism to G. Then it can be applied to G: Applying R consists of adding the conclusion of R to G guided by the homomorphism from H to G. Rules are provided with a logical semantics extending that of BGs, such that graph mechanisms, namely forward chaining and backward chaining, are sound and complete. Let us consider a KB $\mathcal{K}$ composed of a set $\mathcal{F}$ of BGs representing facts and a set $\mathcal{R}$ of rules. A BG is derived from $\mathcal{K}$ if it can be obtained by a sequence of rule applications to the facts. The basic problem considered is thus as follows: Given a KB $\mathcal{K} = (\mathcal{F}, \mathcal{R})$ and a BG Q, is a specialization of Q derivable from $\mathcal{K}$? Due to the soundness and completeness of the graph mechanisms defined, it can be seen as a deduction problem, called $\mathcal{FR}$-DEDUCTION. By enriching BGs with rules, we obtain a computability model. This is an important property but, similar to the fact that no computability model can be the basis for building a universally good programming language, this does not mean that this KR formalism is suitable for all KR domains. Moreover, this high expressivity comes with undecidability of reasoning. That is why properties on rule sets ensuring decidability of reasoning are studied. A simple example is that of rules that do not add unspecified concept nodes (such as the rule in Fig. 1.5, contrary to the rule in Fig. 1.6 which adds an unspecified Gift). In this case, $\mathcal{FR}$-DEDUCTION is NP-complete, thus not more difficult than deduction checking in the BG model, i.e., BG-HOMOMORPHISM.

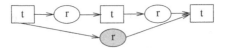

Fig. 1.5 A simple rule

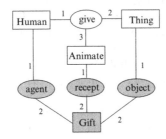

Fig. 1.6 Another rule

Other extensions presented in the book are *type conjunctions*, *nested graphs*, *atomic negation*, and *constraints*, as well as a family of models combining rules and constraints, called the BG-Family.

1.3.3 Several Approaches to CGs

Let us end this section by situating our approach in the CG landscape. Research on CGs can be roughly classified according to three axes: CGs can be seen as a *diagrammatic interface* for other formalisms or as a *diagrammatic calculus of logics*, or as a *graph-based KR formalism*.

Many works are mainly focused on the visual qualities of conceptual graphs. The expressivity and readability of the obtained representations are therefore the main criteria for evaluating a formalism. We shall not forget that a main motivation behind conceptual graphs was the processing of natural language semantics: From this standpoint, the relative easiness in translating a (small) conceptual graph into a natural language sentence it represents is an important criterion. When reasoning mechanisms are associated with representations, they are not part of the CG formalism: They are performed in a procedural way, or rely on another formalism into which the CGs are translated. In this approach, conceptual graphs are seen as a diagrammatical interface for other formalisms (e.g., the Common Logic Standard and

CLIF [CLS07]). They are not a knowledge representation and reasoning formalism *per se*, i.e., in the sense of Sect. 1.1.

Other works can be seen as the continuation of Peirce's work on a diagrammatical system of logics. Conceptual graphs are then diagrams, and reasoning is based on diagrammatic operations. In these works, automated reasoning is not the point and the computational aspect of the formalism is absent.

Finally, the graph-based approach emphasizes the following points (cf. our motto in Sect. 1.3.2), which distinguishes it from the other two approaches:

- CGs are seen as a KR *and reasoning* formalism. Thus, they are provided with their own operations for reasoning.
- Reasoning mechanisms should be *sound and complete* with respect to a formal semantics.
- Reasoning operations should be conducted in an *efficient way*. That is why decidability and complexity studies, as well as the design of efficient algorithms are important issues.
- Reasoning mechanisms are based on graph-theoretic notions, mainly labeled graph homomorphism, as the structure underlying objects is a graph.

We became highly interested in CGs when we proved (cf. [Mug92] and [CM92]) that homomorphism on the BG fragment is complete with respect to the FOL semantics. As Sowa in [Sow84] had proven that homomorphism is sound (with respect to this FOL semantics), the completeness theorem established the graph homomorphism notion as a key reasoning tool. It is agreed that graph homomorphism is a fundamental notion in graph theory and the results we have obtained, as well as results in other fields (e.g., cf. [Jea98] and [FV93]), strengthened our belief that labeled graph homomorphism is a key notion in KR.

Part I
Foundations: Basic and Simple Conceptual Graphs

Chapter 2
Basic Conceptual Graphs

Overview

This chapter presents the kernel of the knowledge representation language, namely basic conceptual graphs. A *basic conceptual graph (BG)* has no meaning independently from a vocabulary, and both are defined in Sect. 2.1. The fundamental notion for reasoning with BGs is the *subsumption* relation. Subsumption is first defined by a *homomorphism*: Given two BGs G and H, G is said to subsume H if there is such a BG homomorphism from G to H. Section 2.2 defines the BG homomorphism, as well as the particular case of BG isomorphism. BGs and subsumption provide a basic query-answering mechanism, as shown at the end of this section. The subsumption properties are studied in Sect. 2.3. Subsumption is not an order as there may be nonisomorphic equivalent BGs, i.e., BGs that subsume each other. However, by suppressing redundant parts, any BG can be transformed into an equivalent *irredundant* BG. Irredundant BGs are unique representatives of all equivalent BGs. Finally, the subsumption relation restricted to irredundant BGs is not only an order but also a lattice. Section 2.4 introduces another way of defining the subsumption relation by sets of elementary graph operations. There is a set of *generalization* operations and the inverse set of *specialization* operations. Given two BGs G and H, G subsumes H if and only if G can be obtained from H by a sequence of generalization operations (or equivalently, H can be obtained from G by a sequence of specialization operations). Section 2.5 introduces the issue of equality, which will be a central topic of the next chapter (Simple Conceptual Graphs). A BG is said to be *normal* if it does not possess two nodes representing the same entity. Normal BGs form the kernel of reasoning based on homomorphism.

In Sect. 2.6, the complexity of BG fundamental problems is studied. In particular, it is proven that checking whether there is a homomorphism between two BGs is an NP-complete problem.

2.1 Definition of Basic Conceptual Graphs (BGs)

2.1.1 Vocabulary

Basic conceptual graphs (BGs) are building blocks for expressing different sorts of knowledge: Assertions or facts, queries or goals, rules describing implicit knowledge, rules describing the evolution of some world, constraints and so on. In this chapter they are used for representing facts and queries, but they will be used for representing more complex knowledge in further chapters.

A fact is an assertion that some entities exist and that these entities are related by some relationships. Any entity has a type (e.g., Car, Person, Toy, etc.), and the set of types is ordered by a *subtyping* relation, also called a *specialization* relation or *a-kind-of* relation (e.g., the type Boy is a kind of the type Person). "The type t is a specialization of the type t'," equivalently "t' is a generalization of t," simply means that any entity of type t is also of type t'. It is assumed that there is a most general type, called the *universal* type, denoted $\top$.

Two kinds of entities are considered. An entity may be either a specific entity (e.g., *that* small red car) or an unidentified entity (e.g., *a* boy). A specific entity is called an *individual* entity and an unidentified entity is called a *generic* entity.

Besides asserting the existence of specific or unidentified typed entities, a fact can assert that relationships hold between these entities (e.g., Paul *possesses* a teddy bear, Paul *is-the-son-of* a person and so on). The relations are structured by a specialization order as the types of entities are. For example, the relation *is-the-son-of* is a specialization of the relation *is-the-child-of*, which is itself a specialization of the relation *is-a-relative-of*. The representation of an entity (of the application domain) is traditionally called a *concept* in the conceptual graphs community and we use this expression hereafter. A basic conceptual graph (BG) is composed of two kinds of nodes, *concept* nodes representing entities that occur in the application domain, and *relation* nodes representing relationships that hold between these entities. The ordered set of concept (entity) types is denoted T_C. An individual concept is referenced by an *individual marker* belonging to a set $\mathcal{I}$ of individual markers, and there is a *generic marker* $*$, which denotes an unspecified entity. The same marker $*$ is used for denoting a generic entity regardless of type. The set of relations is denoted T_R. An element of T_R is called a *relation symbol or a relation type*. These three sets compose the vocabulary used for labeling the two kinds of nodes of a BG: A concept node is labeled by a pair composed of a type and either an individual marker or the generic marker; a relation node is labeled by a relation symbol. A vocabulary is precisely defined as follows:

Definition 2.1 (Vocabulary). A *BG vocabulary*, or simply a vocabulary, is a triple $(T_C, T_R, \mathcal{I})$ where:

- T_C and T_R are finite pairwise disjoint sets.
- T_C, the set of *concept types*, is partially ordered by a relation $\leq$ and has a greatest element denoted $\top$.

- T_R, the set of *relation symbols*, is partially ordered by a relation $\leq$, and is partitioned into subsets $T_R^1, \ldots, T_R^k$ of relation symbols of arity $1, \ldots, k$, respectively. The arity of a relation r is denoted $arity(r)$. Any two relations with different arities are not comparable.
- $\mathcal{I}$ is the set of *individual markers*, which is disjoint from T_C and T_R. Furthermore, $*$ denotes the *generic marker*, $\mathcal{M} = \mathcal{I} \cup \{*\}$ denotes the *set of markers* and $\mathcal{M}$ is ordered as follows: $*$ is greater than any element in $\mathcal{I}$ and elements in $\mathcal{I}$ are pairwise incomparable.

In some works, it is assumed that T_C has a specific structure, such as a tree, a lattice or a semi-lattice. In this book, having a greatest element is the only property required for the ordered set T_C.

A set T_C (resp. T_R) of concept (resp. relation) types is also called a *hierarchy* of concept (resp. relation) types.

The three sets (concept, relation or marker set) play different roles and are assumed to be pairwise disjoint. A specific syntax is used for representing elements of a vocabulary: A concept type begins by an upper case letter, a relation symbol by a lower case letter, and an individual is a proper name or begins by #. This allows the user to quickly determine the role of an identifier and to simply differentiate close identifiers playing different roles. For instance, the word *father* could represent a concept type (e.g., Paul is an individual of type *Father* means that Paul is a father), a binary relation symbol *fatherOf* could represent the binary relation relating a father and one of his children (e.g., $fatherOf(Paul, July)$ means that Paul is the father of July), and if "Father" is the name of the chief of a gang then Father could be used as an individual marker.

Example. Figure 2.1 is a subset of the concept type set corresponding to the children photo example (Sect. 1.1.1). Figure 2.2 is a part of the relation type set for the same example. In this case, for each arity i of relation, there is a greatest element denoted by $\top_i$. $\top_i$ can be seen as representing any relation of arity i.

Type of Individuals and Relation Signatures

A vocabulary can be considered as the representation of a very simple ontology. Different kinds of knowledge can be added to a vocabulary. Two simple extensions are often considered, namely *individual typing* and *relation symbol signatures*.

First, individual markers can be typed. The type of an individual marker represents the most specific information about the category of the individual referenced by this marker. A mapping, say τ, from the set of individuals $\mathcal{I}$ to the set of concept types T_C is thus added with, for an individual m, $\tau(m)$ representing the most specific type of m. Consequently, in any BG relative to this vocabulary, a concept node with marker m will have a type greater than or equal to $\tau(m)$. For instance, let us assume that, in some applications, there is an individual marker $R2$ known to represent a *ToyRobot* (without any other details) i.e., $\tau(R2) = ToyRobot$. Individual concept nodes with marker $R2$ and type *Toy*, or *Robot*, or *MobileEntity* may appear. But, if $Android < ToyRobot$, then no concept node labeled $(Android, R2)$

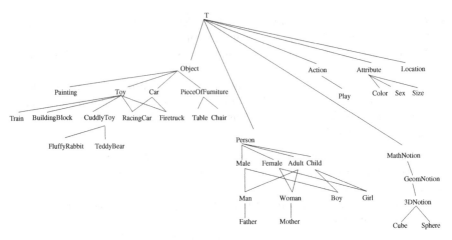

Fig. 2.1 A concept type set

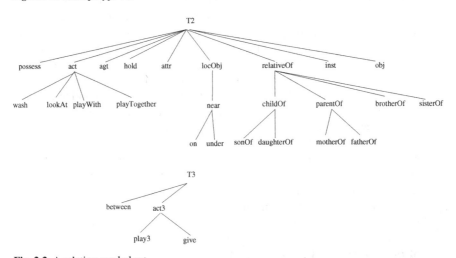

Fig. 2.2 A relation symbol set

can appear because *Android* $< \tau(R2)$. Typing the individuals may be a drawback whenever knowledge about the individuals may evolve. This is typically the case in a knowledge acquisition context. For instance, *R2* may first be known as a *Toy* and later as a *ToyRobot*.

The second extension is the introduction of relation symbol signatures. A relation symbol signature specifies not only the arity of the relation symbol but also the maximal concept type of each of its arguments. For instance, the signature of the relation symbol *possess* could be $(Person, Object)$, which means that the first argument of a relation *possess* is of type *Person* and the second argument is of type *Object*. Relation signatures are formally defined by a mapping σ, which to every relation symbol $r \in T_R^j$, $1 \leq j \leq k$ associates $\sigma(r) \in (T_C)^j$; this mapping also has to fulfill the following condition: $\forall r_1, r_2 \in T_R^j$, $r_1 \leq r_2 \Rightarrow \sigma(r_1) \leq \sigma(r_2)$, i.e., for all $1 \leq i \leq j$, the i-th argument of $\sigma(r_1)$ is less than or equal to the i-th argument of $\sigma(r_2)$. That is, when a relation symbol r_2 is specialized into a relation symbol r_1, its arguments can be specialized but cannot be generalized.

Example. For instance, let us consider the relation symbols $\top_2$, *relativeOf*, *parentOf*, *motherOf*. Signatures for these relations can be: $\top_2(\top, \top)$, *relativeOf(Person, Person)*, *parentOf(Person, Person)*, *motherOf(Mother, Person)*. If the ternary relation *play3* is intended to mean that two persons are playing together with a toy then its signature could be *play3(Person, Person, Toy)*. In the same way, the signature of the ternary relation *give* could be *give(Person, Person, Object)* with the following intuitive meaning: The first person is giving the object to the second person.

In the forthcoming chapters, more complex vocabularies and more complex knowledge, such as rules or constraints, will be studied. Rules and constraints allow the representation, in the conceptual graph model, of "heavyweight" formal ontologies.

2.1.2 Basic Conceptual Graphs

A *basic conceptual graph (BG)* is a bipartite multigraph (cf. Chap. A). "Bipartite" means that a BG has two disjoint sets of nodes and that any edge joins two nodes from different sets (i.e., an edge cannot join two nodes of the same set). "Multigraph" means that a pair of nodes may be linked by several edges. One set of nodes is called the set of concept nodes (representing entities), and the other set is called the set of relation nodes (representing relations between entities). If a concept c is the i-th argument of a relation r then there is an edge between r and c that is labeled i.

Example. Figure 2.3 shows a BG consisting of four concept nodes (the individual *Paul* who is a *Child*, a *Car*, a *Person* and the individual *Small* which is a *Size*) and three relations: one ternary relation, *play3*, whose neighbor list is ([Child: Paul], [Person], [Car]), and two binary relations, *attr* and *possess*, whose neighbor lists are respectively ([Child: Paul], [Car]) and ([Car], [Size: Small]). This graph can be considered as representing the following fact: "There is a car, which is small. This car is possessed by Paul, who is a child. There is a person, and Paul and this person are playing with that car." This way of describing the fact puts the emphasis on the car, but the emphasis could have been put on Paul as the drawing suggests it (e.g., "Paul, who is a child, and a person, are playing with a small car that belongs to Paul") or any subset of entities. A basic conceptual graph has no privileged nodes. Figure 2.4 shows another BG that could represent the assertion that "Paul, who is a child, is washing himself, and he is playing with his mother." Note the parallel edges between the concept [Child:Paul] and the relation (wash), indicating that the agent and the object of the relation (wash) are the same entity, the child Paul. Figure 2.5 shows a BG with more complex cycles, asserting that "a father and his child are playing together on a mat; the mother of the child is looking at them; she is on the sofa near the mat."

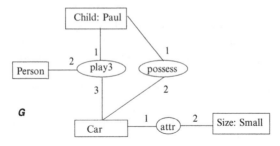

Fig. 2.3 A BG with a ternary relation

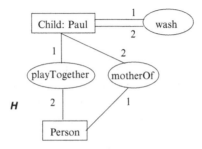

Fig. 2.4 A BG with parallel edges

Every node has a label. A relation node is labeled by a relation symbol and a concept node is labeled by a concept type and by a marker. Thus, a basic conceptual graph is built relative to a vocabulary, and it has to satisfy the constraints enforced by that vocabulary.

Definition 2.2 (Basic Conceptual Graph). A *basic conceptual graph* (BG) defined over a vocabulary $\mathcal{V} = (T_C, T_R, \mathcal{I})$, is a 4-tuple $G = (C, R, E, l)$ satisfying the following conditions:

- (C, R, E) is a finite, undirected and bipartite multigraph called the *underlying graph* of G, denoted $graph(G)$. C is the *concept node* set, R is the *relation node* set (the node set of G is $N = C \cup R$). E is the family of *edges*.
- l is a labeling function of the nodes and edges of $graph(G)$ that satisfies:

    1. A concept node c is labeled by a pair *(type(c), marker(c))*, where *type(c)*$\in T_C$ and *marker(c)*$\in \mathcal{I} \cup \{*\}$,
    2. A relation node r is labeled by $l(r) \in T_R$. $l(r)$ is also called the *type* of r and is denoted by *type(r)*,

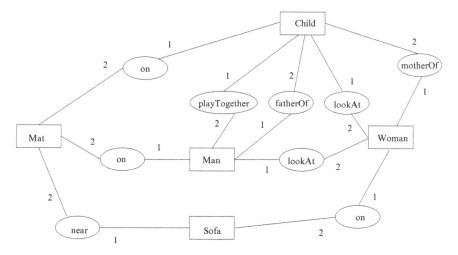

Fig. 2.5 A BG with more complex cycles

3. The degree of a relation node r is equal to the arity of type(r),
4. Edges incident to a relation node r are totally ordered and they are labeled from 1 to $arity(type(r))$.

First note that a BG does not need to be a *connected* graph. It is natural to question what the smallest BGs are. The preceding definition does not prevent a BG from being *empty*, that is to be the tuple $(\emptyset, \emptyset, \emptyset, \emptyset)$. For theoretical purposes, it is sometimes convenient to consider the empty BG. We will denote it by $G_\emptyset$. A BG may be restricted to a single concept node or several concept nodes. We will call them *isolated* concept nodes. But as soon as a BG contains a relation node, it contains at least one concept node, since there are no 0-ary relation symbols. Note also that, as there may be parallel edges, one has to distinguish between the number of edges incident to a relation node (this number is given by the arity of its type) and the number of its neighbors. For instance, the relation (wash) in Fig. 2.4 is incident to two edges but has only one neighbor.

An important kind of BGs consists of BGs having a single relation node.

Definition 2.3 (Star BG). A *star BG* is a BG restricted to a relation node and its neighbors.

With BGs restricted to single concept nodes, star BGs are the elementary building blocks of BGs. This point will be specified in Chap. 8, but let us outline the idea. First, the set of relation symbols of a BG vocabulary can be described by a set of star BGs: A relation symbol r of signature $(t_1, \ldots, t_k)$ is represented by a star BG, the relation node is labeled r and has k neighbors with the i-th being labeled $(t_i, *)$ (cf. Fig. 2.6). Then every non-empty BG definable on this vocabulary can be generated from this set of star graphs by the so-called specialization operations (if the BG has isolated concept nodes, one has to add, to the set of star graphs, the graph restricted to the single concept node $[\top : *]$).

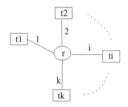

Fig. 2.6 A star BG associated with the relation symbol r with signature $(t_1, \ldots, t_k)$

Notations and Remarks

An edge labeled i between a relation r and a concept c is denoted (r, i, c). The i-th argument of a relation r is also called the i-th neighbor of r and is denoted $r[i]$ i.e., (r, i, c) is in G if and only if $r[i] = c$ in G. The neighbor list $(c_1, \ldots, c_k)$ of a relation r of arity k is the list such that $c_i = r[i]$. It is said that $r(c_1, \ldots, c_k)$ is in G if $(c_1, \ldots, c_k)$ is the neighbor list of the relation r. The same concept node may appear several times in the neighbor list of a relation, so strictly speaking a BG is a multigraph and not a graph.

We will adopt the following classical conventions for BGs. A relation symbol is also called a *relation type* even if a concept type and a relation type have different meanings. A concept type can be considered as a class, i.e., the class of all the entities having this type, but a relation type does not represent a class. Nevertheless, calling "relation type" a relation symbol allows us to simplify notations by considering the type of any node, either concept or relation, in a BG. In a BG drawing, concept nodes are represented by rectangles and relation nodes by ovals. In textual notation, rectangles are replaced by [] and ovals by (). The generic marker is the marker by default, and it is generally omitted. Thus a generic concept of type t, whose label is $(t, *)$, is simply noted [t]. An individual concept with label (t, m) is noted [t:m].

For binary relation nodes, numbers on edges can be replaced by directed edges. A relation node is then incident to exactly one incoming edge and one outgoing edge, linking it to its first and second neighbor, respectively. For example, Fig. 2.7 pictures the same BG as in Fig. 2.3, with arrows on edges incident to binary relations; the arrow from [Car] to (attr) stands for an edge labeled 1, and the arrow from (attr) to [Size : Small] stands for an edge labeled 2.

Relation labels are naturally ordered by the partial order on T_R, and concept labels are ordered as follows:

Definition 2.4 (Ordered set of concept labels). The set of *concept labels*, defined over a given vocabulary, is the set of couples (t, m) such that $t \in T_C$ and $m \in \mathcal{M}$. This set is the cartesian product of the two partially ordered sets T_C and $\mathcal{M}$, thus it is (partially) ordered by $(t, m) \leq (t', m')$ if and only if $t \leq t'$ and $m \leq m'$.

Example. $(Boy, Paul) \leq (Person, Paul) \leq (Person, *)$, but $(Boy, *)$ and $(Person, Paul)$ are not comparable because $Boy < Person$ and $Paul < *$.

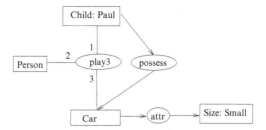

Fig. 2.7 Another drawing of G (cf. Fig. 2.3)

2.1.3 SubBGs and PseudoBGs

We are interested here in subgraphs of a BG that are themselves BGs. We call them *subBGs*. Let us begin with their formal definition before discussing the distinction between a "subgraph of a BG" and a "subBG."

Definition 2.5 (subBG). Let $G = (C, R, E, l)$ be a BG. A *subBG* of G is a BG $G' = (C', R', E', l')$, such that:

- $C' \subseteq C$, $R' \subseteq R$, $E' \subseteq E$,
- l' is the restriction of l to $C' \cup R' \cup E'$.

G' is a *strict* subBG of G if the number of nodes in G' is strictly less than the number of nodes of G.

In graph theory, a subgraph of a graph G is defined as a graph obtained from G by removing nodes (and edges incident to these nodes). A subBG must be a BG, thus it is not possible to remove just any nodes. Since the degree of a relation node has to be equal to the arity of its type, deleting a concept node of G implies deleting all neighboring relation nodes. Similarly, a graph obtained from a BG by removing only edges is not a BG; that is if an edge is removed then its incident relation must be removed too.

Example. G' in Fig. 2.8 is a subgraph of G in Fig. 2.3. G' is not a BG because in G' the relation (play3), which is a ternary relation, has only two neighbors. Thus G' is not a subBG of G.

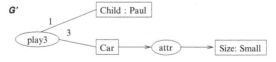

Fig. 2.8 G': a subgraph of G (cf. Fig. 2.4) which is not a subBG

A subBG of G can be obtained from G only by repeatedly deleting a relation or an isolated concept.

Property 2.1. A subgraph G' of a BG G is a subBG if and only if all neighbors in G of any relation r in G' are also neighbors of r in G'.

Definition 2.6 (Boundary node). A *boundary node* of a subBG G' of G is a node of G' that has a neighbor outside G'.

As the arity of a relation type is a constant, a boundary node is necessarily a concept.

If edges or nodes are arbitrarily deleted from a BG, the obtained graph is called a *pseudo-BG*. More generally, a pseudo-BG has concept and relation nodes, and edges connecting concepts to relations, but it might not satisfy other conditions on a BG.

Definition 2.7 (Pseudo-BG). A *pseudo-BG* is obtained from the definition of a BG by removing some conditions from the set of conditions 2.

Example. G' in Fig. 2.8 is a pseudo-BG.

Pseudo-BGs naturally occur during the construction of BGs, using a graph editor for instance. Indeed, the graphs obtained before the whole completion generally are only pseudo-BGs and not BGs. More generally, during knowledge acquisition, any constraint of the model can be a drawback. Pseudo-BGs represent minimal constraints that in our opinion have to be enforced: having two kinds of nodes and preventing edges to connect nodes of the same kind. However, after completion of a knowledge acquisition step, the constraints of the vocabulary (possibly including individual types or relation signatures) are useful validation tools.

2.2 BG Homomorphism

2.2.1 Subsumption and Homomorphism

The *subsumption* relation (denoted $\succeq$) is the fundamental notion for reasoning with BGs. Let G and H be two BGs over the same vocabulary. Intuitively, G subsumes H (noted $G \succeq H$) if the fact—or the information—represented by H entails the fact represented by G, or in other words, if all information contained in G is also contained in H. "G subsumes H" is equivalent to "H is subsumed by G" denoted by $H \preceq G$. Let us keep this intuitive meaning for now; we will come back to the semantic of subsumption later. This subsumption relation can be defined either by a sequence of elementary operations or by the classical homomorphism notion applied to BGs (which essentially is a graph homomorphism, with this point being studied in Chap. 5).

Definition 2.8 (BG homomorphism). Let G and H be two BGs defined over the same vocabulary. A *homomorphism* π from G to H is a mapping from C_G to C_H and from R_G to R_H, which preserves edges and may decrease concept and relation labels, that is:

- $\forall (r,i,c) \in G, (\pi(r), i, \pi(c)) \in H,$
- $\forall e \in C_G \cup R_G, l_H(\pi(e)) \leq l_G(e).$

If there is a homomorphism (say π) from G to H, we say that G *maps* or *projects* to H (by π).

Example. Figure 2.9 G maps to H and K.

Remark

In the conceptual graph community a BG homomorphism is traditionally called a *projection*. We prefer the term BG homomorphism for two reasons: First, it corresponds to the use of "homomorphism" in mathematics, indeed a BG homomorphism is a mapping between BGs preserving the BG structure; secondly, in the relational database model, there is an operation called "projection," which is rather different from a BG homomorphism (cf. Chap. A).

Definition 2.9 (Subsumption relation). Let G and H be two BGs defined over the same vocabulary. The *subsumption* relation $\succeq$ is defined by: $G \succeq H$ if there is a homomorphism from G to H.

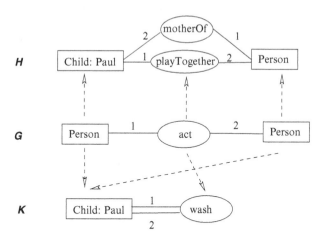

Fig. 2.9 Homomorphisms from G to H and K

Example. Let us consider the BGs G and H in Fig. 2.10. There are two homomorphisms from G to H, namely h_1 and h_2, pictured by dashed and dotted lines, respectively. By h_2, [Child] in G is mapped to [Child:Paul] in H, by h_1 it is mapped to [Child]. The relation (act) is mapped to the appropriate relation (playWith). The other nodes have the same image by both homomorphisms: [Toy] and [Car] are mapped to [FireTruck], the two relations (on) are mapped to the sole relation (on) in H, and each remaining node is mapped to the node with same label in H.

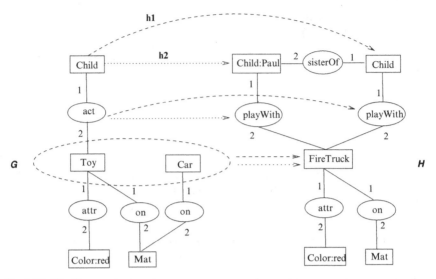

Fig. 2.10 $G \succeq H$

Subsumption emphasizes the fact that a BG is always relative to a given vocabulary and has no meaning independently from it. In other words, the same labeled graph satisfying the properties of Definition 2.2 leads to two different BGs if it is considered relative to two different vocabularies. For instance, let us consider the BG G in Fig. 2.10, which is relative to the vocabulary $\mathcal{V}$ given in Fig. 2.1. If $\mathcal{V}'$ is obtained from $\mathcal{V}$ by deleting the fact that *Firetruck* $<$ *Car*, then there is no homomorphism from G to H considered as BGs over $\mathcal{V}'$, whereas there is a homomorphism from G to H when they are BGs over $\mathcal{V}$.

Definition 2.10 (Image by homomorphism). Let $G = (C_G, R_G, E_G, l_G)$ and $H = (C_H, R_H, E_H, l_H)$ be two BGs and let π be a homomorphism from G to H. The image of G by π is $\pi(G) = (\pi(C_G), \pi(R_G), E', l')$, with E' being equal to the edges of E_H with one extremity in $\pi(C_G)$ and the other in $\pi(R_G)$, and l' being the restriction of l_H. In other words, $\pi(G)$ is the subgraph of H induced by $\pi(C_G) \cup \pi(R_G)$.

It is straightforward to check:

Property 2.2. If G and H are two BGs and π is a homomorphism from G to H, then $\pi(G)$ is a subBG of H.

Let us now observe what happens for homomorphism when BGs have several individual concepts with the same marker. Let us consider the two BGs in Fig. 2.11. There is clearly a homomorphism from G to H but there is no homomorphism in the other direction, i.e., from H to G, even if the two BGs have the same intuitive meaning that is: *The individual m has properties r and s*. We will discuss this point in the section about normal BGs (cf. Sect. 2.5).

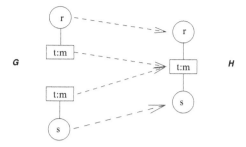

Fig. 2.11 $G \succeq H$ and $H \not\succeq G$

2.2.2 Bijective Homomorphisms and Isomorphisms

The notion of BG isomorphism is naturally defined as follows:

Definition 2.11 (BG isomorphism). An isomorphism from a BG G to a BG H is a bijection π from C_G to C_H and from R_G to R_H, which satisfies:

- $\forall (r,i,c) \in G, (\pi(r),i,\pi(c)) \in H$, and reciprocally, $\forall (r',i,c') \in H$, $(\pi^{-1}(r'),i,\pi^{-1}(c')) \in G$,
- $\forall e \in C_G \cup R_G, l_G(e) = l_H(\pi(e))$.

A bijective BG homomorphism from G to H is an isomorphism from $graph(G)$ to $graph(H)$, but it is a BG isomorphism only if it furthermore preserves labels (i.e., it fullfills condition 2 of the BG isomorphism definition). Indeed, since a relation is mapped to a relation of the same arity, a bijective homomorphism π from G to H always fullfils the property that for all r' and c' in H, for all c and r in G, such that $c' = \pi(c)$ and $r' = \pi(r)$, $(\pi(r),i,\pi(c)) \in H$ implies $(r,i,c) \in G$ (and reciprocally by the homomorphism definition). Thus, the first condition in the BG isomorphism definition is satisfied by a homomorphism as soon as it is bijective.

Examples of bijective BG homomorphisms that are not BG isomorphisms naturally occur when BGs represent data forms to fill in.

A very simple data form schema is represented by DF in Fig. 2.12; F is a partially filled data form corresponding to DF. It is simple to see that the mapping represented by dashed arrows is a bijective homomorphism from DF to F.

Property 2.3. For any BG G, a bijective homomorphism from G to itself is an isomorphism.

Proof. The sets $T_C \times \mathcal{M}$ of concept labels and T_R of relation symbols are partially ordered.

Let us consider two linear extensions of the dual orders of these partially ordered sets, both denoted $\leq_w$, i.e., given two labels l_1 and l_2, $\leq_w$ is a total order and if $l_1 \geq l_2$, then $l_1 \leq_w l_2$.

Assign a weight to each concept and relation, which is the position of its label in $\geq_w$ (starting from 1). It becomes heavier as it becomes more specialized. Then,

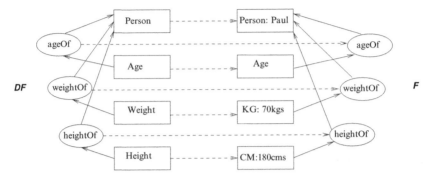

Fig. 2.12 Data form and bijective homomorphism

to a BG, assign a weight (let us denote it w) which is the sum of its relation and concept weights. Note that if there is an injective homomorphism from a BG G_1 to a BG G_2, then $w(G_1) \leq w(G_2)$. Now let π be a bijective homomorphism from G to itself. Assume that π strictly restricts the label of at least one relation or concept of G. Then $w(G) < w(\pi(G))$ which is impossible since $\pi(G) = G$. Thus π is an isomorphism. □

2.2.3 BG Queries and Answers

With BGs and subsumption, we can build a basic query-answering mechanism. Let us consider a KB (knowledge base) B composed of a set of BGs, representing some assertions about a modeled world, e.g., a set of BGs representing facts about the children photo example (Sect. 1.1.1). A *query* made to this base is itself a BG, say Q. Elements answering Q are intuitively defined as the elements in B that entail Q, or equivalently, elements that are specializations of Q, or also, elements that are subsumed by Q.

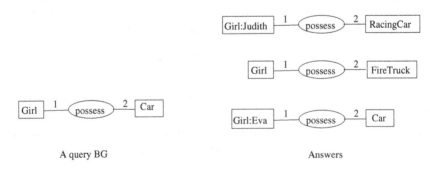

A query BG   Answers

Fig. 2.13 A query and possible answers

More precisely, a query, hence the notion of answer, can be interpreted in different ways:

- A query Q can be seen as a representation of a "yes/no question": *"Is the knowledge represented by Q asserted by the KB?"* For instance, the query in Fig. 2.13 may represent the question, "Is there a girl who owns a car?" or, "Is there a car owned by a girl?" Note that, as a BG does not need to be a connected graph, B can be seen as composing a single BG. The answer is boolean, which is true if and only if Q subsumes B.
- A query Q can be seen as a "pattern" allowing us to extract knowledge from the KB. Generic nodes in the query represent variables to instantiate with individual or generic nodes in the base. The query Q in Fig. 2.13 would become "find all patterns in which a girl owns a car." With this interpretation, each homomorphism from Q to B defines an answer to Q. An answer can be seen as the homomorphism itself, which assigns a node of the base to each node of the query. Or it can be seen as the subgraph of B induced by this homomorphism, i.e., the image of Q (see Fig.2.13 and also Fig. 2.12, where, assuming that the data form DF is used as a query, F can be seen as an answer to this query).

Note that distinct homomorphisms from Q to B may produce the same image graph; thus defining answers as image graphs instead of homomorphisms induces a potential loss of information. In turn, an advantage of considering image graphs is that the set of answers can be seen as a BG. We thus have the property that the results returned by a query are in the same form as the original data. This property would be mandatory if we were interested in processing complex queries, i.e., queries composed of simpler queries; in this context, the answers to a query would be seen as a knowledge base, which could be used to process another query.

A simple extension of this basic mechanism is classically considered. A query is not simply a BG but a BG with distinguished concept nodes defining the part to be considered to define an answer. These nodes are usually distinguished by a question mark (?) added to their marker. Then an answer is given by the part of the homomorphism having for domain the set of marked nodes, or by the associated image graph. With the example in Fig. 2.13, marking the node [Girl] would lead to the query "find all girls who own a car" and, if answers are given as image graphs, would return the graphs [Girl:Judith], [Girl] and [Girl:Eva]. We then have a mechanism equivalent to conjunctive queries in databases, see Chap. 5.

2.3 BG Subsumption Properties

2.3.1 Subsumption Preorder

Properties concerning homomorphisms can be directly proven from the definitions given above. For instance, it is simple to check the following property:

Property 2.4 (Composition of homomorphisms). The composition of two homomorphisms is a homomorphism; thus the subsumption relation $\succeq$ is transitive.

This property is also a direct corollary of two known properties: First, the composition of two graph homomorphisms is a graph homomorphism (the structure is
preserved). Secondly, order relations are transitive relations (the composition of two
decreases of labels is a decrease of labels).

$\succeq$ is also a *reflexive* relation since every BG maps to itself with the identity homomorphism. But $\succeq$ is not *antisymmetric*: Indeed, there are non-isomorphic BGs that
map to each other. For instance, the BGs G and H in Fig. 2.14 are non-isomorphic
whereas $G \succeq H$ and $H \succeq G$ (there is a homomorphism from H to G, which maps
both concept nodes with label l_1 to the same node in G). Figure 2.15 presents a more
complex example: Any BG in the figure subsumes any other.

Definition 2.12 (Hom-equivalence).   Two BGs G and H are said to be *hom-
equivalent* if $G \succeq H$ and $H \succeq G$, also denoted $G \equiv H$.

If $G \succeq H$ and G and H are not hom-equivalent, we say that G is strictly more general
than H and note $G \succ H$ (or $H \prec G$).

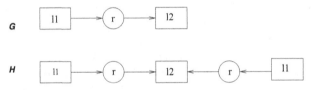

G

H

Fig. 2.14 Two hom-equivalent BGs

An injective homomorphism is a special case of a homomorphism enforcing the
fact that distinct nodes necessarily have distinct images. Let $G \succeq_i H$ denote the fact
that there is an *injective* homomorphism from G to H. In Fig. 2.14 one has $G \succeq_i H$
but not $H \succeq_i G$. It is simple to check that this relation is transitive, reflexive and
antisymmetric; thus $\succeq_i$ is an order relation.

Property 2.5. $\succeq$ is a preorder on the BGs defined over $\mathcal{V}$ (and it is not an order). $\succeq_i$
is an order on the BGs defined over $\mathcal{V}$.

Let $G \succeq_i H$, then there is a bijective homomorphism from G to a subBG of H.
If the injectivity of a homomorphism concerns only the concept nodes, the induced
binary relation is not an order as shown in Fig. 2.16, where $r \leq s$ (however, in this
case, checking "redundancy" becomes simple).

From a knowledge viewpoint, considering $\succeq_i$ instead of $\succeq$ can be pertinent whenever *two different concept nodes always represent two different entities*. From a complexity viewpoint, the decision problems "Is there a homomorphism from G to H,"
"Is there an injective homomorphism from G to H?" and "is there a bijective homomorphism from G to H?" are all NP-complete (cf. Sect. 2.6). Nevertheless, the two
relations $\succeq$ and $\succeq_i$ behave differently from a complexity viewpoint (for instance,

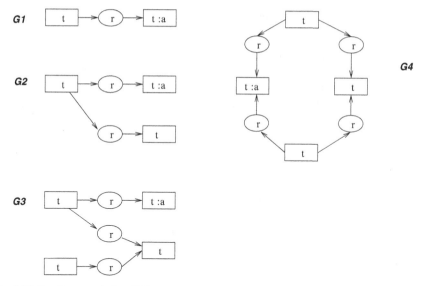

Fig. 2.15 Four hom-equivalent BGs

Fig. 2.16 Injectivity restricted to concept nodes ($r \leq s$)

given two BGs T and H, such that T is acyclic, checking whether $T \succeq H$ is a polynomial problem—see Chap. 7—whereas checking $T \succeq_i H$ remains NP-complete) and, contrary to homomorphism, injective homomorphism is not a complete operation with respect to the formal semantics of BGs studied in Chap. 4.

2.3.2 Irredundant BGs

In this section the equivalence relation associated with the subsumption preorder is studied. We show that any equivalence class has a distinct representative which is the "simplest" BG of the class and that this element is unique (up to isomorphism).

Definition 2.13 (Irredundant and redundant). A BG is called *redundant* if it is hom-equivalent to one of its strict subgraphs. Otherwise it is called *irredundant*.

Let G be a redundant BG. Let us consider a strict subBG H hom-equivalent to G and having a minimum number of nodes. H is irredundant. Thus,

Property 2.6. If a BG G is redundant, then it has an irredundant strict subBG hom-equivalent to it.

Let G be a BG and H a subBG of G. There is a trivial homomorphism from H to G, thus a BG is irredundant if and only if there is no homomorphism from it to one of its strict subBGs. In other words, there is no non-injective (or equivalently there is no non-surjective) homomorphism from G to itself. Equivalently, every homomorphism from G to itself is a bijective homomorphism. We have proven that any such homomorphism is in fact an isomorphism (Property 2.3). Thus we have:

Property 2.7. A BG G is irredundant if and only if every homomorphism from G to itself is an isomorphism (automorphism).

Example. In Fig. 2.14, G is irredundant while H is not; the BGs in Fig. 2.15 are all hom-equivalent, and G_1 is the only irredundant BG.

We now prove that a class of hom-equivalent BGs contains a unique irredundant BG. Thus this irredundant BG can be taken as the representative of the equivalence class. For this we use the following property.

Property 2.8. Let G be any BG and H be an irredundant BG. If G is hom-equivalent to H, then G has a subBG isomorphic to H.

Proof. If G and H are hom-equivalent, there is a homomorphism, say π, from H to G and a homomorphism, say π', from G to H. Now consider the homomorphism $\pi' \circ \pi$ from H to H. $\pi' \circ \pi$ is an isomorphism (by Property 2.7).

Let $G' = \pi(H)$. Let us show that π is an isomorphism from H to G'. π is surjective by definition of G', and π is injective since $\pi' \circ \pi$ is injective. Thus π is a bijective homomorphism from H to G'.

Now, for all concept or relation x in H, $label(x) \geq label(\pi(x)) \geq label(\pi' \circ \pi(x)) = label(x)$ thus $label(x) = label(\pi(x))$. π is thus an isomorphism from H to G', which proves the property.   □

Theorem 2.1. *Each hom-equivalence class contains a unique (up to isomorphism) irredundant BG which is the BG having the smallest number of nodes.*

Proof. Property 2.8 implies that two irredundant BGs are hom-equivalent if and only if they are isomorphic. Thus an equivalence class contains at most one irredundant BG. Now, given an equivalence class, let us take a BG, say G, of minimal size (with the size being the number of concepts and relations). G is irredundant, otherwise it would be hom-equivalent to one of its strict subBGs, which contradicts the hypothesis on G.   □

Example. In Fig. 2.15, each BG contains an irredundant subgraph isomorphic to G_1, to which it is hom-equivalent.

G may contain several irredundant subBGs hom-equivalent to it, but in this case they are all isomorphic to each other. Any of these subBGs can be taken as the *irredundant form* of G. The following property gives more insight into the relationships between a BG and its irredundant form: A redundant BG can be *folded* to its irredundant form.

Definition 2.14 (Folding). Let G be a BG, let π be a homomorphism from G to itself and let $G' = \pi(G)$. π is called a *folding* from G to G' if each node of G' is invariant by π, i.e. $\forall x \in C_{G'} \cup R_{G'}, \pi(x) = x$.

Property 2.9. Let G be a BG and let G' be one of its hom-equivalent irredundant subBGs. Then there is a folding from G to G'.

Proof. By hypothesis there is a homomorphism, say π, from G to G'. Let π' be the homomorphism obtained from π by restricting its domain to G': π' is an isomorphism (from Corollary 2.7). Then $\pi'' = \pi' \circ \pi$ is also a homomorphism from G to G', which is the identity on G'. π'' is thus a folding. $\square$

The folding notion is a direct adaptation for BGs of a classical notion in graph theory. Note that G' being irredundant is essential to this property: A BG cannot always be folded into any of its hom-equivalent subBGs.

If a BG G has duplicate relations, then it is redundant since the subgraph obtained by deleting a duplicate relation is hom-equivalent to it. More generally, if G has two relations with the same neighbors in the same order, such that one relation has a type greater or equal to the type of the other, then the relation with the more general type is redundant. In the absence of redundant relations, G is redundant if and only if there is a homomorphism from G to itself that maps two concepts to the same concept, say c_1 and c_2 to c_2; in this case there is a homomorphism from G to the subBG obtained by deleting c_1 and its neighboring relations. Finally, let us point out that rendering a BG irredundant is generally not a simple task. Indeed, the associated decision problem, namely "Given a BG G, is G redundant?", is an NP-complete problem, as we shall see in Sect. 2.6.

Considering irredundant BGs only, the subsumption relation becomes a special partial order—it is a lattice.

Property 2.10. Let $\mathcal{G}$ be the set of all irredundant BGs definable over a given vocabulary. Then $(\mathcal{G}, \preceq)$ is a lattice.

Proof. The set of BGs over a given vocabulary admits a greatest element: the empty BG. Should we exclude the empty BG, there is still a greatest element, the BG restricted to one generic concept with universal type. Now let us consider G and H two (irredundant) BGs and their disjoint sum $G + H$. G and H subsume $G + H$. And every BG K which is subsumed both by G and H is also subsumed by $G + H$ (the union of two homomorphisms from G to K and from H to K defines a homomorphism from $G + H$ to K). Thus the irredundant BG hom-equivalent to $G + H$ is the greatest lower bound of G and H. The set of irredundant BGs is thus an inf-semi-lattice, and since it possesses a greatest element, it is a lattice. $\square$

Note that the previous property is true even if the set of concept node labels is not a lattice. The only property required, if the empty BG is not admitted, is that the concept type set has a greatest element. There is no condition on the relation symbol set.

Example. In Fig. 2.17, the vocabulary given on the left is restricted to a concept type set, and it should be noted that this set is not a lattice. The lattice on the right represents all irredundant BGs constructible on this vocabulary, except for the empty BG.

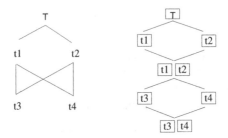

Fig. 2.17 A lattice of irredundant graphs

2.4 Generalization and Specialization Operations

Transforming a BG into its image by a homomorphism is a global operation between BGs that can be decomposed into simpler operations. Sets of elementary operations "equivalent" to homomorphism—in the sense that there is a homomorphism from *G* to *H* if and only there is a sequence of elementary operations transforming *G* into *H*—are presented in this section. Some of these elementary operations are frequently used in applications (e.g., the join in natural language processing). More complex operations can be defined using these elementary operations (e.g., extended join and maximal join defined in Chap. 8). Finally, these elementary operations are used in order to inductively define BGs (cf. Chap. 8).

There are five elementary generalization operations and five inverse operations, called elementary specialization operations.

2.4.1 Elementary Generalization Operations for BGs

Any generalization operation is a "unary" operation, i.e it has a BG (and some elements of it) as input and has a BG as output.

Definition 2.15 (Generalization operations). The five elementary generalization operations are:

- **Copy.** Create a disjoint copy of a BG *G*. More precisely, given a BG *G*, $copy(G)$ is a BG which is disjoint from *G* and isomorphic to *G*.
- **Relation duplicate.** Given a BG *G* and a relation *r* of *G*, $relationDuplicate(G, r)$ is a BG obtained from *G* by adding a new relation node *r′* having the same type

and the same list of arguments as r, i.e., the same neighbors in the same order. Two such relations of the same type and having exactly the same neighbors in the same order are called *twin relations*.

- **Increase.** Increase the label of a node (concept or relation). More precisely, given a BG G, a node x of G, and a label $L \geq l(x)$ $increase(G, x, L)$ is the BG obtained from G by increasing the label of x up to L, that is its type if x is a relation, its type and/or its marker if x is a concept.
- **Detach.** Split a concept into two concepts. More precisely, let c be a concept node of G and $\{A_1, A_2\}$ be a partition of the edges incident to c, $detach(G, c, A_1, A_2)$ is the BG obtained from G by deleting c, creating two new concept nodes c_1 and c_2, having the same label as c, and by attaching A_1 to c_1, and A_2 to c_2 (A_1 or A_2 may be empty).
- **Substract.** Given a BG G, and a set of connected components $C_1, \ldots, C_k$ of G, $substract(G, C_1, \ldots, C_k)$ is the BG obtained from G by deleting $C_1, \ldots, C_k$ (the result may be the empty graph).

Remark

Copy and relation duplicate are actually equivalence operations, i.e., they do not change the semantics of the graph (cf. Chap. 4).

The *copy* operation is in fact not needed in generalization operations since a copy of a graph can be obtained by a Relation Duplicate sequence followed by a Detach sequence ended by a Substract. It is considered as a generalization operation for symmetry reasons with specialization operations.

The copy operation is also needed because, in some situations, we do not consider BGs up to isomorphism. For instance, in implementation and drawings two copies of the same BG must be considered as two different BGs.

The substract operation is defined as the deletion of a set of connected components, and not the deletion of only one component, in order to be the inverse of the disjoint sum, which is an elementary specialization defined later.

Example. The *relation duplicate* operation is illustrated in Fig. 2.18. The *detach* operation is illustrated in Fig. 2.19.

It is now possible to precisely define what *a BG G is a generalization of a BG H* means:

Definition 2.16 (Generalization). A BG G is a generalization of a BG H if there is a sequence of BGs $G_0 = (H), G_1, \ldots, G_n = (G)$, and, for all $i = 1, \ldots, n$, G_i is obtained from G_{i-1} by a generalization operation.

Example.

Figure 2.20 presents some of the graphs occurring in a generalization sequence from H to G. H_1 is obtained from H by splitting the nodes [Child:Paul] and [FireTruck] (Detach operation), then deleting the connected component K pictured in the dashed rectangle (Substract operation). H_2 is obtained from H_1 by duplicating the (on) relation (Relation Duplicate operation). H_3 is obtained from H_2 by splitting

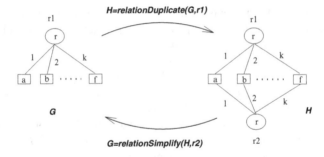

Fig. 2.18 Relation duplicate and its inverse Relation simplify

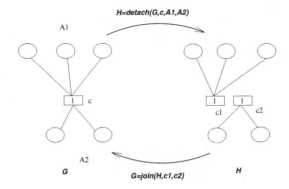

Fig. 2.19 Detach and its inverse Join

the node [FireTruck] (Detach operation). Finally, H_4 isomorphic to G is obtained by increasing some labels (Increase operation). This generalization sequence can be associated with the homomorphism h_1 in Fig. 2.10. The equivalence between a homomorphism and a generalization sequence is stated in next properties (Property 2.12 and Property 2.13).

Property 2.11. A subBG G' of a BG G is a generalization of G.

Proof. Any boundary concept of G' can be split into c_1 and c_2 in the following way: The neighbors of c_1 are the neighbors of c, which belong to G', and the neighbors of c_2 are the neighbors of c, which do not belong to G'. Then, all connected components having a node outside G' are removed, and the remaining BG, which has been obtained from G by generalization, is precisely G'. □

2.4.2 Generalization and Homomorphism

In this section, we prove that G subsumes H if and only if G is a generalization of H. Thus, homomorphism, which is a global notion (it is a mapping between sets),

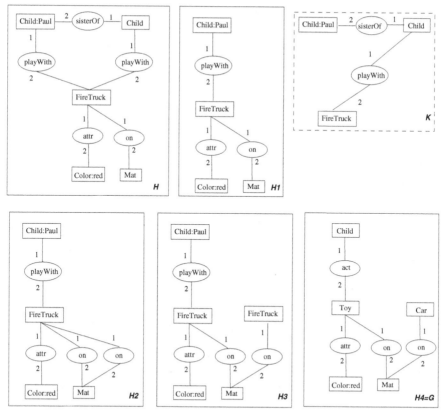

Fig. 2.20 Generalization from H to G

can be replaced by a sequence of simple operations: Copy and substract are simple to understand, and the three others are local. Furthermore, the five operations are simple to draw.

Property 2.12. If a BG G is a generalization of a BG H, then there is a homomorphism from G to H ($G \succeq H$).

Proof. One first proves that if G is obtained from H by an elementary generalization operation then there is a homomorphism from G to H. It is trivial for the copy operation since an isomorphism is a homomorphism.

- *Relation duplicate.* Let r' be a relation duplicate of a relation r. The mapping π defined as follows is a homomorphism:
  $\pi(r') = r$ and $\forall x \neq r$, x node of G, $\pi(x) = x$.

- *Increase.* Let $G = increase(H, x, L)$. The mapping which associates to each node of G its corresponding node in H is a homomorphism (the only difference between G and H is that $l_G(x) \geq l_H(x)$).

- *Detach.* Let c_1 and c_2 be the concepts resulting from the split of c. The mapping π defined as follows is a homomorphism:
  $\pi(c_1) = \pi(c_2) = c$ and $\forall x \in C_G \cup R_G \setminus \{c_1, c_2\}$, $\pi(x) = x$.

- *Substract.* G is equal to a subBG of H and the identity is a homomorphism.

As the composition of two homomorphisms is a homomorphism, one concludes by recurrence on the length of a generalization sequence. □

Property 2.13. If there is a homomorphism from a BG G to a BG H, then G is a generalization of H.

Proof. Let π be a homomorphism from G to H. We build a sequence of generalization operations from H to G as follows.

1. We first build a generalization sequence from H to $\pi(G)$. As $\pi(G)$ is a subBG of H, property 2.11 can be applied: The sequence is composed of a sequence of Detach on the boundary nodes of $\pi(G)$ followed by a Substract.
2. for every relation r such that $|\pi^{-1}(r)| = 1$, say $\pi^{-1}(r) = \{r'\}$, the label of r is increased to that of r'.
3. for every relation r such that $\pi^{-1}(r) = \{r_1, \ldots, r_k\}$, with $k \geq 2$, r is duplicated $k - 1$ times into $s_1(= r), \ldots, s_k$, and for any i, the label of s_i is increased to that of r_i.
4. for every concept c such that $|\pi^{-1}(c)| = 1$, say $\pi^{-1}(c) = \{c'\}$, the label of c is increased to that of c'.
5. for every concept c such that $\pi^{-1}(c) = \{c_1, \ldots, c_k\}$, with $k \geq 2$, c is split by a Detach into k concepts $d_1, \ldots, d_k$, and, for any i, the edges incident to d_i are the images by π of the edges incident to c_i. Furthermore, for any i, the label of d_i is increased to $l_G(c_i)$. We obtain (a BG isomorphic to) G.
   □

The two previous properties give the following theorem, which justifies the term *generalization* for the existence of both a homomorphism and a sequence of elementary generalization operations.

Theorem 2.2 (Homomorphism and generalization). *There is a homomorphism from a BG G to a BG H ($G \succeq H$) if and only if G is a generalization of H.*

Example. In Fig. 2.20 some graphs of a generalization sequence from a BG H to a BG G are pictured. This sequence corresponds to the homomorphism h_1 from G to H presented in Fig. 2.10. It follows the proof steps of Property 2.13, except that all labels are increased at the last step instead of during the construction. More specifically, H_1 is equal to $h_1(G)$; its construction thus corresponds to step 1 of the proof. Step 2 is delayed. The construction of H_2 corresponds to step 3, with the increase in labels being delayed. Similarly, the construction of H_3 corresponds to step 4, with the increase in labels being delayed. Finally, the construction of H_4 gathers all label increases.

2.4.3 Elementary Specialization Operations

The elementary specialization operations defined in this section are inverse (up to an abuse of language explained hereafter) operations of the elementary generalization operation defined previously. Besides the copy operation, there are three unary operations and one binary operation.

Definition 2.17 (Elementary specialization operations). The five elementary specialization operations are:

- **Copy** (already defined as a generalization operation).
- **Relation simplify.** Given a BG G, and two twin relations r and r' (relation with the same type and the same list of neighbors), $relationSimplify(G, r')$ is the BG obtained from G by deleting r'.
- **Restrict.** Given a BG G, a node x of G, and a label $l \leq l(x)$ $restrict(G, x, l)$ is the BG obtained from G by decreasing the label of x to l, which is its type if x is a relation, its type and/or its marker if x is a concept.
- **Join.** Given a BG G, and two concepts c_1 and c_2 of G with the same label, $join(G, c_1, c_2)$ is the BG obtained from G by merging c_1 and c_2 in a new node c, i.e. the edges incident to c are all edges incident to c_1 and to c_2 and c_1 and c_2 are deleted.
- **Disjoint sum.** Given two disjoint BGs $G = (C, R, E, l)$ and $H = (C', R', E', l')$, $G + H$ is the union of G and H that is the BG $(C \cup C', R \cup R', E \cup E', l \cup l')$. G or H may be empty.

It is noted in the definition of the *disjoint sum* operation that G or H may be empty. This specific case is needed in order to obtain any BG from the empty BG $G_\emptyset$, with $G_\emptyset$ being more general than any BG. The Relation simplify and Join operations are pictured in Fig. 2.18 and Fig. 2.19, together with their inverse generalization operations, respectively Relation duplicate and Detach.

Example. In reading Fig. 2.20 from G to H, one obtains the sketch of a specialization sequence. This sequence corresponds to the homomorphism h_1 from G to H presented in Fig. 2.10. More specifically, H_3 is obtained from G by four restrict operations. H_2 is obtained from H_3 by a join of the concept nodes with label $(FireTruck, *)$. A relation simplification of a relation with label (on) in H_2 yields $H_1 = h_1(G)$. Finally, a disjoint sum of H_1 and K followed by two joins on the concept nodes with labels, respectively $(Child, Paul)$ and $(FireTruck, *)$, produces H.

It is straightforward to check the following relationships between elementary generalization operations and elementary specialization operations:

- If H is obtained from G by the deletion of a relation r', twin of a relation r, then G is isomorphic to a BG obtained from H by duplicating r, and conversely.
- If H is obtained from G by restricting to b the label a of a node x, then G can be obtained from H by increasing the label of x from b to a, and conversely.

- If H is obtained from G by joining two nodes x and y into a new node xy, then G is isomorphic to a BG obtained from H by detaching xy into two nodes x' and y' having the same neighboring as x and y in G, and conversely.
- If H is equal to the disjoint sum $G + K$, then G can be obtained from H by a substract operation.

Thanks to these relationships, we will make an abuse of language whereby we say that elementary specialization operations are inverse of elementary generalization operations.

The direct definition of a specialization notion inverse to the definition of a generalization (cf. Definition 2.16) for specialization operations is not so simple. This is due to the operation disjoint sum which is a binary operation which has one new input BG. Thus, instead of a *sequence*, we will first consider a *tree*, as illustrated in Fig. 2.21 and defined below.

Definition 2.18 (Specialization tree). A specialization tree of a BG G is an anti-rooted tree T whose nodes are labeled by BGs in such a way that:

- T has an anti-root labeled by G.
- Let x be any node with label H, then

  − x is a leaf (i.e., x has no predecessors), or
  − x has exactly one predecessor labeled K, and H is obtained from K by one of the unary specialization operations, or
  − x has exactly two predecessors, respectively labeled K_1 and K_2, and $H = K_1 + K_2$.

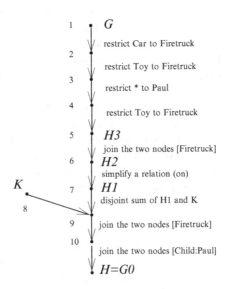

Fig. 2.21 The specialization tree corresponding to Fig. 2.20

The transitive closure of an anti-rooted tree is a partial order, and any total ordering of the nodes containing that partial order is called a linear extension of the partial order.

Definition 2.19 (Specialization). Given a specialization tree of a BG G, the sequence of BGs labeling any linear extension of the tree is called a specialization sequence of G. G is said to be a *specialization* of any BG H appearing in a specialization sequence of G.

Example. In Fig. 2.21, the vertex set of the specialization tree is $\{1, ..., 11\}$. There are several linear extensions of this tree which differ only by the rank of the vertex 8 labeled by K. In any case, this node has to appear before node 9. For instance the linear extension $(1 ...7\ 8\ 9\ 10\ 11)$ yields a specialization sequence of form $(G = H_4, ..., H_3, , H_2, H_1, K, ..., H)$.

Property 2.14. If G and H are two BGs, then H is a specialization of G if and only if G is a generalization of H

Proof. Let us consider a specialization sequence from G to H. Then, it is possible to transform this sequence into a generalization sequence from H to G by replacing each elementary specialization operation by an elementary generalization operation. Conversely, any generalization sequence from H to G can be transformed into a specialization sequence from G to H. □

In some situations, especially when one wants to actually build a BG by a sequence of elementary specialization operations, it is interesting to consider a specialization graph and not only a tree. Let us assume, for instance, that there is a set of BGs $\mathcal{B}$ used as building blocks (cf. Chap. 8). In other words, the leaves of the specialization tree can be labeled only by elements of $\mathcal{B}$. Then imagine that, to build a certain BG, one needs twice the "same" subBG G' (e.g., this BG has two isomorphic disjoint subgraphs). Assume that a first subBG G' has been built and G'' isomorphic to G' is needed. If a specialization tree is considered, a second graph has to be built from the building blocks in the same way as G'. In a specialization graph, G'' can be simply obtained by copying G'.

Definition 2.20 (Specialization graph). A specialization graph of a BG H is a directed graph without circuits D whose nodes are labeled by BGs in such a way that:

- D has an anti-root labeled by H.
- Let x be any node with label G, then

    - x is a source (i.e., x has no predecessors), or x has exactly one predecessor labeled K, and G is obtained from K by one of the unary specialization operations, or x has exactly two predecessors, respectively labeled K_1 and K_2, and $G = K_1 + K_2$,
    - x may have several successors, but at most one does not correspond to a copy operation.

Property 2.15. Given any specialization graph D of H, there is a specialization tree T of H with the same set of leaf labels.

Proof. Let us consider a specialization graph D with its anti-root labeled by H and the set of its leaf labels equal to $\{L_1,\dots,L_p\}$. Let us prove the property by recurrence on the number of nodes having at least two successors. If D has no node with two or more successors, then D is a tree. Otherwise, let us consider a node x of D having at least two successors such that all the predecessors of x have exactly one successor. Let the successor set of x be $\{y_1,\dots,y_k\}$, $k \geq 2$, with $y_2,\dots,y_k$ being obtained from x by a copy operation. Let us denote by D_x the subgraph of D induced by all the ascendants of x (x included). The graph D' obtained from D by creating $k-1$ disjoint copies of D_x with anti-roots $x_2,\dots,x_k$, respectively, and by replacing each arc from x to y_i by the arc joining x_i to y_i, for $i=2,\dots,k$, is a specialization graph with its anti-root labeled by H and its leaf label set being equal to $\{L_1,\dots,L_p\}$. Furthermore, the number of nodes of D' having at least two successors is equal to the number of nodes of D having at least two successors minus one. One concludes by using the recurrence hypothesis. $\square$

2.4.4 Specialization and Homomorphism

From Property 2.14 and Theorem 2.2, one obtains:

Theorem 2.3 (Homomorphism and Spec./Gen.). *Let G and H be two BGs. The three following propositions are equivalent:*

1. *G is a generalization of H*
2. *H is a specialization of G*
3. *there is a homomorphism from G to H, i.e. $G \succeq H$*

Let us consider a homomorphism π from G to H. We have shown how it is possible to build a sequence of elementary generalization operations transforming H into G (cf. proof of property 2.13). Using the duality between generalization and specialization operations, it is possible to transform this sequence into a sequence of elementary specialization operations transforming G into H (Property 2.14). For the sake of completeness, we briefly indicate how a specialization sequence transforming G into H can be directly built from π (i.e., without considering a generalization sequence from H to G). Note that this sequence is not the exact inverse of the generalization sequence given in the proof of Property 2.13 because the order in which label modifications naturally occur are not the same.

1. For every concept or relation node x in G, the label of x is restricted to the label of $\pi(x)$. We obtain G_1. Now, all nodes with the same image by π have the same label.
2. All concept nodes in G_1 with the same image in H are joined. More precisely, for every concept node c' in H such that $\pi^{-1}(c') = \{c_1,\dots,c_k\}$, $k \geq 2$, all the c_i

are joined. We obtain G_2. Now, all relations with the same image by π are twin relations.

3. A sequence of relation simplifications is performed on G_2 to keep only one relation per set of relations with the same image in H. More precisely, for every relation r' in H such that $\pi^{-1}(r') = \{r_1, \ldots, r_k\}$, $k \geq 2$, $(k-1)$ relation simplications on r_2, ..., r_k (for instance) are performed. The graph obtained, i.e., G_3, is isomorphic to $\pi(G)$.
4. Let us consider the BG K obtained from H by deleting all nodes of $\pi(G)$ that are not boundary nodes. A disjoint sum of G_3 and K yields G_4.
5. Finally, a sequence of join operations are applied on G_4: Each boundary node of G_3 is joined to its corresponding node in K. The BG obtained is (isomorphic to) H.

Example. Let us consider Fig. 2.20, which sketches a generalization sequence from H to G guided by the homomorphism h_1 in Fig. 2.10. Reading this figure in the inverse direction, one obtains the sketch of a specialization sequence from G to H built from h_1 as above. More specifically, H_3, H_2 and H_1 respectively correspond to the BG G_1, G_2 and G_3 built at the above steps 1, 2 and 3. The union of H_1 and K corresponds to G_4. Finally, H is obtained.

2.5 Normal BGs

2.5.1 Definition of Normal BGs

Two concept nodes having the same individual marker represent the same entity. This can be seen as the simplest kind of equality. Introducing the equality between any kind of concept nodes (i.e., not only individual concepts but also generic concepts) is the main objective of the next chapter, and the present section can be considered as an introduction to this chapter.

If a BG has several individual concept nodes with the same marker, we would like to be able to merge these nodes into a single individual concept, while keeping the same intuitive meaning (and the same formal semantics, see Chap. 4). Note first that, even when there is an obvious way of merging the concepts, the obtained BG is generally not hom-equivalent to the original BG. This point is illustrated by Fig. 2.22. Let us consider the BG G, in which the individual marker m appears twice. The nodes to be merged have exactly the same label, thus merging them simply consists of performing a Join operation. Let H be the resulting BG. G and H are obviously semantically equivalent, but H is strictly more specific than G for the subsumption relation (indeed there is a homomorphism from G to H but no homomorphism from H to G). Now, assume for instance, that G represents an assertion and let Q be the query pictured in the figure. There is a homomorphism from Q to H but not to G, thus H answers Q while G does not.

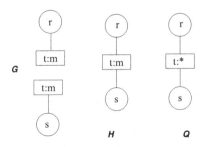

Fig. 2.22 Homomorphism and normality

Thus, BGs having several individual concept nodes with the same marker do not have a *normal* behavior with respect to the subsumption relation, and this naturally leads to the following definition of normal BGs:

Definition 2.21 (Normal BG). A BG is called *normal* if there is at most one individual concept with a given marker, i.e., all individual concepts have distinct markers.

Example. H and Q in Fig. 2.22 are normal while G is not. G in Fig. 2.23 is not normal because the individual *Paul* occurs in two concept nodes, and $norm(G)$ is a normal BG semantically equivalent to G.

A second issue concerns the conditions under which any BG can be transformed into a semantically equivalent normal BG. Let P_c denote the partition of the set of concept nodes of a BG defined as follows: Any generic concept defines a class restricted to itself, and all the individual concepts with the same marker define a class. A BG is normal if and only if its partition P_c is the discrete partition. If a BG is not normal, then one can consider its quotient graph G/P_c defined as follows:

Definition 2.22 (G/P_c). Let $G = (C, R, E, l)$ be a BG. G/P_c is the graph defined as follows:

- its set of concept nodes is in bijection with P_c,
- its set of relation nodes is in bijection with R,
- for any (r, i, c) in G, (r', i, c') is in G/P_c, where r' is the relation node corresponding to r, and c' corresponds to the class containing c.

In order to transform the graph G/P_c into a BG, a label has to be defined for the classes of P_c. The label of a node corresponding to a trivial class of P_c is the label of the unique node of that class. The marker associated with a non trivial class is obviously the individual marker appearing in all nodes of this class. But, if no conditions are enforced on the types of individual concepts composing the class, there is no garantee that a type preserving semantic equivalence can be defined.

We shall now enumerate some simple conditions, but, as we shall see, none is completely satisfactory from a knowledge representation viewpoint.

- **Unicity condition**
  One can impose that all concepts with the same individual marker have the same

type. Then the label of the node of G/P_c corresponding to a non-trivial class is the label of any concept in this class. In this case, G and G/P_c have the same intuitive meaning.

Example. By merging c_1 and c_2 in G in Fig. 2.23, the normal BG $norm(G)$ is obtained.

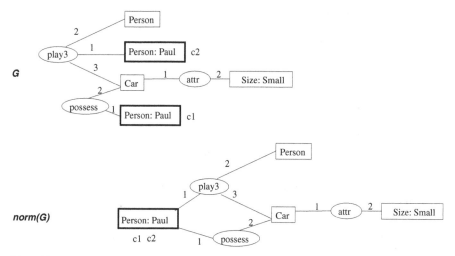

Fig. 2.23 Normalization of a BG

In such cases, G/P_c is a normal BG that is called the *normal* form of G. The notation $norm(G)$ is also used. A canonical homomorphism from G to G/P_c can be defined. This homomorphism associates to any concept node c the node corresponding to the class containing this node, and to any relation its corresponding relation.

But allowing individual concepts with the same marker and different types could be useful. For instance, let us consider the following situation where #123 is a pick-up car. One would like to represent specific characteristics of the pickup #123 (e.g., number of kilometers in the bush, composition of first-aid kit, state of the tires) by an individual concept of type *Pickup* and marker #123, and to represent administrative data of the car #123 (e.g., names and addresses of the successive owners) by another individual concept of type Car and the same individual marker. In this situation, as *Pickup* is a subtype of *Car*, these two individual concepts could be merged into a single individual concept of type *Pickup* while keeping the same semantics. This leads to the following condition.

- **Minimality condition**

  One can relax the unicity condition of types and replace it by a minimality condition, i.e., for any class A of P_c the set T_A of types of concepts in A has a minimal element $min(T_A)$ (within T_A). In this case, G and G/P_c have the same intuitive meaning (with the type of concept in G/P_c associated with the class A being equal to $min(T_A)$). But this minimality condition is still too strong. Indeed, let us consider a BG having a class $A = \{c, c'\}$ of P_c with labels $l(c) = (t, m)$ and $l(c') = (t', m)$, and t and t' incomparable. The minimality condition is not respected, and the two individual concepts cannot be merged, even if m actually represents an entity of type t'' less than t and t'. But what happens if $\{t, t'\}$ has several minimal elements (within T_C)?

This leads to the following condition.

- **Greatest lower bound condition**

  This condition consists of requiring that the set of all concept types of any class of P_c has a greatest lower bound (glb) in T_C. Hence, as P_c is relative to a specific BG, generalizing this glb condition to any BG leads us to assume that T_C is an inf-semi-lattice. But in this case, if one wants to keep the BG semantic, one must assume the following property: If m is a t and m is a t', then m is a $glb(t, t')$. This lattice-theoretic interpretation of the concept type set is discussed in the next chapter.

- **Typing of individual markers**

  Let us recall that a typing of the individual markers is a mapping τ from the set of individuals $\mathcal{I}$ to the set of concept types T_C. For an individual m, all occurrences of individual concept nodes with marker m must have a type greater than or equal to $\tau(m)$. Having a typing of individual markers comes down to the unicity condition. Indeed, before any computation with a BG, all types of individual concepts with marker m can be replaced by $\tau(m)$.

  In the Simple Conceptual Graphs model presented in the next chapter, we will see how this problem is tackled and how it is possible to explicitly express that several concepts, either generic or individual, represent the same entity. Thus the simple conceptual graph model can be considered as the basic conceptual model enriched by equality between concepts.

2.5.2 Elementary Operations for Normal BGs

The subsumption relation, defined through the homomorphism notion, is the fundamental reasoning tool in conceptual graphs. Thus, normal BGs, which have a nice behavior relative to homomorphism, are the fundamental class of BGs (and of the simple conceptual graphs studied in the next chapter). Previous generalization and specialization operations correspond to homomorphisms between BGs. But can we take these operations as reference operations on normal BGs? The only flaw is that, as such, they are not internal operations on normal BGs since a non-normal BG can be produced from a normal BG. Moreover, given two normal BGs G and H with $G \succeq H$, there may be no way of deriving one from the other without producing a non-normal BG somewhere in the derivation (cf. Fig. 2.24).

If we consider generalization, the guilty operation is the operation *detach* (when it splits an individual node). Concerning specialization, there are two guilty operations, *restrict* (when it restricts a generic concept to an individual concept, whereas

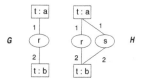

Fig. 2.24 Every specialization sequence from G to H contains a non-normal BG

the individual marker already appears elsewhere in the BG) and *disjoint sum* (when there are individual concepts with the same label in the two BGs).

Let us consider a situation where it is possible to merge individual concepts having the same marker. Then the operations *detach*, *restrict* and *disjoint sum* can be slightly transformed as follows:

- **detach$_{norm}$.** Modify *detach* as follows: If c is an individual concept, replace the marker of c_2 by a generic marker (which involves an increase).
- **restrict$_{norm}$.** Modify *restrict* as follows: Let x be a generic concept restricted to an individual node; if there is already a concept with the same individual marker, merge these concepts (which involves a restrict of the greater type to the smaller, if the types of the concepts are not equal, followed by a join).
- **disjoint-sum$_{norm}$.** Modify *disjoint sum* as follows: A disjoint sum operation is followed by the merging of the individual concepts having the same marker.

Note that the modified generalization and specialization operations are internal operations on normal BGs. It remains to be shown that they define the same relationship between normal BGs as the original ones.

Property 2.16. Let G and H be two *normal* BGs. G is a generalization of H using detach$_{norm}$ instead of detach (notation $G \succeq_{norm} H$) if and only if there is a homomorphism from G to H.

Proof. Similar to the proofs of Property 2.12 and Property 2.13 for BGs. Each generalization operation yielding G_j from G_i defines a homomorphism from G_j to G_i, thus by composition of all homomorphisms associated with a derivation from H to G, one obtains a homomorphism from G to H. Reciprocally, let us consider the proof of Property 2.13: detach$_{norm}$ preserves the property that any subBG of H is a generalization of H (Property 2.11), thus the proof still holds for $\succeq_{norm}$. $\square$

Property 2.17. Let G and H be two *normal* BGs. H is a specialization of G (notation: $H \preceq_{norm} G$) using restrict$_{norm}$ instead of restrict and disjoint-sum$_{norm}$ instead of disjoint sum if and only if $G \succeq_{norm} H$.

Proof. Similar to the proof of Property 2.14.

If G_j is obtained from G_i by a relation duplicate operation, then G_i is obtained from G_j by a relation simplify operation.

If G_j is obtained from G_i by a detach$_{norm}$ and the marker m of the individual node c_2 becomes a generic marker, then G_i can be obtained from G_j by a restrict$_{norm}$ of

c_2 to the marker m.

If G_j is obtained from G_i by an increase operation, then G_i is obtained from G_j by a restrict operation, therefore by a restrict$_{norm}$ operation.

If G_j is obtained from G_i by a substract operation, then G_i is obtained from G_j by a disjoint sum, therefore by a disjoint-sum$_{norm}$ operation.

It is straightforward to check the other direction. □

Corollary 2.1. *Let G and H be two* normal *BGs. $H \preceq_{norm} G$ if and only if $G \succeq_{norm} H$.*

Theorem 2.4 ($\preceq_{norm}$) and homomorphism). *Let G and H be two* normal *BGs. The three following propositions are equivalent:*

1. $H \preceq_{norm} G$
2. $G \succeq_{norm} H$
3. there is a homomorphism from G to H.

2.6 Complexity of Basic Problems

Let us consider the following three basic problems:

- BG-HOMOMORPHISM: given two BGs G and H, is there a homomorphism from G (the source) to H (the target)?
- HOM-EQUIVALENCE: given two BGs, are they equivalent?
- REDUNDANCY: given a BG, is it redundant?

We will show that they are all NP-complete. Let us first stress two points. First, the fact that these problems are NP-complete does not mean that they cannot be efficiently solved in practice. Chapter 6 is devoted to algorithmic techniques for finding homomorphisms. Chapter 7 is devoted to particular cases in which BG-HOMOMORPHISM can be solved in polynomial time.

Secondly, the really fundamental problem is BG-HOMOMORPHISM, and that explains why we focus on it throughout this book. Indeed, let us assume that we have an algorithm for solving BG-HOMOMORPHISM; checking whether two BGs are hom-equivalent can be done with two calls to this algorithm, and checking whether a graph G is redundant can be done with a number of calls to this algorithm, which is linear in the size of G. In the trivial case where G has a redundant relation node (i.e., a relation node with type t such that there is another relation node with the same argument list and type $t' \leq t$), it is redundant. Otherwise, G is redundant if and only if it can be mapped to one of its subgraphs $G - \{x\}$ obtained by deleting a concept node x and all relation nodes incident to it. There are at most $|C_G|$ such subgraphs. Thus, an efficient algorithm for BG-HOMOMORPHISM directly yields an efficient algorithm for HOM-EQUIVALENCE and REDUNDANCY.

In practice, we are interested in computing the irredundant form of a graph, and not only in checking whether it is redundant. Assume that we have an algorithm able to exhibit a homomorphism from G to H if any. A simple algorithm for computing the irredundant form of G is as follows: First, delete redundant relation nodes in G; secondly, check if there is a homomorphism from G to one of its subgraphs $G - \{x\}$; if no, G is its own irredundant form; if yes, let π be a homomorphism from G to $G - \{x\}$: The irredundant form of G is the irredundant form of $\pi(G)$.

Theorem 2.5. BG-HOMOMORPHISM *is NP-complete.*

Proof. BG-HOMOMORPHISM is obviously in NP. To prove the completeness we build a reduction from the well-known 3-SAT problem. The input of 3-SAT is a propositional formula f in 3-conjunctive normal form (3-CNF), i.e., a conjunction of disjunctions (or clauses), each with three literals, and the question is whether there is a truth assignment of the variables in f such that f is true.

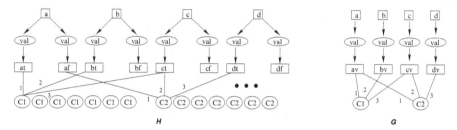

Fig. 2.25 Example of transformation from 3-SAT to BG-HOMOMORPHISM

Let $f = C_1 \wedge \ldots \wedge C_k$ be an instance of 3-SAT. Without loss of generality, then we suppose that a variable appears at most once in a clause. Let us create four concept types for each variable x: x, xf, xt and xv. We also create one relation type C_i for each clause C_i, and a relation type val. Each concept type xv is greater than xt and xf, these are the only possible comparisons between distinct types.

We build the graph $H(f)$ as follows: For every variable x in f, we have three concept nodes [x], [xt] and [xf] in $H(f)$ and two relation nodes typed val linking the first to the latter ones (intuitively, this means that the variable x can be valuated by true or false). Let us say that the truth value true (resp. false) is associated with [xt] (resp. [xf]). Then for every clause $C_i = (l_x \vee l_y \vee l_z)$ in f (where l_x, l_y and l_z are literals over variables x, y and z), we add the 7 relation nodes typed C_i, having as first argument [xt] or [xf], as second argument [yt] or [yf], and as third argument [zt] or [zf], that correspond to an evaluation of the clause to true (more precisely, if we replace, in the clause C_i, each positive (resp. negative) literal l_j, $1 \le j \le 3$, by the truth value (resp. the negation of the truth value) associated with the *jth* neighbor of the relation node, C_i is evaluated to true).

In graph $G(f)$, two concept nodes [x] and [xv] are created for each variable x and they are linked by a binary relation (val). For each clause $C_i = (l_x \vee l_y \vee l_z)$, there is one relation (C_i) linked to [xv], [yv] and [zv]. This question means "Is there a valuation of variables such that all clauses evaluate to true?"

This transformation from the 3-SAT formula $(a \vee b \vee \neg c) \wedge (\neg a \vee c \vee \neg d)$ is illustrated in Fig. 2.25. In graph H, not all edges issued from the clauses have been drawn, for readability reasons. It is immediate to check that, for a formula f, there is a valuation of its variables such that each clause is evaluated to true if and only if $G(f)$ can be mapped into $H(f)$. $\square$

Chapter 5 provides other polynomial reductions to BG-HOMOMORPHISM, in particular from GRAPH-HOMOMORPHISM, CONJUNCTIVE-QUERY-CONTAINMENT, CONJUNCTIVE-QUERY-EVALUATION and CSP (Constraint Satisfaction Problem).

Deciding whether two BGs G and H are hom-equivalent or checking whether a graph is redundant could be problems that are simpler than BG-HOMOMORPHISM, but they are not:

Theorem 2.6. HOM-EQUIVALENCE *and* REDUNDANCY *are NP-complete.*

Proof. Let (G,H) be an instance of BG-HOMOMORPHISM. It is easily checked that there is a homomorphism from G to H if and only if H is hom-equivalent to $G+H$ (the disjoint sum of G and H), hence we have an immediate reduction from BG-HOMOMORPHISM to HOM-EQUIVALENCE. Let us now build a reduction from BG-HOMOMORPHISM to REDUNDANCY. Without loss of generality we assume that G and H have no redundant relation nodes. We build G' and H' from G and H, respectively, by adding some "gadgets" such that (1) G' and H' are irredundant, (2) there is no homomorphism from H' to G', and (3) there is a homomorphism from G to H if and only if there is a homomorphism from G' to H'. For instance, let r and s be binary relation types that do not occur in G and H and that are incomparable to all other relation types. We obtain G' from G by adding a relation node of type r between any two distinct concept nodes x and y in G. We obtain H' from H by adding a relation node of type r between all concept nodes x and y in H, where x and y may be the same node, and a relation node of type s between all distinct x and y in H. Check that G' and H' satisfy conditions (1), (2) and (3). Because of (1) and (2), their disjoint sum $G' + H'$ is redundant if and only if there is a homomorphism from G' to H', thus, from (3), if and only if there is a homomorphism from G to H. $\square$

2.7 Bibliographic Notes

In his seminal book [Sow84], Sowa first settled a simple model and progressively added more complex notions. This simple model corresponds to the "basic conceptual graphs" studied in this chapter. In [CM92] a first study of its properties was conducted. This basis has evolved over the years. The definitions and results of this chapter integrate this evolution, but essentially rely on [Sow84] and [CM92].

The "vocabulary" defined in Sect. 2.1.1 is an evolution of the "support" in [CM92], itself based on the semantic network in [Sow84]. Originally, the concept type set was a lattice and relation symbols were not ordered. In addition, a "conformity relation" enforced constraints on labels of individual concept nodes. It is equivalent to the individual typing mapping τ mentioned at the end of Sect. 2.1.1.

In Sowa's book, basic conceptual graphs were connected graphs. Specialization operations, called canonical formation rules, were introduced. It was shown that a mapping π from G to H, called a projection, could be associated with every specialization sequence from G to H. In [CM92] projection was identified as a graph homomorphism and its equivalence with a specialization sequence was stated, by proving the reciprocal property: To every projection a specialization sequence can be associated. Generalization operations dual of specialization operations were also introduced. Equivalence between projection/homomorphism, generalization and specialization was thus proved (Theorem 2.2 and Theorem 2.3). Specialization and generalization rules of the present book basically come from the extension of [CM92] to non-connected graphs in [MC96] (research report [CM95] for an English version). Other sets of rules have been proposed, we will comment on them in the bibliographical notes in the Simple Conceptual Graphs chapter, as these rules consider graphs with equality. The fact that this subsumption is not an order (contrary to the claim in [Sow84]), but only a preorder had been several times noted (e.g., [Jac88]). [CM92] introduced irredundancy and proved that each BG equivalence class contains a unique irredundant graph (Theorem 2.1). The additional property 2.9 is from [CG95]. In the same paper, an alternative proof of Property 2.3 is provided. Normal graphs were independently introduced in [MC96] and [GW95]. The NP-completeness of BG-HOMOMORPHISM, HOM-EQUIVALENCE and REDUNDANCY was proven in [CM92] (with different proofs).

Chapter 3
Simple Conceptual Graphs

Overview

Concept nodes representing the same entity are said to be *coreferent* nodes. In basic conceptual graphs (BGs) only individual concept nodes may be coreferent. *Simple Conceptual Graphs (SGs)* enrich BGs with unrestricted coreference. The introduction of this chapter develops the discussion concerning equality started in the previous chapter and presents the *conjunctive type* and *coreference* notions. A new definition of a vocabulary extending the previous one with conjunctive types is given in Sect. 3.2. Section 3.3 defines SGs. An SG is simply a BG plus a coreference relation. The coreference relation is an equivalence relation over the concept node set with the following meaning: All concept nodes in a given equivalence class represent the same entity.

SG homomorphisms naturally extend BG homomorphisms. In Sect. 3.4, generalization and specialization operations defined on BGs are extended to SGs. Normal SGs are introduced in Sect. 3.5. An SG is *normal* if its coreference relation is the identity relation, i.e., each node is solely coreferent with itself. A normal SG can be associated with any SG. In fact, normal SGs and normal BGs can be identified, which emphasizes the importance of normal BGs. The notion of *coref-homomorphism*, which is specific to SGs, is introduced in Sect. 3.6. Instead of mapping concept nodes onto concept nodes as for a homomorphism, a coref-homomorphism maps *coreference classes* onto coreference classes. Relationships between homomorphisms and coref-homomorphisms are studied, and it is shown that, in the presence of a coreference, the intuitive meaning of generalization or specialization operations is better captured by coref-homomorphism than by homomorphism. The normal form of an SG is in a sense the most compact form of an SG. The *antinormal* form studied in the last section can be considered as the most scattered form of an SG: Each relation node is separated from any other relation node, and there are no multiple edges.

3.1 Introduction

The essential difference between *basic* conceptual graphs (BGs) and *simple* conceptual graphs (SGs) is that in SGs several concept nodes *not necessarily individual* can represent the same entity. Nodes representing the same entity are said to be *coreferent*, i.e., they refer to the same entity. Then the first question is: Can any pair of concepts be coreferent?

This question is closely related to the structure and the interpretation of the concept type set. Besides the structure of the ordered concept type set (e.g., a semilattice or lattice), an essential point is how the type hierarchy is interpreted. When a type is interpreted as a set of entities, the order over the hierarchy is interpreted as the inclusion. Then, the greatest lower bound (*glb*) of two types, when it exists, can be *lattice-theoretically* or *order-theoretically* interpreted. In the order-theoretic interpretation there is nothing more than the interpretation of the subtype relation by set inclusion. In the lattice interpretation, the interpretation of the glb of two types is then interpreted as the intersection of these sets. E.g., let *Building* and *OldThing* be two incomparable types, and let *HistoricLandmark* be their glb (see Fig. 3.1 left). With both interpretations, every entity of type *HistoricLandmark* is also of types *Building* and *OldThing*. With the lattice interpretation, every entity of both types *Building* and *OldThing* is necessarily of type *HistoricLandmark*. With the order-theoretic interpretation, the glb of two types is simply a subtype, thus a subset of their intersection; an entity of both types *Building* and *OldThing* is not necessarily a *HistoricLandmark* (cf. Fig. 3.1 right).

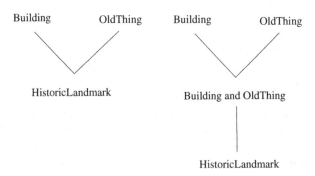

Fig. 3.1 Interpretation of types

The way the concept type set is interpreted has an effect on reasoning.

For instance, let Q be the query [HistoricLandmark:*] ("is there a historic landmark?") and let G be a fact containing the coreferent concept nodes [Building:a] and [OldThing:a], thus indicating that a is an entity which is a *Building* and an *OldThing*. With the lattice interpretation, G should provide an answer to Q while in the order-theoretic interpretation it should not.

The lattice-theoretic interpretation explicitly requires building all types of intersections (for instance, if an entity of types *Building* and *OldThing* is not necessarily of type *HistoricLandmark* one has to build their common subtype *BuildingandOldThing*, which is a supertype of *HistoricLandmark*, see Fig. 3.1 right). Thus it leads to a number of new and artificial types which can be exponential in the number of useful types.

On the other hand, the drawback of the order-theoretic interpretation is that we lose the possibility of expressing conjunctions of types, e.g., that *HistoricLandmark* is exactly the intersection of *OldThing* and *Building*. This property could be expressed by a rule (cf. Chap. 10), stating that every entity of type *Building* and of type *OldThing* is also of type *HistoricLandmark*. However, coding the conjunction of types in the hierarchy itself is more efficient than using rules, exactly as coding the subtype relation in a type hierarchy is more efficient than considering a flat set of types and representing the hierarchy by rules.

Conjunctive Type

A concept node represents an entity, but an entity can be represented by several coreferent concept nodes. A way of gathering all or some concept nodes representing the same entity is desirable to ease the building and understanding of an SG by a human being.

A simple way to gather these nodes is to merge them into one node. This operation has to keep the meaning of the original SG. Thus if an entity is represented by a node of type t and by a node of type t' there must be a type representing exactly the entities which are of both types t and t'. More generally, we need to express that a type is the conjunction of several types. Any subset of (incomparable) primitive types defines a conjunctive type. The set of conjunctive types is partially ordered by an order extending the order defined on primitive types.

However, not all conjunctions of types have a meaning. Thus we need a way of expressing that the conjunction of two (or more) types is *banned*, or equivalently that they cannot have a common subtype. A classical way of doing this is to add a special *Absurd* type below disjoint types. But this technique is not always precise enough. For instance, t_1, t_2, t_3 being direct supertypes of *Absurd* means that the intersection of any two of them is empty. We cannot express that t_1 and t_2 are disjoint as well as t_1 and t_3, but that there can be an entity of type t_2 and t_3 (e.g. $t_1 = Animal$, $t_2 = Ship$, $t_3 = Robot$, with all being subtypes of *MobileEntity*).

Giving all acceptable (i.e., non-banned) types in extension is not conceivable in practice, so we define the set of acceptable conjunctive types by the primitive type set and the (maximal) banned conjunctive types. In the previous example, the types $\{t_1, t_2\}$ and $\{t_1, t_3\}$ would be banned. Theoretically, the number of banned conjunctive types can be exponential in the number of primitive types but, practically, considering banned types of cardinality two seems sufficient and their number is at most quadratic in the number of primitive types.

Coreference

An individual marker is generally considered as a surrogate or an identifier (in the programming meaning) of an entity. According to the Unique Name Assumption (UNA) common in knowledge representation, it is thus assumed that two distinct individual markers represent distinct entities. Hence nodes with distinct individual markers cannot belong to the same coreference set.

Let us consider two concepts with incomparable types t_1 and t_2. Can these nodes be coreferent without any condition on t_1 and t_2? The answer is positive in many works. As already said, in our opinion important properties required are, first, that *(a)* coreferent concepts can always be merged into a single node, secondly, that *(b)* the meaning of the obtained graph is the same as that of the original one. With a lattice-interpretation of concept types these properties are fulfilled if the types of the coreferent nodes have a glb (distinct from the *Absurd* type), which becomes the type of the new node. With the order-theoretic assumption, and in the absence of conjunctive types, the only way to fulfill these properties is to require that the set of types of the coreferent nodes possesses a minimal element in this set itself. Indeed, let us suppose for instance that there are two coreferent concepts with types *Building* and *OldThing*, respectively. Either these concepts cannot be merged and *(a)* is not satisfied, or they can and the type of the new node is a subtype of *Building* and *OldThing* and the obtained graph is (in general) strictly more specialized than the original one. That is why in works where properties *(a)* and *(b)* are required, it is required that coreferent nodes have the same type. In the framework of conjunctive types proposed here, *(a)* and *(b)* are obtained as soon as the conjunction of the types of the coreferent nodes is not a banned type. Then every SG can be easily transformed into an equivalent normal SG, which is called its *normal form*.

Whether a coreference set contains both generic and individual nodes is less important. Indeed, making a generic node and an individual node coreferent can always be performed by restricting the marker of the generic concept to the individual marker of the other node without changing the meaning of the graph.

3.2 Vocabulary

In a *conjunctive vocabulary*, the set T_C of concept types is built with three components: a set of primitive concept types, an operation of type conjunction and a set of banned conjunctive types. A set of primitive concept types has the same structure as a set of concept types in a vocabulary as defined in Chap. 2, i.e., it is a partially ordered set with a greatest element.

A conjunctive type is a set of n incomparable primitive types $E = \{t_1, \ldots, t_n\}$. A conjunctive type is either acceptable or banned. Thus, T_C is the set of acceptable conjunctive types. Let $\{t_1, \ldots, t_n\}$ be an acceptable conjunctive type, then every subset of this set is also an acceptable conjunction. Let $\{t_1, \ldots, t_n\}$ be a banned conjunctive type, which does not mean that any subset of this conjunction is banned but that there is no entity with all these types.

Let us assume that incomparability of types in a conjunctive type is not required. Let $\{t_1,...,t_n\}$ be an acceptable conjunctive type with, for instance, $t_i \leq t_j$. Then the conjunction of t_i and t_j has the same meaning as t_i since every entity of type t_i is of type t_j, hence t_j, the greatest type, is useless. For example, let us consider $t = \{Boy, Person, SmallEntity\}$, as $Boy < Person$, t has the same meaning as $t' = \{Boy, SmallEntity\}$. If the conjunction of $\{t_1,...,t_n\}$ is banned, then t_j is useless too, because if there cannot be an entity with types $\{t_1,...,t_n\}$, there cannot be an entity with types $\{t_1,...,t_n\} \setminus \{t_j\}$. For example if $t = \{RacingCar, Animal, Car\}$ is banned, then as $RacingCar < Car$, t has the same meaning has $t' = \{RacingCar, Animal\}$ which is banned too.

Thus we restrict conjunctive types to the set of minimal types of a type set, and if $\{t_1,...,t_n\}$ is a set of types, then its associated conjunctive type is denoted $min\{t_1,...,t_n\}$ or $t_1 \wedge ... \wedge t_n$.

Due to the associativity of the conjunction operator, defined as the minimal operator, a conjunction of conjunctions of primitive types is a conjunction of primitive types. Therefore, we consider only conjunctions of primitive types.

The set of all acceptable conjunctions of primitive types can be exponentially bigger than the primitive type set. This is why the set of concept types is not defined in extension but by means of the primitive type set and a set of assertions stating which types are banned. Let us define these notions more formally.

Definition 3.1 (Primitive concept type set). A *primitive concept type set* is an ordered set $(T, \leq)$ with a greatest element denoted by $\top$.

This definition is the same as the concept type set definition (cf. Definition 2.1).

Definition 3.2 (Conjunctive concept type). A *conjunctive concept type* is given by a (non-empty) set of incomparable types $\{t_1,...,t_n\}$. This type is denoted by the set itself or by $t_1 \wedge ... \wedge t_n$. Let A be any (non-empty) subset of T, then the conjunctive type associated with A is defined as the set, $min(A)$, of minimal elements of A (a minimal element of A is an element t of A such that $\forall t' \in A, t' \not< t$).

A primitive type t is identified with the conjunctive type $\{t\}$. Conjunctive types are provided with a natural partial order that extends the order defined between primitive types: Given two conjunctive types t and s, t is a specialization of s if every primitive type of s has a specialization (possibly equal to itself) in t.

Definition 3.3 (Conjunctive concept type set). Let T be a set of primitive concept types. $T^{\sqcap}$ denotes the set of all conjunctive types over T. It is provided with the following partial order, which extends the partial order on T: Given two types $t = \{t_1,...,t_n\}$ and $s = \{s_1,...,s_p\}$, $t \leq s$ if for every $s_j \in s$, $1 \leq j \leq p$, there is a $t_i \in t$, $1 \leq i \leq n$, such that $t_i \leq s_j$.

Example. Let $t = BuidingBlock \wedge Cube \wedge Small$ and $s = Toy \wedge Cube$. $BuidingBlock \leq Toy$ and $Cube \leq Cube$ then $t \leq s$.

Property 3.1. $(T^\sqcap, \leq)$ is a lattice.

The previous property relies on basic results in order theory. In order theory, a subset of incomparable elements is called an *antichain*, and it is well-known that the set of antichains provided with the previous partial order is a lattice: Each pair of antichains has a greatest lower bound that is the set of minimal elements of the union of the antichains, and a least upper bound which is the set of minimal elements of the intersection of the filters generated by the two antichains. Furthermore, the inf-irreducible elements of that lattice are exactly the primitive types (cf. Appendix).
Example. Figure 3.2 represents an ordered set and the lattice of its antichains; the first set can be seen as the set T of primitive types and the lattice is then the set $T^\sqcap$ of conjunctive types over T.

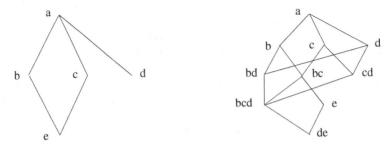

Fig. 3.2 An ordered set and the lattice of its antichains

The property of being a banned conjunctive type is hereditary: If a conjunctive type A is banned, all conjunctive types less than A are also banned. For instance, if $A = \{Car, Animal\}$ is banned then $\{RacingCar, Animal\}$ is banned too.

Definition 3.4 (Banned type set). Let B denote a set of conjunctive types. An element of $T^\sqcap$ is said to be *banned* with respect to B if it is less than or equal to (at least) an element of B. B^* denotes the set of all banned types: $B^* = \{t \in T^\sqcap \mid \exists t' \in B, t \leq t'\}$.

In the ordered sets terminology, B^* is the union of the ideals generated by the banned types.
Example. Figure 3.3 presents examples of banned and non-banned types. $\{Building-Block, Cube, Firetruck\}$ is banned because it is less than $\{BuildingBlock, Car\}$, which is itself banned.

The set of non-banned types, i.e., of acceptable types, is obtained from $T^\sqcap$ by removing B^*:

Definition 3.5 (Concept type hierarchy). A concept type hierarchy $T_C(T, B)$, is given by a couple (T, B), where:

- T, the set of primitive concept types, is partially ordered by $\leq$, with a greatest element, denoted by $\top$ and called the universal type,

Banned types   **Acceptable types**

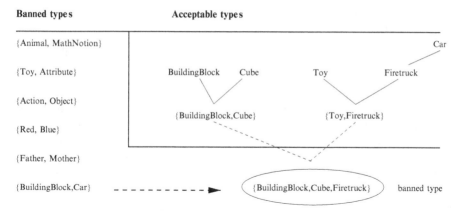

Fig. 3.3 Examples of banned and acceptable concept types

- $\mathcal{B}$, the set of basic banned conjunctive types, is composed of conjunctive types over T,
- $\mathcal{B}$ complies with T, i.e., for all $b \in \mathcal{B}$, there is no type $t \in T$ with $t \leq b$ i.e., $\mathcal{B}^* \cap T = \emptyset$.

$T_C(T, \mathcal{B})$ is defined as the set $T^{\sqcap} \setminus \mathcal{B}^*$. $T^{\sqcap}$ is thus partitioned into the acceptable types $T_C(T, \mathcal{B})$ and the banned types $\mathcal{B}^*$. Whenever there is no risk of ambiguity, $T_C(T, \mathcal{B})$ is simply denoted T_C.

An important particular concept type hierarchy is the case where there are no banned types, i.e., $\mathcal{B}$ is empty. In this case $T_C = T^{\sqcap}$ and T_C is a lattice. In general case, T_C is a sup-semi-lattice, i.e., each pair of types has a least upper bound.

A conjunctive vocabulary contains three parts: A hierarchy of concept types, a hierarchy of relation types, and a set of markers.

Definition 3.6 (Vocabulary). A *conjunctive vocabulary* is a triple $(T_C, T_R, \mathcal{I})$.

- $T_C, T_R, \mathcal{I}$ are pairwise disjoint sets.
- T_C is the *concept type hierarchy* defined by a set T of primitive concept types and a set $\mathcal{B}$ of banned conjunctions.
- T_R, the set of relation symbols, is ordered by a relation $\leq$, and is partitioned into subsets $T_R^1, \ldots, T_R^k$ of relation symbols of arity $1, \ldots, k$ respectively. The arity of a relation r is denoted $arity(r)$. Furthermore, any two relations with different arities are not comparable.
- $\mathcal{I}$ is the set of individual markers.

BGs (cf. Definition 2.2) can be defined on a conjunctive vocabulary and hereafter a conjunctive vocabulary is simply called a vocabulary.

The set of ordered concept labels is defined in the same way as in Chap. 2:

Definition 3.7 (Ordered set of concept labels). The ordered set of *concept labels* is the cartesian product of the ordered sets T_C and $\mathcal{I} \cup \{*\}$.

The introduction of conjunctive types gives a semi-lattice structure to the set of concept labels:

Property 3.2. The ordered set of concept labels on a conjunctive vocabulary is a sup-semi-lattice.

Proof. T_C is a sup-semi lattice since it is obtained from a lattice by removing elements while keeping the least upper bound of all remaining elements (if t and t' are acceptable, then $t \vee t'$ is acceptable too). $\mathcal{I} \cup \{*\}$ is a sup-semi-lattice. One concludes using a (simple) semi-lattice property: The product of two sup-semi-lattices is a sup-semi-lattice (indeed, let (t, m) and (t', m') be two concept labels; it is easy to check that $(lub(t, t'), lub(m, m'))$ is equal to $lub((t, m), (t', m')))$. Thus $T_C \times \{\mathcal{I} \cup \{*\}\}$ is a sup-semi-lattice. $\square$

A set of labels therefore has a least upper bound. An entity of the application domain can be represented by several concepts. In this case the labels of these concepts have to satisfy two conditions. The first condition, already seen in Chap. 2, corresponds to the Unique Name Assumption: Two distinct individual markers cannot represent the same entity. The second condition requires that the conjunction of the types of the concepts representing an entity does not yield a banned type. These two conditions are fulfilled in the following definition.

Definition 3.8 (Compatible concept labels). The concept labels $(t_1, m_1), \ldots, (t_k, m_k)$ are *compatible* if

- there is at most one individual marker, i.e., $|\mathcal{I} \cap \{m_1, \ldots, m_k\}| \leq 1$,
- the conjunctive type $t' = min\{t_1 \cup \ldots \cup t_k\}$ is in T_C.

Property 3.3. The concept labels $(t_1, m_1), \ldots, (t_k, m_k)$ are compatible if and only if these labels possess a greatest lower bound.

The greatest lower bound of concept labels $(t_1, m_1), \ldots, (t_k, m_k)$ is the concept label (t', m') with $t' = min\{t_1 \cup \ldots \cup t_k\}$ and $m' = min\{m_1, \ldots, m_k\}$, i.e., m' is the generic marker if all m_i are generic markers, otherwise m is the only individual marker appearing in the m_i.
Example. $(Object, *), (Cube, C3), (Toy \wedge Small, *)$ are compatible. Assume that $\{Young, Mother\}$ is not banned then $(Young, Judy)$ and $(Mother, *)$ are compatible, but $(Young, Judy)$ and $(Mother, Mary)$ are not compatible because Judy and Mary are different individuals.

3.3 Simple Conceptual Graphs (SGs)

Roughly said, an SG is a BG plus a coreference relation. A coreference relation is an equivalence relation over the concept set and two equivalent concepts are called coreferent concepts. The important properties to fulfill are, first, that coreferent concepts can always be merged, secondly, that the formal semantics of the obtained graph is equivalent to those of the original one (cf. Chap. 4).

Definition 3.9 (Compatible concept nodes). A set of *compatible concept nodes* is a set of concept nodes such that the set of their labels is compatible, i.e., these labels possess a greatest lower bound. The *label of a compatible concept set X* is defined as $l(X) = glb(\{l(x) \mid x \in X\})$.

Definition 3.10 (Simple conceptual graph). A *simple conceptual graph* (in short SG) over a vocabulary $\mathcal{V}$ is a 5-tuple $(C, R, E, l, coref)$, such that:

- (C, R, E, l) is a basic conceptual graph over $\mathcal{V}$.
- $coref$ is an equivalence relation over C, such that:

    any coref class is compatible,
    if two individual concepts have the same marker then they are in the same coref class.

In graph drawings, a coreference relation is usually only *partially* represented by coreference links (cf. Fig. 3.4). A *coreference link* connects two coreferent concepts. Coreference links are usually represented by dashed lines. Note that the coreference relation is the reflexive and transitive closure of the relation, which is the union of the set of coreference links and the relation linking two individual concept nodes having the same marker. When it is totally represented in a graph drawing, the coreference links form a "clique" on concepts belonging to the same coreference class, i.e., each pair of concepts of a coreference class is connected by a coreference link.

Example. An SG is presented in Fig. 3.4. Its coreference classes are: $\{c_1, c_2, c_3\}$, $\{d_1, d_2\}$, and the trivial classes $\{a\}$, $\{b\}$ and $\{c\}$. The coreference link between c_1 and c_3 is not represented, it is obtained by transitive closure.

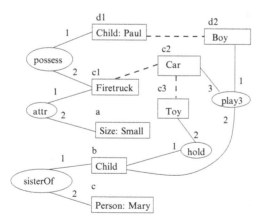

Fig. 3.4 A Simple conceptual graph

What are coreference links useful for? A first feature is that they can be exploited in *user interfaces*. Indeed, one advantage of BGs is their readability, a quality that is lost when the BG becomes too big (e.g., too big to be entirely drawn on a screen) or too complex. A BG that is too big can be split into pieces, with each piece being displayable on a screen. The split can also be performed to decompose a BG into semantic modules, or to enhance a particular substructure. Coreference links enable the representation of a BG by several pieces, while keeping its global structure. A user can build a BG piece by piece, then connect them with coreference links. Another important case is when a conceptual graph basis has been constructed by different people or at different times. Finally, coreference links can also be used to decompose a graph into a simpler structure that can be exploited in *algorithms*. See for instance the tree covering notion in Chap. 7.

We will see that it is easy to construct a BG (i.e., to get rid of the coreference relation) that is semantically equivalent to an SG. Thus, even if coreference is a useful representation tool, it does not add expressivity to BGs. The reasoning mechanisms defined for BGs can still be used, with the SGs being transformed into BGs before any computation. Let us point out, however, that coreference will play a more substantial role in more complex graphs such as nested conceptual graphs (Chap. 9) and rules (Chap. 10). It will also be a fundamental element in full conceptual graphs (Chap. 12).

Definitions concerning BGs can be directly adapted to SGs. Indeed, an SG is a BG plus a coreference relation on the set of concept nodes. Thus, let P be a notion concerning BGs, then an extension of this notion to SGs can be obtained as follows. Let $G = (C, R, E, l, coref)$ be an SG: The notion P is considered for the BG (C, R, E, l) and a condition taking the coreference relation into account is added. Let us give some examples.

Definition 3.11 (subSG). Let $G = (C, R, E, l, coref)$ be an SG. A *subSG* of G is an SG $G' = (C', R', E', l', coref')$, such that:

- (C', R', E', l') is a subBG of (C, R, E, l).
- $coref'$ is the restriction of $coref$ to C' (i.e., a class of $coref'$ is the intersection of a class of $coref$ with C').

The homomorphism notion for BGs can also be easily extended to SGs. Two coreferent concepts must have coreferent images: either the same image (since coref is a reflexive relation) or distinct coreferent images.

Definition 3.12 (SG homomorphism and $\succeq$). Let $G = (C_G, R_G, E_G, l_G, coref_G)$ and $H = (C_H, R_H, E_H, l_H, coref_H)$ be two SGs defined on the same vocabulary. A *homomorphism* π from G to H is a homomorphism from the BG (C_G, R_G, E_G, l_G) to the BG (C_H, R_H, E_H, l_H) that preserves the coreference, i.e. $\forall x, y \in C_G, coref_G(x, y) \rightarrow coref_H(\pi(x), \pi(y))$.
$G \succeq H$ means that there is a homomorphism from G to H.

An SG isomorphism is naturally defined as follows:

Definition 3.13 (SG isomorphism). Let $G = (C, R, E, l, coref)$ and $G' = (C', R', E', l', coref')$ be two SGs. G and G' are *isomorphic* if first, there is a BG isomorphism π from (C, R, E, l) to (C', R', E', l'); secondly, π preserves the coref classes, that is x and y are coreferent in G if and only if $\pi(x)$ and $\pi(y)$ are coreferent in G'.

Before studying the SG homomorphism in more detail it is useful to define generalization and specialization operations.

3.4 Generalization and Specialization Operations

In this section, the elementary generalization and specialization operations defined in Chap. 2 are extended to SGs, i.e., the coreference relation is taken into account.

We will group operations into three clusters: Equivalence operations, (plain) generalization operations and (plain) specialization operations. By "plain" we mean that these operations generally do not produce an equivalent SG although they may be equivalent in some cases. Note that when operations have the same name as for BGs, they are the same operation, i.e., the coreference relation is not changed by the operation.

Definition 3.14 (SG Equivalence operations). The elementary equivalence operations are:

- **Copy.** Create a disjoint copy of an SG G. More precisely, given an SG G, $copy(G)$ is an SG which is disjoint from G and isomorphic to G.
- **Relation duplicate.** Given an SG G and a relation r of G, $relDuplicate(G, r)$ is the SG obtained from G by adding a new relation node r' having the same type and the same list of arguments as r.
- **Relation simplify.** Given an SG G, and two twin relations r and r' (relation with the same type and the same list of neighbors), $relSimplify(G, r')$ is the SG obtained from G by deleting r'.
- **Concept split.** Split a concept into coreferent concepts. Let c be a concept node of G and $\{A_1, A_2\}$ be a partition of the edges incident to c. $split(G, c, A_1, A_2)$ is the SG obtained from G by: creating two new concept nodes c_1 and c_2 with the same label as c, attaching A_1 to c_1 and A_2 to c_2 (A_1 or A_2 may be empty), adding c_1 and c_2 to the coreference class of c, and finally deleting c.
- **Coreferent nodes merge.** Given an SG G, and two coreferent concepts c_1 and c_2 of G, merge c_1 and c_2, i.e., create a node c with the label being the greatest lower bound of c_1 and c_2 labels, attach to c all edges incident to c_1 or c_2 (replace every edge (r, i, c_1) or (r, i, c_2) with an edge (r, i, c)), add c in the coreference class of c_1 and c_2 and delete these nodes.
- **Concept label modify.** Let c be a concept node in a coreference class X of an SG $G = (C, R, E, l, coref)$. Change $l(c) = L$ into L' in such a way that the label of X is unchanged; otherwise said, the modification of the label of c does not change the glb of the label of concepts in X.

Definition 3.15 ((plain) SG Generalization operations). The elementary generalization operations are:

- **Increase.** Increase the label of a node (concept or relation). More precisely, given an SG G, a node x of G, and a label $L \geq l(x)$ $increase(G, x, L)$ is the SG obtained from G by increasing the label of x up to L, i.e., its type if x is a relation, its type and/or its marker if x is a concept.

- **Coreference delete.** Split a coreference class into two (non-empty) classes.
- **Substract.** Given an SG G, and a set of connected components $C_1, \ldots, C_k$ of G, $substract(G, C_1, \ldots, C_k)$ is the SG obtained from G by deleting $C_1, \ldots, C_k$ (the result may be the empty graph).

Definition 3.16 ((plain) SG Specialization operations). The elementary specialization operations are:

> **Restrict.** Given an SG G, a node x of G, and a label $l \leq l(x)$ $restrict(G,x,l)$ is the SG obtained from G by decreasing the label of x to l that is its type if x is a relation, its type and/or its marker if x is a concept. The compatibility condition has to be preserved.
>
> **Coreference add up.** Make the union of two coreference classes of G provided that their union is a compatible set.
>
> **Disjoint sum.** Given two disjoint SGs $G = (C, R, E, l, coref)$ and $H = (C', R', E', l', coref')$, $G + H = (C \cup C', R \cup R', E \cup E', l \cup l', coref \cup coref')$ (G or H may be empty).

Generalization and specialization of SGs are defined similar to BGs as a sequence of elementary operations.

Definition 3.17 (SG generalization and specialization operations). The set of generalization operations is composed of equivalence operations and plain generalization operations. The set of specialization operations is composed of equivalence operations and plain specialization operations. Given two SGs G and H, G is a generalization (resp. specialization) of H if there is a generalization (resp. specialization) sequence from G to H.

The disjoint sum operation is not the inverse of the substract operation, but the inverse of a substract operation can be performed by a disjoint sum plus a coreference addition, thus:

Property 3.4. Given two SGs G and H, G is a generalization of H if and only if H is a specialization of G.

Definition 3.18 (Gen-Equivalence). Given two SGs G and H they are called *gen-equivalent* if G is a generalization (or specialization) of H and reciprocally.

Operations on BGs are naturally included in operations on SGs, either directly under the same name, or they can be performed by a combination of operations of the same category on SGs. Indeed, a detach operation for SGs can be obtained by a split into coreferent nodes followed by a coreference deletion, which are both generalization operations, and a join for SGs can be obtained by a co-reference addition followed by a merging of co-referent nodes, which are both specialization operations.

3.5 Standard and Normal SGs

A *normal* SG is such that *coref* is the identity relation, i.e., each node is uniquely
coreferent with itself. In the notation of a normal SG, *coref* is usually omitted; a
normal SG is thus simply denoted by (C,R,E,l), as a (normal) BG. This is a consistent notation because a BG can be considered as an SG with the following implicit
coreference relation: Each generic concept constitutes a class and all individual concepts with the same marker are coreferent.

   Thus, normal SGs and normal BGs can be *identified*. Furthermore, a normal SG
can be associated with any SG G by considering the SG $G/coref$, obtained by merging all concepts belonging to a coreference class. *Merging* a coreference class X
consists of two successive sequences of elementary specialization operations: First,
all concept labels in X are restricted to $l(X) = glb(\{l(c) \mid c \in X\})$, then the class is
condensed to a single concept by a merge coreferent nodes operation. The first step
is called the *standardization* of an SG G. The result is the *standard form* of G.

Definition 3.19 (Standard SG). A *standard* SG is an SG such that all coreferent
nodes have the same label. The *standard form* of an SG G, denoted $stand(G)$, is
obtained from G by restricting the label of each concept in a coreference class X to
$l(X) = glb(\{l(x) \mid x \in X\})$.

Example. Figure 3.5 presents the standard form of the SG in Fig. 3.4.

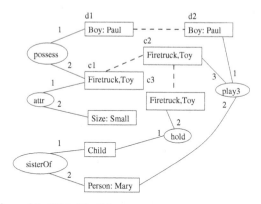

Fig. 3.5 Standard form of the SG in Fig. 3.4

   The normal form of an SG G is indifferently denoted $norm(G)$ or $G/coref$. The
latter notation shows that the normal form can be computed by first computing the
(ordinary) quotient graph $graph(G)/coref'$, where $coref'$ is obtained from $coref$
by adding trivial classes corresponding to the relation nodes of G (each relation node
constitutes a class), then by giving the correct labels to the nodes.

Definition 3.20 ($G/coref$ and $norm(G)$). Let $G = (C,R,E,l,coref)$ be an SG.
$G/coref$ is the normal SG obtained as follows: for any coref class $X = \{c_1,\dots,c_k\}$,

all c_i are merged into one concept X. $G/coref$ is also called the *normal form* of G, and denoted by $norm(G)$.

The normal form of the SG G in Fig. 3.4 is the SG in Fig. 3.6. Note that the standard and normal forms can be computed in linear time (relative to the size of the original graph) if the computation of the glb of k concept labels is in $O(k)$.

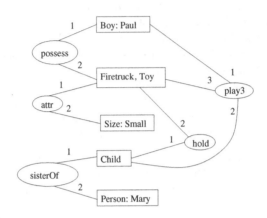

Fig. 3.6 $G/coref$

The three SGs G, $stand(G)$ and $norm(G)$ are gen-equivalent in the sense of Definition 3.18.

Property 3.5. For any SG G, the three SGs G, $stand(G)$, and $norm(G)$ are gen-equivalent.

Proof. One can see that there are sequences of equivalence operations for transforming any SG in $\{G, stand(G), norm(G)\}$ into any other. $stand(G)$ is obtained from G by a sequence of *concept label modify* operations. $norm(G)$ is obtained from $stand(G)$ by a sequence of *coreferent node merge* operations, finally G is obtained from $norm(G)$ by a sequence of *concept split* and *concept label modify* operations. □

Intuitively, G, $stand(G)$ and $norm(G)$ have the same meaning. We will see later that their formal semantics are indeed equivalent.

3.6 Coref-Homomorphism

Let us now study the relationships between an SG G, its standard form $stand(G)$ and its normal form $norm(G)$ with respect to SG homomorphism. The following property is immediate:

Property 3.6. Let G be an SG. There is a homomorphism from G to $stand(G)$ and a homomorphism from $stand(G)$ to $norm(G)$. Thus: $G \succeq stand(G) \succeq norm(G)$.

The homomorphisms associated with the sequence of equivalence operations described above yielding $stand(G)$ and $norm(G)$ are called canonical homomorphisms from G to $stand(G)$ and $norm(G)$. The former is restricted to the identity (it is a bijective homomorphism) and the latter is the surjective mapping naturally associated with the quotient operation $G/coref$. Thus, as natural as it may seem, the SG homomorphism notion is not entirely satisfactory.

The converse homomorphisms do not generally occur (unless G is standard or normal).

We are faced with the problem we had foreseen on BGs: Due to the presence of coreferent nodes, graphs with the same intuitive semantic may not be equivalent for homomorphism. Let us go into further detail.

Property 3.7. Given SGs G and H, there is a bijective mapping between the set of homomorphisms from G to $norm(H)$ and the set of homomorphisms from $norm(G)$ to $norm(H)$.

Proof. Let π_G be the canonical homomorphism from G to $norm(G)$. Let P be the mapping that associates the homomorphism $\pi' = \pi \circ \pi_G$ from G to $norm(H)$ with any homomorphism π from $norm(G)$ to $norm(H)$.
P is injective because, given two distinct homomorphisms π_1 and π_2 from $norm(G)$ to $norm(H)$ and any node x of $norm(G)$, if $\pi_1(x) \neq \pi_2(x)$, then for any node y with $\pi_G(y) = x$, $\pi_1(\pi_G(y)) \neq \pi_2(\pi_G(y))$.
Let us prove that P is surjective. Let π' be any homomorphism from G to $norm(H)$, then we prove that it can be decomposed into $\pi \circ \pi_G$, where π is a homomorphism from $norm(G)$ to $norm(H)$. Indeed, π is built as follows: For every node x in $norm(G)$, if x is a node of G, in particular if x is a relation node, then $\pi(x) = \pi'(x)$. Otherwise let $c_1...c_n$ be the nodes of G merged into x in $norm(G)$; by definition of π_G, for all these nodes c_i, $\pi_G(c_i) = x$, and since π' is a homomorphism and $norm(H)$ is normal, all c_i have the same image by π', then, given any c_i, one takes $\pi(x) = \pi'(c_i)$. By definition of coreference and homomorphism, for any $i = 1,...,n$ $label(c_i) \geq label(\pi'(c_i)) = label(\pi(x))$, so $label(x) = glb(\{label(c_i) \mid i = 1,...,n\}) \geq label(\pi(x))$. Furthermore, let (r,i,x) be in $norm(G)$ then $\pi(r) = \pi'(r)$. If x belongs to G then $\pi(x) = \pi'(x)$. As π' is a homomorphism, $(\pi'(r),i,\pi'(x)) = (\pi(r),i,\pi(x))$ is in $norm(H)$. Otherwise, x is obtained from $c_1...c_n$ and there is a c_j, $1 \leq j \leq n$, with (r,i,c_j) in G. As π' is a homomorphism, $(\pi'(r),i,\pi'(x)) = (\pi(r),i,\pi(x))$ is in $norm(H)$. Thus, π is a homomorphism. Check that $\pi \circ \pi_G = \pi'$.
P is injective and surjective so it is bijective. $\quad\square$

Figure 3.7 shows Property 3.7 as a commutative diagram.

Thus, in order to obtain a "good" SG homomorphism behavior, one only has to ensure that the target graph H is in normal form, regardless of whether the source graph G is normal or not.

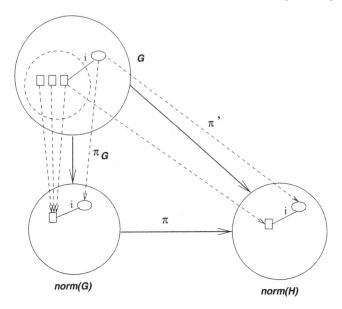

Fig. 3.7 Illustration of Property 3.7

The normal form of an SG always exists and is easily computable, however in some cases it may be important to keep SGs exactly as they are. For example, consider a set of SGs over the same vocabulary built by different users. If a query is made on this base the whole base has to be considered to compute the answers, but the SGs must not be changed. Another example is that of a base distributed on several sites.

The question now is the following: Why does homomorphism not deal correctly with coreference? The reason is that it is first of all *a mapping on concepts*; if we modify it so that it maps coreference classes onto coreference classes, we obtain the desired notion.

Definition 3.21 (coref-homomorphism). Let $G=(C_G,R_G,E_G,l_G,corefg_G)$ and $H=(C_H,R_H,E_H,l_H,coref_H)$ be two SGs defined on the same vocabulary. A *coref-homomorphism* from G to H is a mapping π from $coref_G$ to $coref_H$ and from R_G to R_H, such that:

1. $\forall(r,i,c) \in G$, let C be the coreference class of c, then there is a concept $c' \in \pi(C)$ such that $(\pi(r),i,c') \in H$,
2. $\forall C \in coref_G$, let $l_H(\pi(C))$ be the greatest lower bound of the concept labels in $\pi(C)$ then, for all $c \in C$, $l_G(c) \geq l_H(\pi(C))$,
3. $\forall r \in R_G$, $l_G(r) \geq l_H(\pi(r))$.

Each homomorphism defines a coref-homomorphism, since all coreferent nodes have their images in the same coreference class. The following property specifies the relationships between both notions.

Property 3.8. Given SGs G and H, there is a bijective mapping between the set of coref-homomorphisms from G to H and the set of homomorphisms from $norm(G)$ to $norm(H)$.

Proof. There is a bijection, say b, between concept nodes of $norm(G)$ (resp. $norm(H)$) and coreference classes in G (resp. H). Let us show that b defines the desired mapping between the set of coref-homomorphisms from G to H and the set of homomorphisms from $norm(G)$ to $norm(H)$. Let π be a coref-homomorphism from G to H. Let π' be the induced mapping from nodes of $norm(G)$ to nodes of $norm(H)$: For each concept c in $norm(G)$, $\pi'(c) = b^{-1}(\pi(b(c)))$ and for each relation r of $norm(G)$, $\pi'(r) = \pi(r)$.

This correspondence is injective. Indeed, let π_1 and π_2 be two coref-homomorphisms from G to H. There is a concept c in $norm(G)$ with $\pi_1(c) \neq \pi_2(c)$, then $\pi'_1(c) \neq \pi'_2(c)$.

We check that π' is a homomorphism from $norm(G)$ to $norm(H)$: For every edge (r,i,c) in $norm(G)$, there is an edge (r,i,d) in G where d is in the coreference class $b(c)$, thus, due to condition 1 of the coref-homomorphism, there is an edge $(\pi'(r),i,d')$ in H where d' is in the coreference class $\pi(b(c))$. Thus an edge $(\pi'(r),i,\pi'(c))$ since $\pi'(c) = b^{-1}(\pi(b(c)))$. For every concept c in $norm(G)$, condition 2 of coref-homomorphism ensures that for all $c_i \in b(c)$, $l(c_i) \geq l(\pi'(c))$; thus by definition of a greatest lower bound, $l(c) \geq l(\pi'(c))$. In the same way, we check that any homomorphism π' from $norm(G)$ to $norm(H)$ yields a coref-homomorphism π from G to H using bijection b. For every edge (r,i,c_j) in G, let C be the coreference class of c_j, there is an edge (r,i,c) in $norm(G)$, where $C = b(c)$, hence an edge $(r,i,\pi'(c))$ in $norm(H)$, thus an edge (r,i,c'_j) in H such that c'_j belongs to the coreference class $b(\pi'(c))$, which is exactly $\pi(C)$. For every coreference class C in G, for every c in C, $l(c) \geq l(\pi'(b^{-1}(C)))$, which is exactly $l(\pi(C))$, thus $l(c) \geq l(\pi(C))$.

Thus, the correspondence between coref-homomorphisms from G to H and homomorphisms from $norm(G)$ to $norm(H)$ is surjective and injective. $\square$

The following theorem summarizes the "safe" ways of comparing two SGs.

Theorem 3.1. *Let G and H be two SGs. There is a bijective mapping between:*

- *coref-homomorphisms from G to H,*
- *homomorphisms from $norm(G)$ to $norm(H)$,*
- *homomorphisms from G to $norm(H)$.*

The intuitive meaning of generalization or specialization operations is better captured by coref-homomorphism than by homomorphism. This is shown in the following theorem for SGs, which is similar to the Theorem 2.3 for BGs where homomorphism is replaced by coref-homomorphism:

Theorem 3.2. *Let G and H be two SGs. The three following propositions are equivalent:*

1. H is a specialization of G,
2. G is a generalization of H,

3. there is a coref-homomorphism from G to H.

Proof. The equivalence between (1) and (2) has already been stated (cf. Sect. 3.4). (1) $\Rightarrow$ (3): in a specialization sequence from G to H, each specialization step, say from H_i to H_{i+1}, defines a coref-homomorphism from H_i to H_{i+1}. By composition of these coref-homomorphisms, one obtains a coref-homomorphism from G to H. (3) $\Rightarrow$ (1): Let π be a coref-homomorphism from G to H. Let us show that H is a specialization of G. From G, one builds $norm(G)$ using the equivalence operation *coreferent nodes merge*. By Property 3.8, there is a homomorphism from $norm(G)$ to $norm(H)$. Thus by Theorem 2.3, there is a BG specialization sequence from $norm(G)$ to $norm(H)$. This BG specialization sequence can be rewritten as an SG specialisation sequence, with the join operation being replaced either by a *coreferent nodes merge* if the joined nodes are individual nodes, or decomposed into a coreference addition followed by a *coreferent nodes merge* if the joined nodes are generic nodes. By Property 3.5, there is a specialization sequence from $norm(H)$ to H. $\square$

3.7 Antinormal Form and Homomorphism

The normal form of an SG is, in a sense, the most compact form of an SG. The antinormal form studied in this section can be considered as the most scattered form of an SG: Each relation node is disconnected from the other relation nodes and there are no multiple edges. Thus, a relation node and its neighbors can be considered as a tuple. The antinormal form of an SG can be considered as a representation of an SG by tables in the database relational model. Indeed, gathering all tuples corresponding to relation nodes with the same label is equivalent to building a table. Then usual database systems and operations can be used. The correspondence between conceptual graphs and relational structures is studied in Chap. 5. In this section, it is assumed that an SG has no isolated concept node, and it is straightforward to extend all the results if isolated concepts exist.

Definition 3.22 (Antinormal SG). An SG is called *antinormal* if:

- any concept node is incident to exactly one edge,
- it is standard, i.e., all the concept labels in any coref class are identical.

For any SG G *antinorm(G)* is the SG obtained from G as follows: First it is standardized; then each concept node c with $k > 1$ incident edges, say $e_1, \ldots, e_k$, is split into k coreferent nodes $c_1, \ldots, c_k$, with each edge $e_i = (r, j, c)$ becoming an edge (r, j, c_i). *antinorm(G)* is obtained from G by a sequence of specialization operations, moreover:

Property 3.9. Let G be an SG. There is a homomorphism from *antinorm(G)* to *stand(G)*, i.e., *antinorm(G)* $\succeq$ *stand(G)*.

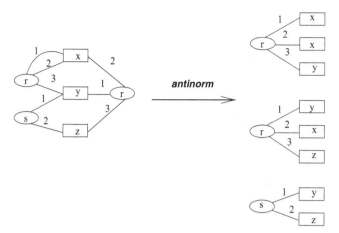

Fig. 3.8 An SG and its antinormal form

Example. Figure 3.8 presents an SG and its antinormal form. In this figure, the concepts labeled by *x* (or by *y* or by *z*) are coreferent.

Figure 3.9 shows that *antinorm(G)* does not generally map to *G* if *G* is not standard.

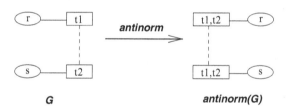

Fig. 3.9 antinorm(G) does not map to $\succeq$ G

Note that the relation node sets of *antinorm(G)* and *G* are equal, even if *G* is not standard.

If an SG is not antinormal then it does not generally map to its antinormal form by a homomorphism (cf. example given in Fig. 3.8).

The subsumption relation behaves in the same way for the normal forms and antinormal forms.

Property 3.10. Given two SGs *G* and *H*, there is a bijection from the set of homomorphisms from *norm(G)* to *norm(H)* to the set of homomorphisms from *antinorm(G)* to *antinorm(H)*.

Proof. Let π be a homomorphism from *norm(G)* to *norm(H)* and ϕ_H be the canonical homomorphism from *antinorm(H)* to *norm(H)*. A homomorphism π' from *antinorm(G)* to *antinorm(H)* can be constructed as follows. For any relation node,

$\pi'(r) = \pi(r)$ since the relation nodes of an SG and its normal form are the same. Let c be a concept node of *antinorm*(G). c is in the coreference class of a concept d of *norm*(G) and is incident to a unique edge (r, i, c). There is a unique concept c' of *antinorm*(H), which corresponds to the edge $(\pi(r), i, \pi(d))$ in *norm*(H). One takes $\pi'(c) = c'$ (cf. Fig. 3.10). If one considers two different homomorphisms from *norm*(G) to *norm*(H) the associated homomorphisms from *antinorm*(G) to *antinorm*(H) are also different. Reciprocally, let π' be a homomorphism from *antinorm*(G) to *antinorm*(H). The mapping π defined as follows is a homomorphism from *norm*(G) to *norm*(H). To any relation node r of *norm*(G) $\pi(r) = \pi'(r)$. Let d be a concept of *norm*(G). For any (r, i, d) in *norm*(G) corresponds a single c in *antinorm*(G) with (r, i, c) in *antinorm*(G). $(\pi'(r), i, \pi'(c))$ is in *antinorm*(H). One takes $\pi(d) = \phi_H(\pi'(c))$.

This does not depend on the choice of the edge (r, i, d) incident to d since for any other edge (r', j, d) in *norm*(G) corresponding with the node c' in *antinorm*(G), we have $\phi_H(\pi'(c)) = \phi_H(\pi'(c'))$ by the preservation of coreference by π'.

Figure 3.10 illustrates this part of the proof. Indeed, one can say that the diagram in Fig. 3.10 commutes. If one considers two different homomorphisms from *antinorm*(G) to *antinorm*(H), the associated homomorphisms from *norm*(G) to *norm*(H) are also different. The construction is injective in the two directions, so there is a bijection between the two homomorphism sets. $\square$

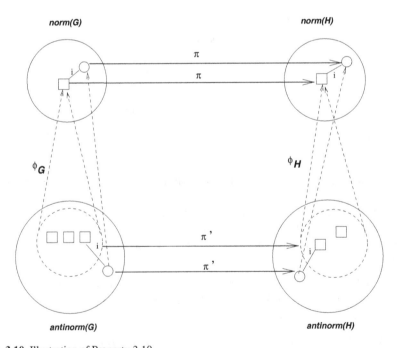

Fig. 3.10 Illustration of Property 3.10

Let us use an example to illustrate the importance for Property 3.10 of the last condition in Definition 3.22 of an antinormal SG. Figure 3.11 presents two non-standard SGs G_1 and G_2, where the relation types r and s are not comparable. These SGs satisfy the first antinormality condition since they are totally scattered. One has $norm(G_1) = norm(G_2) = G$, but there is neither a homomorphism from G_1 to G_2 nor from G_2 to G_1.

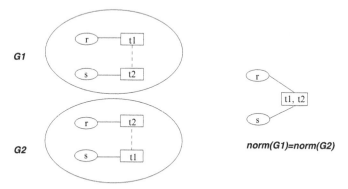

Fig. 3.11 G_1 and G_2 are not antinormal SGs

The following property shows the difference between the behavior of the homomorphism for the normal form and the antinormal form (cf. Property 3.7).

Property 3.11. Let G and H be two SGs. If $antinorm(G) \succeq H$, then $antinorm(G) \succeq antinorm(H)$. When H is standard, the converse is true: If $antinorm(G) \succeq antinorm(H)$ then $antinorm(G) \succeq H$.

Proof. Let π be a homomorphism from $antinorm(G)$ to H. A homomorphism π' from $antinorm(G)$ to $antinorm(H)$ can be constructed as follows. For any relation node, $\pi'(r) = \pi(r)$ since the relation nodes of an SG and its antinormal form are the same. Let (r,i,c) in $antinorm(G)$. $(\pi(r),i,\pi(c))$ is in H and the label of c is greater than or equal to the label of $\pi(c)$. There is a single concept d in $antinorm(H)$ which is adjacent to an edge numbered i to $\pi(r)$. The label of $\pi(c)$ is greater than or equal to the label of d and one takes $\pi'(c) = d$ (cf. Fig. 3.12). Let us assume that H is standard. Property 3.9 and the transitivity of $\succeq$ yield the second part of the property. □

Example. Let us consider an SG H isomorphic to the SG G on the left side in Fig. 3.9). There is a (trivial) homomorphism from $antinorm(G)$ to $antinorm(H) = antinorm(G)$, but there is no homomorphism from $antinorm(G)$ to H.

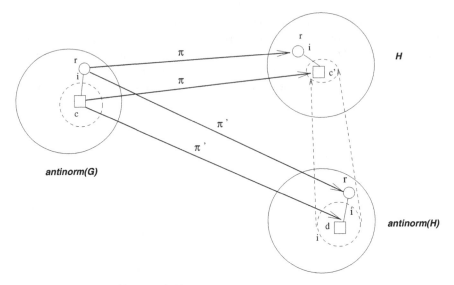

Fig. 3.12 Illustration of Property 3.11

3.8 Bibliographic Notes

The notion of a *coreference link* was introduced in [Sow84], mainly as a means of representing an anaphoric reference in the context of natural language processing.

In [Sow84] the concept type set is a *lattice*. In later works, it becomes a partially ordered set with a greatest element, thus a less constrained structure. The distinction between lattice-theoretical and order-theoretical interpretations is from [BHP+92]), and its effect on reasoning with CGs was pointed out in [WL94] and [CCW97].

The framework presented in this chapter is based on [CM04], which itself gathers and generalizes several previous works. Let us mention [CMS98], in which coreference (called co-identity) was defined as an equivalence relation on concept nodes. In this work, coreferent nodes were either individual nodes with the same label, or generic nodes with the same type, with these restrictions being related to properties (*a*) and (*b*) discussed in the introduction to this chapter. The extension of generalization and specialization operations to this restricted coreference relation was presented in [Mug00]. Type conjunction in conceptual graphs was first considered in the context of fuzzy types [CCW97] (and later [TBH03a] [TBH03b]) or in relationship with description logics [BMT99]. In [Bag01], the set of conjunctive concept types was defined in intension by a hierarchy of primitive concept types; type conjunction was also defined for relations, on the condition that these relations have the same arity. [CM04] mainly added the notion of banned conjunctive type and coref-homomorphism. Note that if we allow to name conjunctive types, we obtain a simple type definition mechanism, with conjunction as the only constructor. More complex type definition mechanisms are mentioned in Chap. 8.

According to the unique name assumption, different individual markers represent different entities. An exception to this common assumption is the framework of [Dau03a] which allows coreference between concept nodes with different individual markers. In other words, there are aliases among the individual markers. This case could be included in our framework with slight modifications.

The notion of antinormal form of an SG was introduced in [GW95]. The definition did not include a standardization step. It was thus well-suited to the restricted case where all coreferent nodes have the same label but not to the general case.

Chapter 4
Formal Semantics of SGs

Overview

The first section of this chapter presents a *model semantic* (or set semantic) for SGs. First, a model of a vocabulary is defined. It consists of a set, the set of entities also called a universe, upon which the concept types, the relation types and the individuals are interpreted. A concept type is interpreted as a subset of the universe, a relation type is interpreted as a set of tuples of elements of the universe and an individual is interpreted as an element of the universe. Secondly, a model of an SG over a vocabulary $\mathcal{V}$ is defined. It is a model of $\mathcal{V}$ enriched by an interpretation of the concept nodes as elements of the universe. Then, an entailment relation is defined, i.e., what it means that an SG H entails an SG G, or equivalently, what means that G is a consequence of H. The canonical model of a SG G, which plays a specific role in the characterization of the SGs consequences of G, is defined. This first section ends by the fundamental theorem stating that if there is a homomorphism from G to H then H entails G and if $H/coref$ entails G then there is a homomorphism from $H/coref$ to G (i.e., a soundness and completeness theorem of SG homomorphism with respect to entailment).

The second section concerns the presentation of a *first order logical semantic* for SGs. This semantic is defined through a mapping from SGs to FOL formulas called Φ. First, a FOL language corresponding to the elements of a vocabulary $\mathcal{V}$ is defined and a set of FOL formulas, denoted $\Phi(\mathcal{V})$, corresponding to the order relations over $\mathcal{V}$ is defined. Secondly, for any SG G a FOL formula $\Phi(G)$, built on the language associated with $\mathcal{V}$, is defined. Thirdly, it is shown that the classical model theory of FOL is equivalent to the model semantic presented in the first section.

In the third section relationships between SGs and the positive, conjunctive and existential fragment of FOL are studied. First, we introduce the $\mathcal{L}$-substitution notion and we use it to give another proof of the soundness and completeness theorem. Secondly, we prove the "equivalence" between the SGs and the positive, conjunctive and existential fragment of FOL. Finally, we present a second FOL semantic, denoted Ψ, which is less intuitive than Φ but has interesting technical properties.

FOL is used to give a semantic to SGs, but not to reason with them. According to our claim that SGs have good computational properties, we will build graph-based deduction algorithms that are not translation of logical procedures. This is developed in Chap. 6 and Chap. 7.

The final section is a note on the relationships between description logics and conceptual graphs.

4.1 Model Semantic

4.1.1 Models of SGs

In order to provide SGs with a set semantic, one first has to define a model of a vocabulary. A model of a vocabulary consists of a non-empty set D, called the universe or the domain of the model, and a mapping that associates with any concept type a subset of D, with any relation type of arity k a k-ary relation over D, i.e., a subset of D^k, and with any individual i an element of D. A model is also called an *interpretation*. Primitive type concepts are interpreted by subsets of D and conjunctive types respect the set intersection.

Definition 4.1 (Model of a vocabulary). Let us consider a vocabulary $\mathcal{V}=(T_C(T,\mathcal{B}), T_R, \mathcal{I})$. A *model* of $\mathcal{V}$ is a pair $M = (D, \delta)$, where D is a non-empty set called the domain of M, and δ is a mapping such that:

- $\forall p \in T \cup T_R, \delta(p) \subseteq D^{arity(p)}$,
- $\forall p, q \in T \cup T_R$ if $p \leq q$ then $\delta(p) \subseteq \delta(q)$,
- $\forall i \in \mathcal{I}, \delta(i) \in D$.
  The mapping δ is extended to $T_C(T,\mathcal{B})$ as follows:
  $\forall t = \{t_1, \ldots, t_k\} \in T_C, \delta(t)$ is defined by $\delta(t) = \delta(t_1) \cap \ldots \cap \delta(t_k)$.

Given a model (D, δ) of a vocabulary $\mathcal{V}$, in order to interpret an SG one must supplement (D, δ) by a mapping from the concepts of G to D. Therefore, each concept of G is interpreted by an element of the domain, and each k-ary relation of G is interpreted by the k-tuples of interpretations of its neighbors.

Definition 4.2 (Model of an SG). Let $G = (C, R, E, l, coref)$ be an SG over a vocabulary $\mathcal{V}$. A *model* of G is a triple (D, δ, α) such that: (D, δ) is a model of $\mathcal{V}$ and α, called an *assignment*, is a mapping from C to D, such that $\forall c_1, c_2 \in C$ if $coref(c_1, c_2)$ then $\alpha(c_1) = \alpha(c_2)$.

A model of an SG G satisfies G if it respects the structure and the labeling of G. More precisely,

Definition 4.3 (Model satisfying a SG). Let $G = (C, R, E, l, coref)$ be an SG over $\mathcal{V}$, and (D, δ, α) be a model of G.
(D, δ, α) is a *model satisfying* G, noted $(D, \delta, \alpha) \models_{sg} G$, if α is an assignment such that:

- for any individual concept c with marker i: $\alpha(c) = \delta(i)$,
- $\forall c \in C$, $\alpha(c) \in \delta(type(c))$,
- $\forall r \in R$, if $(c_1, \ldots, c_k)$ is the list of neighbors of r, then $(\alpha(c_1), \ldots, \alpha(c_k)) \in \delta(type(r))$.

Note that if $\delta(\top) \neq D$ then all the elements that do not belong to $\delta(\top)$ are useless and can be removed, since they cannot appear in any model satisfying an SG over $\mathcal{V}$.

Example. Let G be the SG in Fig. 4.1. Let us build a model (D, δ, α) satisfying G. D contains toys, persons, colors, sizes, etc.

$\delta(Car)$ is the subset of D containing all the cars.

$\delta(Person)$ is the subset of D containing all the persons, and so on for the different concept types.

$\delta(possess)$ is the subset of $D \times D$ containing all the couples (d, d') such that d is a person who possesses the object d'.

$\delta(play3)$ is the subset of $D \times D \times D$ containing all the triples (d, d', d'') such that d and d' are persons who play with the toy d'', and so on for the different relation types.

$\delta(Paul)$ is the element of D representing Paul, $\delta(Small)$ is the element of D representing the size small, and so on for all other individuals

$\alpha(a_1) = \alpha(a_2) = \delta(Paul)$, $\alpha(b_1) = \alpha(b_2) = \alpha(b_3) = u_1 \in D$, $\alpha(c) = \delta(Small)$, $\alpha(d) = u_2 \in D$, $\alpha(e) = \delta(Mary)$

$(\delta(Paul), u_1) \in \delta(possess)$, $(\delta(Paul), u_2, u_1) \in \delta(play3)$, $(u_2, \delta(Mary)) \in \delta(daughterOf)$, $(u_1, \delta(Small)) \in \delta(attr)$, etc.

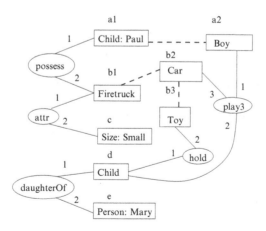

Fig. 4.1 An SG

Using the notion of an SG model it is now possible to define an entailment relation between SGs as follows:

Definition 4.4 (Entailment and equivalence). Let G and H be two SGs over $\mathcal{V}$.

- H *entails* G, or G is a *consequence* of H, notation $H \models_{sg} G$, if for any (D, δ) model of V and for any α such that $(D, \delta, \alpha) \models_{sg} H$, then there is an assignment β of the concept nodes in G such that $(D, \delta, \beta) \models_{sg} G$.
- G and H are *model-equivalent*, notation $G \equiv_{sg} H$, if $G \models_{sg} H$ and $H \models_{sg} G$.

The previous definitions ensure that an SG and its normal form have the same semantic, i.e., are model-equivalent.

Property 4.1. Let G be an SG, then $G \equiv_{sg} G/coref$.

Proof. Let (D, δ, α) be a model satisfying G. If f denotes the mapping from C_G to $C_{G/coref}$ induced by the construction of $G/coref$, then the relation $\beta = \alpha \circ f^{-1}$ is a mapping from $C_{G/coref}$ to D. Indeed, let us consider two coreferent concepts $c_1, c_2 \in C_G$ belonging to a coref class X, then $\beta(X) = \alpha(c_1) = \alpha(c_2)$. Let us check that (D, δ, β) is a model satisfying $G/coref$.

1. If X is an individual concept of $G/coref$ with marker i, then $\exists c \in C_G$ with $marker(c) = i$ and $\beta(X) = \alpha(f^{-1}(X)) = \alpha(c) = \delta(i)$.
2. If X is a generic concept of $G/coref$, then $f^{-1}(X) = \{c_1, \ldots, c_k\}$. Let $t_i = type(c_i)$, then $type(X) = min\{t_1 \cup \ldots \cup t_k\}$. Furthermore, $\forall i = 1, \ldots, k$, $\alpha(c_i) \in \delta(t_i)$, therefore $\forall i = 1, \ldots, k$, $\alpha(c_i) \in \delta(t_1) \cap \ldots \cap \delta(t_k) = \delta(min\{t_1 \cup \ldots \cup t_k\})$ and $\beta(X) = \alpha(c_i) \in \delta(type(X))$.
3. If r is a relation in $G/coref$ with a list of neighbors $X_1, \ldots, X_k$, then there is a relation s in G, with $type(r) = type(s) = t$ and with a list of neighbors $c_1 \ldots c_k$ such that for all $i = 1 \ldots k$, $X_i = f(c_i)$. By hypothesis $(\alpha(c_1), \ldots, \alpha(c_k)) \in \delta(t)$, thus $(\beta(X_1), \ldots, \beta(X_k)) \in \delta(t)$.

Conversely, if (D, δ, β) is a model satisfying $G/coref$, then it is simple to check that (D, δ, α), with $\alpha = \beta \circ f$, is a model satisfying G. □

The canonical model of a SG G plays a specific role in the characterization of the SGs consequences of G. We first define the canonical model of an SG, then we show that the canonical model of G is indeed a model satisfying G.

Definition 4.5 (Canonical model of normal SGs). Let $G = (C, R, E, l)$ be a normal SG over a vocabulary V. The *canonical model* of G is the triple $M_G = (D, \kappa, id)$ defined as follows.

- Domain. D is the set of concept nodes of G supplemented by a new element, i.e. $D = C \cup \{z\}$, where $z \notin C$.
- Interpretation of individual markers. For any c individual concept of G with marker i, $\kappa(i) = c$, and for any individual marker i which does not appear in G, $\kappa(i) = z$.
- Interpretation of concept types. For any $t \in T_C$, $\kappa(t) = \{c \in C | type(c) \leq t\}$.
- Interpretation of relation types. For any relation $p \in T_R$ with $arity(p) = k$, $\kappa(p) = \{(c_1, \ldots, c_k) | \exists r(c_1, \ldots, c_k) \in G, type(r) \leq p\}$.
- id is the identity over C.

Example. The SG in Fig. 4.2 is the normal form of the SG in Fig. 4.1. Its canonical model is the following:

$D = \{a,b,c,d,e\}$
$\kappa(Paul) = a$, $\kappa(Small) = c$, $\kappa(\{Firetruck, Toy\}) = \{b\}$, $\kappa(Child) = \{a,d\}$,
$\kappa(Size) = \{c\}$, $\kappa(Person) = \{a,d,e\}$, $\kappa(play3) = \{(a,d,b)\}$, $\kappa(possess) = \{(a,b)\}$,
$\kappa(attr) = \{(b,c)\}$, $\kappa(daughterOf) = \{(d,e)\}$.

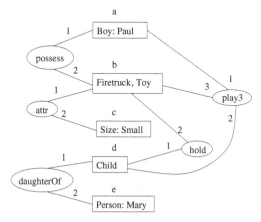

Fig. 4.2 A normal SG

The next property gathers facts simple to check:

Property 4.2 (Properties of canonical models). Let $G = (C,R,E,l)$ be a normal SG, (C,κ,id) the canonical model of G, satisfies:

1. $(C,\kappa,id) \models_{sg} G$,
2. let t be a concept type and $c \in \kappa(t)$, then c is a concept node of G with $type(c) \le t$,
3. let p be a relation type and $(c_1,\ldots,c_k) \in \kappa(p)$, then $c_1 \ldots c_k$ is the neighbor list of some relation node r of G such that $type(r) \le p$.

4.1.2 Soundness and Completeness of (coref) Homomorphism

We now prove a soundness and completeness theorem of SG homomorphism with respect to entailment.

Lemma 4.1. *Let G and H be two SGs over the same vocabulary. If there is a homomorphism from G to H, then $H \models_{sg} G$.*

Proof. Let π be a homomorphism from $G = (C_G, R_G, E_G, l_G, coref_G)$ to $H = (C_H, R_H, E_H, l_H, coref_H)$, and (D,δ,α) be a model satisfying H. Let us check that

$(D, \delta, \beta = \alpha \circ \pi)$ is a model satisfying G. First, if c_1 and c_2 are coreferent concepts of G, then their images by π are coreferent concepts of H, therefore β is an assignment of G. Let us now check that the three conditions of definition 4.3 are satisfied:

- If c is an individual concept of G with marker i, then its image by a homomorphism is an individual concept with the same marker i, thus $\beta(c) = \alpha(\pi(c)) = \delta(i)$.
- Let c be a concept of G with $\pi(c) = c'$, then $type(c') \leq type(c)$, and, since (D, δ) is a model of $\mathcal{V}$, it holds that $\delta(type(c')) \subseteq \delta(type(c))$. (D, δ, α) is a model satisfying H, thus $\alpha(c') \in \delta(type(c'))$, and one has $\beta(c) = \alpha(\pi(c)) \in \delta(type(c'))$, therefore $\beta(c) \in \delta(type(c))$.
- Let $r(c_1, \ldots, c_k)$ be in G. $(\alpha(\pi(c_1)), \ldots, \alpha(\pi(c_k)))$ is an element of $\delta(type(\pi(r)))$ and $type(\pi(r)) \leq type(r)$, thus $(\beta(c_1), \ldots, \beta(c_k)) \in \delta(type(r))$. $\square$

Lemma 4.2. *Let G and H be two SGs over the same vocabulary. The following properties are equivalent.*

1. *$H \models_{sg} G$.*
2. *There is an assignment α such that $(C_{H/coref}, \kappa, \alpha) \models_{sg} G$, where $(C_{H/coref}, \kappa, id)$ is the canonical model of $H/coref$.*
3. *There is a homomorphism from G to $H/coref$.*

Proof.
$1 \Rightarrow 2$
Since $(C_{H/coref}, \kappa, id) \models_{sg} H/coref$ and $H/coref \models_{sg} H$, there is β such that $(C_{H/coref}, \kappa, \beta) \models_{sg} H$. Thus, if $H \models_{sg} G$, there is α such that $(C_{H/coref}, \kappa, \alpha) \models_{sg} G$.
$2 \Rightarrow 3$
Let H' denote $H/coref$. By hypothesis, there is α such that $(C_{H'}, \kappa, \alpha)$ is a model satisfying G. We first show that α is a mapping from C_G to $C_{H'}$ that respects the condition of a homomorphism, and then, we show that there is a mapping β from R_G to $R_{H'}$, such that (α, β) is a homomorphism from G to H'.

- If c is a concept in G, then $\alpha(c) \in \kappa(type_G(c))$. Thus, $\alpha(c)$ is a concept in H' with $type_{H'}(\alpha(c)) \leq type_G(c)$. Let c be an individual concept of G with marker i. Then, H' contains an individual concept with marker i otherwise $\kappa(i) \notin C_{H'}$ and $(C_{H'}, \kappa, \alpha)$ does not satisfy G. H' is normal, thus $\alpha(c) = \kappa(i)$, which is the only concept of H' with marker i.
- Let us now show that α can be extended to a homomorphism from G to H. Let us consider any relation r of G and its neighbors list $c_1 \ldots c_k$. $(C_{H'}, \kappa, \alpha) \models_{sg} G$ implies that $(\alpha(c_1), \ldots, \alpha(c_k)) \in \kappa(type_G(r))$. Therefore, according to the statement 3 of Property 4.2, there is a relation s of H' such that $\alpha(c_1) \ldots \alpha(c_k)$ is the neighbor list of s with $type_{H'}(s) \leq type_G(r)$. Let us define the mapping β from R_G to $R_{H'}$ by $\beta(r) = s$. (α, β) is a homomorphism from G to H'.

$2 \Rightarrow 3$
From Lemma 4.1 and Property 4.1. $\square$

Note that $1 \Leftrightarrow 2$ in the previous lemma shows that deciding whether an SG G is a consequence of an SG H can be done by considering the canonical model of $H/coref$ instead of all models satisfying H.

Lemmas 4.1 and 4.2 yield the following soundness and completeness result:

Theorem 4.1. *Let G and H be two SGs over the same vocabulary. If there is a homomorphism from G to H, then $H \models_{sg} G$. If $H/coref \models_{sg} G$, then there is a homomorphism from G to $H/coref$.*

Using the previous theorem and theorem 3.1 one obtains:

Theorem 4.2. *Let G and H be two SGs over the same vocabulary. There is a coref-homomorphism from G to H if and only if $H \models_{sg} G$.*

4.2 Logical Semantic

In this section a logical semantic for the SGs is defined as follows: First, a FOL language is naturally associated with any vocabulary $\mathcal{V}$ and a set of FOL formulas, denoted $\Phi(\mathcal{V})$, corresponding to the order relations existing over $\mathcal{V}$, is defined. Secondly, for any SG G over $\mathcal{V}$, a FOL formula $\Phi(G)$, built on the language associated with $\mathcal{V}$, is defined. Thirdly, it is shown that the classical model theory of FOL is equivalent to the model semantic presented in the first section. Therefore, a corollary of the previous soundness and completeness theorem states that SG homomorphism is sound and complete with respect to deduction in FOL.

4.2.1 The FOL Semantic Φ

Logical Semantic of a Vocabulary

A FOL language $\mathcal{L}_\mathcal{V}$ can be naturally associated with any SG vocabulary $\mathcal{V} = (T_C, T_R, \mathcal{I})$. $\mathcal{L}_\mathcal{V}$ is defined as follows: A constant is assigned to each individual marker, a n-ary predicate is assigned to each n-ary relation type, and a unary predicate is assigned to each *primitive* concept type.

For simplicity, we consider that each logical constant or predicate has the same name as the element of the vocabulary it is assigned to, i.e., the logical constant associated with the individual marker i is also denoted by i, and the predicate associated with the type t is also denoted by t.

The ordered facet of the vocabulary $\mathcal{V}$ is represented by the following set $\Phi(\mathcal{V})$ of FOL formulas.

Definition 4.6 ($\Phi(\mathcal{V})$).

$\Phi(\mathcal{V})$ is equal to the following set of formulas: $\forall x_1 \ldots x_k (t_2(x_1, \ldots, x_k) \rightarrow t_1(x_1, \ldots, x_k))$, for all t_1 and t_2 primitive concept types or relation types of $\mathcal{V}$ such that $t_2 < t_1$, k being the arity of t_1 and t_2.

Example. Let *wash* and *act* be two binary relation types, with *wash* < *act*. $\Phi(\mathcal{V})$ contains the formula $\forall x \forall y\ (wash(x,y) \rightarrow act(x,y))$. Let *Boy*, *Child* and *Person* be three concept types with *Boy* < *Child* < *Person*. $\Phi(\mathcal{V})$ contains the formulas $\forall x\ (Boy(x) \rightarrow Child(x))$ and $\forall x\ (Child(x) \rightarrow Person(x))$. It also contains the formula $\forall x\ (Boy(x) \rightarrow Person(x))$, but this formula can be obtained by transitivity from the two first ones. More generally, the covering relation of the order could be interpreted instead of < itself. However, as reasonings are performed on the graphs themselves, and not on the associated logical formulas, this question is irrelevant.

Let us consider two conjunctive types $t = \{t_1,\ldots,t_n\}$ and $s = \{s_1,\ldots,s_p\}$. Let us recall that $t \le s$ if and only if for every $s_j \in s$, $1 \le j \le p$, there is a $t_i \in t$, $1 \le i \le n$, such that $t_i \le s_j$. It is straightforward to check the following property:

Property 4.3. Let $t = \{t_1,\ \ldots,\ t_n\}$ and $s = \{s_1,\ \ldots,\ s_p\}$ be two conjunctive types with $t \le s$, then:
$\forall x(t_1(x) \wedge \ldots \wedge t_n(x) \rightarrow s_1(x) \wedge \ldots \wedge s_p(x))$ is a consequence of $\Phi(\mathcal{V})$.

The previous property explains why only primitive concept types are considered in Definition 4.6. The formulas representing the order relation on the non-primitive types are not explicitly needed since they are consequences of the formulas representing the order relation on the primitive types.

Logical Semantic of an SG

The mapping Φ transforms any SG over $\mathcal{V}$ into a formula of the existential conjunctive and positive fragment of FOL over the language $\mathcal{L}_\mathcal{V}$. This fragment is denoted $FOL(\wedge, \exists, \mathcal{L}_\mathcal{V})$. Roughly said, concepts are translated into terms—variables or constants—and relations are transformed into atoms (containing these terms).

Since the meaning of an SG cannot be given independently from its vocabulary, deduction in the logical fragment associated with SGs is relative to $\Phi(\mathcal{V})$. That is, given two formulas f and g of this fragment, the basic question is not whether f is deducible from g, but rather whether f is deducible from g *and* $\Phi(\mathcal{V})$.

Definition 4.7 ($\Phi(G)$). Given any SG G, the formula $\Phi(G)$ is built as follows.

- A term $term(c)$ is assigned to each concept c in the following way. First, a term $term(X)$ is assigned to each class X of coref. If X contains an individual marker i, then $term(X) = i$. Otherwise, $term(X)$ is a variable, and distinct variables are associated with different classes. Secondly, if c belongs to the coref class X, then $term(c) = term(X)$.
- Let c be a concept node with type $\{t_1,\ldots,t_n\}$. The formula $t_1(term(c)) \wedge \ldots \wedge t_n(term(c))$ is assigned to c.
- The atom $r(term(c_1),\ldots,term(c_k))$ is assigned to each relation node x, where r is its type, k the arity of r, and c_i denotes the i-th neighbor of x.
- Let $\phi(G)$ be the conjunction of the formulas associated with all the concept and the relation nodes.
- Finally, $\Phi(G)$ is the existential closure of $\phi(G)$.

Note that if *term* is one-to-one over the set of coref classes of some SG G, it is not necessarily the case over the concept set of G. More precisely, *term* is one-to-one over the concept set of G if and only if G is normal.

Also note that $\Phi(G) = \Phi(G/coref)$.

Example. Let G be the SG in Fig. 4.2. Let us build $\Phi(G)$. As G is normal, each coreference class contains exactly one concept node. We assign to the nodes a, b, c, d and e, respectively, the terms *Paul*, x, *Small*, y and *Mary*, where x and y are the sole variables. The conjunction of the formulas assigned to the concept nodes is:
$A = Boy(Paul) \wedge Firetruck(x) \wedge Toy(x) \wedge Size(Small) \wedge Child(y) \wedge Person(Mary)$.
The conjunction of the formulas assigned to the relation nodes is:
$B = possess(Paul,x) \wedge attr(y,Small) \wedge play3(Paul,y,x) \wedge hold(y,x) \wedge$
$daughterOf(y,Mary)$. Finally, $\Phi(G) = \exists x \exists y (A \wedge B)$, that is:
$\Phi(G) = \exists x \exists y (Boy(Paul) \wedge Firetruck(x) \wedge Toy(x) \wedge Size(Small) \wedge Child(y) \wedge$
$Person(Mary) \wedge possess(Paul,x) \wedge attr(y,Small) \wedge play3(Paul,y,x) \wedge hold(y,x) \wedge$
$daughterOf(y,Mary))$.

Now, let us consider the non-normal *SG* in Fig. 4.1, say G'. The normal form of G' is G. The coreference classes of G' are $\{a_1,a_2\}$, $\{b_1,b_2,b_3\}$, $\{c\}$, $\{d\}$ and $\{e\}$. Let us assign to these classes *Paul*, x, *Small*, y and *Mary*, i.e., the same terms as for the corresponding nodes in G. The conjunction of the formulas assigned to the concept nodes in G' is $A' = Child(Paul) \wedge Boy(Paul) \wedge Firetruck(x) \wedge Toy(x) \wedge$
$Car(x) \wedge Size(Small) \wedge Child(y) \wedge Person(Mary)$. $\Phi(G') = \exists x \exists y (A' \wedge B)$, where B is the same as in $\Phi(G)$. Note that $\Phi(\mathcal{V})$ contains $\forall x \ (Boy(x) \rightarrow Child(x))$ and $\forall x \ (FireTruck(x) \rightarrow Car(x))$, thus the atoms $Child(Paul)$ and $Car(x)$ are redundant in $\Phi(G')$. More precisely, from $\Phi(\mathcal{V})$ it can be deduced that $\Phi(G')$ and $\Phi(G)$ are equivalent.

4.2.2 Model Semantic and Φ

The set semantic proposed in Sect. 4.1 is equivalent to the model theory for the FOL semantic Φ.

In the following property, a (SG) model satisfying an SG G over a vocabulary $\mathcal{V}$ (cf. Definition 4.3) is compared to a (FOL) model of $\mathcal{L}_\mathcal{V}$ satisfying the formula $\Phi(G)$.

Property 4.4 (FOL and SG models). Let $\mathcal{V}$ be an SG vocabulary, and G and H be two SGs over $\mathcal{V}$.

1. Let $M = (D,\delta)$ be a FOL model of $\mathcal{L}_\mathcal{V}$. M is a FOL model satisfying $\Phi(\mathcal{V})$ if and only if M is an SG model of $\mathcal{V}$.
2. (D,δ,α) is an SG model satisfying G if and only if (D,δ) is a FOL model satisfying $\Phi(G) \wedge \Phi(\mathcal{V})$.
3. $H \models_{sg} G$ if and only if $\Phi(\mathcal{V}), \Phi(H) \models \Phi(G)$.

Proof. 1. It is immediate to check that an SG model of a vocabulary $\mathcal{V}$ (see Definition 4.1) is a FOL model of $\mathcal{L}_\mathcal{V}$ (cf. Definition A.23).

Let $M = (D, \delta)$ be a FOL model of $\mathcal{L}_\mathcal{V}$ satisfying $\Phi(\mathcal{V})$. If $p, q \in T \cup T_R$ with $p \leq q$ and $\delta(p) \not\subseteq \delta(q)$, then the formula $\forall x(p(x) \rightarrow q(x))$, which is in $\Phi(\mathcal{V})$, is false (consider an element $d \in \delta(p) \smallsetminus \delta(q)$). $M = (D, \delta)$ is thus an SG model of $\mathcal{V}$. Conversely, if $M = (D, \delta)$ is an SG model of $\mathcal{V}$, then $\forall p, q \in T \cup T_R$ with $p \leq q$ one has $\delta(p) \subseteq \delta(q)$, and $M = (D, \delta)$ satisfies $\Phi(\mathcal{V})$.

2. If (D, δ, α) is an SG model satisfying G, then (D, δ) is a FOL model of $\mathcal{L}_\mathcal{V}$ and is a FOL model satisfying $\Phi(\mathcal{V})$. Furthermore, the properties of α show that values in D can be assigned to the existentially quantified variables of $\Phi(G)$ in such a way that $\Phi(G)$ is true. Reciprocally, let us consider a FOL model (D, δ) satisfying $\Phi(G) \wedge \Phi(\mathcal{V})$. Then, as (D, δ) is a FOL model satisfying $\Phi(\mathcal{V})$ it is a model of $\mathcal{V}$. The assignment α is built as follows: if c is a concept coreferent to an individual concept with marker i, then $\alpha(c) = \delta(i)$; if c is generic, $\alpha(c)$ is equal to the element of D that can be assigned to the variable associated with c in the FOL model.

3. The previous result leads to "$H \models_{sg} G$ iff any FOL model satisfying $\Phi(H) \wedge \Phi(\mathcal{V})$ is a FOL model satisfying $\Phi(G)$".

$\square$

Theorem 4.1 and the previous property gives the following theorem:

Theorem 4.3 (Soundness and completeness of homomorphism with respect to Φ). *Let G and H be two SGs over a vocabulary $\mathcal{V}$ with H normal. There is a homomorphism from G to H iff $\Phi(\mathcal{V}), \Phi(H) \models \Phi(G)$.*

4.3 Positive, Conjunctive, and Existential Fragment of FOL

Let $\mathcal{L}$ be a FOL language composed of a partially ordered set P of predicate symbols of arity ≥ 1 and a set of constants C. Two comparable predicates must have the same arity. $A(\mathcal{L})$ denotes the set of atoms built over $\mathcal{L}$, and $FOL(\wedge, \exists, \mathcal{L})$ denotes the existential conjunctive and positive fragment of FOL over $\mathcal{L}$. Let f be a wff in $FOL(\wedge, \exists, \mathcal{L})$, $atoms(f)$ (resp. $terms(f)$, $vars(f)$, $consts(f)$) denotes the set of atoms (resp. terms, variables, constants) of f. For any f, $terms(f) = vars(f) \cup consts(f)$. Any f in $FOL(\wedge, \exists, \mathcal{L})$ is equivalent to the existential closure of the conjunction of $atoms(f)$. Hereafter, we assume that any wff in $FOL(\wedge, \exists, \mathcal{L})$ respects this form, i.e., is in prenex form.

An $\mathcal{L}$-*substitution* from a subset of $A(\mathcal{L})$ to another one is defined in Sect. 4.3.1. Then, the $\mathcal{L}$-substitution lemma states that $\mathcal{L}$-substitution is, in a way that will be specified, equivalent to logical entailment. Remark that an ordered FOL language $\mathcal{L}_\mathcal{V}$ can be canonically associated with an SG vocabulary $\mathcal{V}$, in such a way that if G is an SG built over $\mathcal{V}$ then $\Phi(G)$ is a wff in $FOL(\wedge, \exists, \mathcal{L}_\mathcal{V})$.

The $\mathcal{L}$-substitution lemma is used in Sect. 4.3.2 to give another proof of the soundness and completeness theorem 4.3. More generally, the $\mathcal{L}$-substitution lemma can be used to study different FOL semantics of different conceptual graph sorts. We claimed in the introduction that SGs are "equivalent" to the positive, conjunctive and existential fragment of FOL without functions. The precise meaning of that claim is explained in Sect. 4.3.3. Finally, in Sect. 4.3.4 a FOL semantic representing the whole structure of an SG is presented.

4.3.1 Ordered Language and $\mathcal{L}$-Substitution

Definition 4.8 ($\mathcal{L}$-substitution). Let g and h be two wffs in $FOL(\wedge, \exists, \mathcal{L})$, with $vars(g) \cap vars(h) = \emptyset$. A $\mathcal{L}$-substitution from g to h is a substitution σ such as: for any $p(\vec{e})$ in $atoms(g)$, there is an atom $q(\sigma(\vec{e}))$ in $atoms(h)$ with $q \leq p$.

Let $\Phi(\mathcal{L})$ be the set of formulas which are the universal closures of $q(\vec{e}) \rightarrow p(\vec{e})$, where $\vec{e}$ is a list of distinct variables, for all $q \leq p$ in $\mathcal{L}$. Let us give some definitions about FOL models adapted for $FOL(\wedge, \exists, \mathcal{L})$ over an ordered language.

Definition 4.9 ($\mathcal{L}$-Models). Let (D, δ) be a (FOL) model of $\mathcal{L}$.

- An *assignment* is a mapping α from variables and constants to D such that for any constant a, $\alpha(a) = \delta(a)$.
- A model (D, δ) *satisfies* a wff f if there is an assignment α such that for any $p(\vec{e}) \in atoms(f)$, $\alpha(\vec{e})$ is in $\delta(p)$.
- A $\mathcal{L}$-*model* is a model (D, δ) of $\mathcal{L}$ respecting the partial order, i.e., if $\vec{d} \in \delta(q)$ and $q \leq p$ then $\vec{d} \in \delta(p)$.

One has:

Property 4.5. (D, δ) is a $\mathcal{L}$-model of $\mathcal{L}$ if and only if $(D, \delta) \models \Phi(\mathcal{L})$.

The relationships between $\mathcal{L}$-substitution and logical entailment is stated as follows:

Lemma 4.3 ($\mathcal{L}$-substitution lemma). *Let g and h be two wffs in $FOL(\wedge, \exists, \mathcal{L})$, with $vars(g) \cap vars(h) = \emptyset$. There is an $\mathcal{L}$-substitution from g to h iff $\Phi(\mathcal{L}), h \models g$.*

Proof. Let σ be a $\mathcal{L}$-substitution from g to h. Let us consider a model (D, δ) satisfying h and $\Phi(\mathcal{L})$. For any $p(\vec{e}) \in atoms(g)$, there is an atom $q(\sigma(\vec{e})) \in atoms(h)$ with $q \leq p$. As (D, δ) satisfies h, there is an assignment α, such that for any constant a, $\alpha \circ \sigma(a) = \alpha(a) = \delta(a)$, and for any atom $q(\vec{d}) \in atoms(h)$, $\alpha(\vec{d})$ is in $\delta(q)$, thus $\alpha(\sigma(\vec{e}))$ is in $\delta(q)$. As (D, δ) also satisfies $\Phi(\mathcal{L})$, (D, δ) is an $\mathcal{L}$-model, therefore $\alpha(\sigma(\vec{e}))$ is in $\delta(p)$. Thus $\alpha \circ \sigma$ is a mapping from $terms(g)$ to D such that for any atom $p(\vec{e}) \in atoms(g)$, $\alpha \circ \sigma(\vec{e})$ is in $\delta(p)$, therefore (D, δ) is a model of g. Conversely, let us assume that $\Phi(\mathcal{L}), h \models g$, any model (D, δ) satisfying h and $\Phi(\mathcal{L})$, is an $\mathcal{L}$-model, and it also satisfies g. Let us take $D = terms(h) \cup \{constants\}$, $\delta(a) = a$ for any constant a, and for any predicate p, $\delta(p) = \{\vec{e} \mid$ there is $q(\vec{e}) \in atoms(h)$ and $q \leq p\}$. There is an assignment α such that for any constant a $\alpha(a) = \delta(a) = a$, and for any $p(\vec{e}) \in atoms(g)$, $\alpha(\vec{e})$ is in $\delta(p)$. By definition of δ, as $\alpha(\vec{e}) \in \delta(p)$, there is $q \leq p$ with $q(\alpha(\vec{e})) \in atoms(h)$, and α is a $\mathcal{L}$-substitution from g to h. $\square$

One can also simply prove the $\mathcal{L}$-substitution lemma by using the SLD resolution. Indeed, the clausal forms of $h, \Phi(\mathcal{L})$, are composed of Horn clauses, and $\neg g$ is a negative Horn clause.

4.3.2 Soundness and Completeness Revisited

In this section we do not use the set semantic for studying the logical semantic Φ as in Sect. 4.2.1, but we directly deal with FOL. More precisely, the $\mathcal{L}$-substitution lemma shows that the existence of an $\mathcal{L}$-substitution from a formula A to a formula B is equivalent to the logical entailment from B and the formulas representing the order over $\mathcal{L}$ to A. In order to state a soundness and completeness theorem (of homomorphism with respect to logical entailment) we establish relationships between the existence of a homomorphism from G to H and the existence of an $\mathcal{L}$-substitution from $\Phi(G)$ to $\Phi(H)$. Such a technique can be used for any sort of conceptual graphs equipped with a semantic in $FOL(\wedge, \exists)$ (e.g., it is used in Chap. 9).

Property 4.6. Let G and H be two SGs defined on $\mathcal{V}$ and $\mathcal{L}$ the ordered FOL language associated with $\mathcal{V}$. If there is a homomorphism from G to H, then there is an $\mathcal{L}$-substitution from $\Phi(G)$ to $\Phi(H)$.

Proof. Let π be a homomorphism from G to H, and let $g = \Phi(G)$ and $h = \Phi(H)$, with $vars(g) \cap vars(h) = \emptyset$. From π, we build a mapping σ from $terms(g)$ to $terms(h)$ in the following way: for any term e in $terms(g)$, let c be any concept such that $e = term(c)$, then $\sigma(e) = term(\pi(c))$. Note that $\sigma(e)$ is uniquely defined whatever the chosen c is. Indeed, if $term(c) = a$ then the image of c is an individual concept with marker a, and if c and c' are two coreferent concept nodes in G then the same term is associated with their images in H. Let us show that σ is an $\mathcal{L}$-substitution from $\Phi(G)$ to $\Phi(H)$.

Let $A = t(term(c))$ be the atom in g associated with the concept c of G. Let $c' = \pi(c)$ and t' be the type of c'. $t'(term(c')) = t'(\sigma(c))$ is an atom in h, and by definition of homomorphism $t' \leq t$. Let $A = t(term(c_1), \ldots, term(c_k))$ be the atom in g associated with the relation r in G. $t'(term(c'_1), \ldots, term(c'_k))$ is an atom in h, where for all i, $1 \leq i \leq k$, $c'_i = \pi(c_i)$, and t' is the type of $\pi(r)$. $t'(term(c'_1), \ldots, term(c'_k)) = t'(\sigma(c_1), \ldots, \sigma(c_k))$ and $t' \leq t$. $\square$

The converse property is not true in general: If there is an $\mathcal{L}$-substitution from $\Phi(G)$ to $\Phi(H)$ there is not necessarily a homomorphism from G to H in the case where H is not normal. The reason has been previously seen: The mapping Φ does not translate the whole structure of a BG. If several concepts c_1 and c_2 have the same individual marker m, i.e., if a BG is not normal, a relational atom $r(\ldots, m, \ldots)$ of its logical translation does not keep the information about which concept m comes from.

Example. Let G and H be the graphs Fig. 4.3.

$g = \Phi(G) = t(m) \wedge r(m) \wedge s(m)$ and $h = \Phi(H) = t(m) \wedge t(m) \wedge r(m) \wedge s(m)$. The identity mapping is an $\mathcal{L}$-substitution from g to h (since $atoms(g) = atoms(h)$ and

$terms(g) = terms(h))$ but there is no homomorphism from G to H: A homomorphism is a mapping from the set of nodes of G to the set of nodes H; therefore the individual concept of G cannot be mapped onto the two different individual concepts of H.

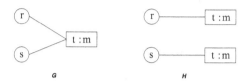

Fig. 4.3 Normality and $\mathcal{L}$-substitution

Property 4.7. Let G and H be two SGs, with H being normal. If ρ is an $\mathcal{L}$-substitution from $\Phi(G)$ to $\Phi(H)$, then there is a homomorphism π from G to H. Moreover, for any concept node c in G associated with the term u in $\Phi(G)$, $\pi(c)$ is the concept node in H associated with the term $\rho(u)$ in $\Phi(H)$.

Proof. Let ρ be an $\mathcal{L}$-substitution from $\Phi(G)$ to $\Phi(H)$. From ρ we build a mapping π from G to H. First, let us consider the terms of $\Phi(G)$. Since H is normal, there is a bijection from the concept set of H to the terms of $\Phi(H)$. All concepts of G associated with a specific term t of $\Phi(G)$ have for image by π the unique concept of H that is associated with $\rho(t)$. The property of ρ about the atoms corresponding to the concept types ensures that π satisfies the homomorphism property for the concepts. Let us now consider the atoms associated with the relations of G. If $r'(\rho(t_1),\ldots,\rho(t_k))$ is the atom associated with $r(t_1,\ldots,t_k)$ by ρ, then all (duplicate) relations in G with atom $r(t_1,\ldots,t_k)$ have for image by π any relation of H with atom $r'(\rho(t_1),\ldots,\rho(t_k))$. Then, π is a homomorphism from G to H. □

The following theorem gathers Property 4.6 and Property 4.7.

Theorem 4.4 (Homomorphism and $\mathcal{L}$-substitution). *Let G and H be two SGs defined on $\mathcal{V}$ and $\mathcal{L}$ the ordered FOL language associated with $\mathcal{V}$. If there is a homomorphism from G to H, then there is an $\mathcal{L}$-substitution from $\Phi(G)$ to $\Phi(H)$. If H is normal and if there is an $\mathcal{L}$-substitution from $\Phi(G)$ to $\Phi(H)$, then there is a homomorphism from G to H.*

The soundness and completeness theorem can now be obtained as a corollary of the previous theorem and of the $\mathcal{L}$-substitution lemma.

4.3.3 Relationships Between SGs and FOL($\wedge$, $\exists$)

The precise meaning of the "equivalence" between SGs and FOL($\wedge$, $\exists$) is now given.

Let $\mathcal{L}$ be a (classical) FOL language, i.e., unordered, without functions and 0-ary predicates. First, we show that the set of formulas in FOL($\wedge$, $\exists$, $\mathcal{L}$) can be translated into the set SG($\mathcal{V}_{\mathcal{L}}$) of SGs over a vocabulary $\mathcal{V}_{\mathcal{L}}$, in such a way that deduction is translated into homomorphism. Then, we show that the set of SGs over a vocabulary $\mathcal{V}$ can be translated into the set of formulas in FOL($\wedge$, $\exists$, $\mathcal{L}_{flat(\mathcal{L})}$)), in such a way that homomorphism is translated into deduction.

4.3.3.1 From FOL Formulas to SGs

In order to associate SGs with FOL formulas, the first step is to associate an SG vocabulary with a FOL language. Let $\mathcal{L}$ be a FOL language (P,C), where P is the set of predicates and C is the set of individual constants. $\mathcal{V}_{\mathcal{L}}$ is the vocabulary $(\{\top\},P,C)$, i.e., the vocabulary with a unique concept type $\top$, with the relation type set being equal to the predicate set of $\mathcal{L}$ and the order on P being the discrete order, and with the individual marker set being equal to the individual constant set of $\mathcal{L}$. Now, $f2G$ is the mapping from FOL($\wedge$, $\exists$, $\mathcal{L}$) to the set SG($\mathcal{V}_{\mathcal{L}}$), defined as follows: Any formula f is mapped onto an SG G whose concept node set is in bijection with the set of terms occurring in f (one node [$\top$: ∗] for each variable and one node [$\top$: a] for each constant a), and whose relation node set is in bijection with the set of atoms of f (for each atom $p(e_1, ..., e_k)$ one node (p) whose i-th neighbor is the node assigned to e_i). Figure 4.4 pictures the graph $f2G(f)$ assigned to the formula $f = \exists x(Car(x) \wedge Toy(x) \wedge Boy(Paul) \wedge possess(Paul,x))$.

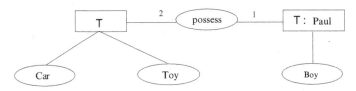

Fig. 4.4 The mapping f2G

The following property states that $f2G$ transforms deduction into homomorphism.

Property 4.8. Let g and h be two formulas of FOL($\wedge$, $\exists$, $\mathcal{L}$). The mapping $f2G$ associates with g and h two (normal) SGs $G = f2G(g)$ and $H = f2G(h)$ over the vocabulary $\mathcal{V}_{\mathcal{L}}$ such that $h \vDash g$ if and only if there is a homomorphism from G to H.

Proof. Let $g' = \Phi(G)$ and $h' = \Phi(H)$. Check that $h \vDash g$ if and only if $h' \vDash g'$. We conclude using soundness and completeness of homomorphism (Theorem 4.3), and noticing that $\Phi(\mathcal{V}_{\mathcal{L}})$ is empty.   □

Note that, for any formula f the SG $f2G(f)$ is a normal SG. Furthermore, $f2G$ is a bijection. More precisely,

Property 4.9. $f2G$ is a bijection from the set $\text{FOL}(\wedge, \exists, \mathcal{L})$ (to within variable renaming) to the set of normal SGs without isolated concept nodes (to within isomorphism) on the vocabulary $\mathcal{V}_{\mathcal{L}}$.

Proof. Let G be a normal SG without isolated concept nodes on $\mathcal{V}_{\mathcal{L}}$, and let $\Phi'(G)$ denote the existential closure of the conjunction of all the atoms associated by Φ with all relation nodes of G. One can check that $f2G \circ \Phi'$ (and $\Phi' \circ f2G$) is the identity mapping on its domain. □

Remark that, if we restrict ourselves to normal SGs over $\mathcal{V}_{\mathcal{L}}$, the only difference between $f2G^{-1}$ and Φ is that $f2G^{-1}$ does not associate atoms with concept nodes. If we associate with $\mathcal{V}_{\mathcal{L}}$ the logical interpretation $\Phi(\mathcal{V}_{\mathcal{L}}) = \{\forall x \top(x)\}$ instead of the empty set, then for any graph G over this vocabulary, one has $\Phi(\mathcal{V}_{\mathcal{L}}), f2G^{-1}(G) \vDash \Phi(G)$ (and of course $\Phi(G) \vDash f2G^{-1}(G)$).

4.3.3.2 From SGs to FOL Formulas

For translation in the other direction, the apparent problem is that formulas assigned to the vocabulary by Φ are universally quantified and are used in the deduction process. However, we can do without them, replacing the vocabulary by a vocabulary with only one concept type, and with the order on relation type set being the discrete order (i.e., it is reduced to the equality). We will define a mapping $G2f$, which is a variant of Φ. It maps an SG to an existential formula integrating the needed knowledge about the vocabulary.

Definition 4.10 (Flat vocabulary, $flat(\mathcal{V})$ and $flat(G)$). An SG vocabulary is *flat* if its concept type set is reduced to $\{\top\}$ and the order on its relation type set is the discrete order.

- Let $\mathcal{V} = (T_C, T_R, \mathcal{I})$ be an SG vocabulary. $flat(\mathcal{V})$ is the flat vocabulary obtained from $\mathcal{V}$ as follows: The new relation type set is the union of T_R and $T_C \setminus \{\top\}$, with the concept types becoming unary relation types ordered by the discrete order; the new concept type set has a single element $\top$; $\mathcal{I}$ is unchanged.
- Let G be an SG on $\mathcal{V}$. $flat(G)$ is the SG over $flat(\mathcal{V})$ obtained from G as follows: For each concept node c of G of type t_c, the new type of c is $\top$, and for each primitive concept type greater or equal to t_c and distinct from $\top$, one adds a relation node with type t_c and neighbor c; for each relation node r of type t_r in G, for each relation type $t_{r'}$ in $\mathcal{V}$ strictly greater than t_r, a relation node of type $t_{r'}$ with the same neighbors as r is added.

It is simple to check that, given two SGs G and H on $\mathcal{V}$, there is a homomorphism from G to H if and only if there is a homomorphism from $flat(G)$ to $flat(H)$, both relative to $flat(\mathcal{V})$.

Example. Let $\mathcal{V} = (T_C, T_R, \mathcal{I})$, with: $T_C = \{\top, Person, Boy, Object, Car, FireTruck\}$ and $Person, Object < \top$, $Boy < Person$, $Car < Object$, $FireTruck < Car$; $T_R = \{\top_2, possess\}$, with $possess < \top_2$; $\mathcal{I} = \{Paul\}$. $flat(\mathcal{V}) = (\{\top\}, T_R', \mathcal{I})$ with $T_R' =$

{*Person, Boy, Object, Car, FireTruck*, $\top_2$, *possess*}. Figure 4.5 illustrates the translation of a graph on $\mathcal{V}$ to a graph on $flat(\mathcal{V})$.

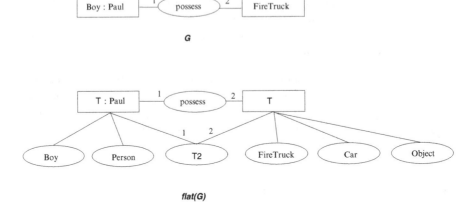

Fig. 4.5 Flattening an SG

Now, let us define $G2f(G) = \Phi(flat(G))$. For instance, the formula assigned to the graph G in Fig. 4.5 is :
$\exists x (\top(x) \wedge FireTruck(x) \wedge Car(x) \wedge Object(x) \wedge \top(Paul) \wedge Person(Paul) \wedge Boy(Paul) \wedge possess(Paul, x) \wedge \top_2(Paul, x))$.

$G2f$ transforms homomorphism into deduction.

Property 4.10. Let G and H be two normal SGs on $\mathcal{V}$. The mapping $G2f$ associates with G and H two formulas of $FOL(\wedge, \exists, \mathcal{L}_{flat(\mathcal{V})})$, $g = G2f(G)$ and $h = G2f(H)$, such that there is a homomorphism from G to H if and only if $h \vDash g$.

Proof. One has $g = G2f(G) = \Phi(flat(G))$ and $h = G2f(H) = \Phi(flat(H))$. Using soundness and completeness of homomorphism, $h \vDash g$ if and only if $flat(G) \succeq flat(H)$, since $\Phi(\mathcal{L}_{flat(\mathcal{V})}) = \emptyset$ and $flat(G)$ and $flat(H)$ are normal graphs. It remains to check that $flat(G) \succeq flat(H)$ iff $G \succeq H$.   □

If G and H are two non-isomorphic normal SGs, then $flat(G)$ and $flat(H)$ are also non-isomorphic, and the two associated formulas are different, thus:

Property 4.11. $G2f$ is an injective application from the set of normal SGs (to within isomorphism) defined on $\mathcal{V}$ to the set $FOL(\wedge, \exists, \mathcal{L}_{flat(\mathcal{V})})$ (to within variable renaming).

4.3.4 Another FOL Semantic: Ψ

When the FOL semantic Φ is considered, the normality condition for SG homomorphism occurs because $\Phi(G)$ exactly represents G only when G is normal. Another

FOL semantic, denoted Ψ, can be considered which represents the whole structure of an SG. In this semantic, two terms are associated with a concept node c. The first term, $\bar{c}$ represents the coreference class of c as in Φ; it is a constant if c is coreferent to an individual concept, otherwise it is a variable. The second term, x_c, is a variable representing the node itself. All these variables are pairwise distinct.

If $\mathcal{V}$ is a vocabulary, $\Psi(\mathcal{V})$ is defined as $\Phi(\mathcal{V})$, except that predicates associated with concept types become binary. If G is an SG, $\Psi(G)$ is built as follows. Assign the atom $t(\bar{c}, x_c)$ to each concept node c with type t, and assign the atom $r(x_{c_1}, \ldots, x_{c_k})$ to each relation node having r for relation type and $c_1 \ldots c_k$ for neighbor list. $\Psi(G)$ is the existential closure of the conjunction of these atoms.

Example. The formula $\Psi(G)$ associated with the graph G in Fig. 4.2 is obtained as follows:

The variables corresponding to the five concept nodes a, b, c, d, e are x_1, x_2, x_3, x_4, x_5, respectively. The variables corresponding to the two generic concept nodes b, d are y_1, y_2, respectively. The atoms associated with the concepts are:

$Boy(Paul, x_1)$, $Toy(y_1, x_2)$, $Firetruck(y_1, x_2)$, $Size(Small, x_3)$, $Child(y_2, x_4)$, $Person$ $(Mary, x_5)$.

The atoms associated with the relations are:

$possess(x_1, x_2)$, $attr(x_2, x_3)$, $daughterOf(x_4, x_5)$, $hold(x_4, x_2)$, $play3(x_1, x_4, x_2)$.

$\Psi(G)$ is the existential closure of the conjunction of the atoms associated with the concepts and of the atoms associated with the relations.

Now, if an SG is not normal, then it is not logically equivalent to its normal form anymore. Consider for instance the graphs G and H in Fig. 4.3, G being the normal form of H. $\Psi(G) = \exists x \ (t(m,x) \wedge r(x) \wedge s(x))$ and $\Psi(H) = \exists x_1 \exists x_2 \ (t(m,x_1) \wedge t(m,x_2) \wedge r(x_1) \wedge s(x_2))$. $\Psi(H)$ can be deduced from $\Psi(G)$, but $\Psi(G)$ cannot be deduced from $\Psi(H)$, this is in accordance with the fact that there is a homomorphism from H to G, but not in the opposite direction.

One can show that the homomorphism is sound and complete with respect to the semantic Ψ without the normality condition.

Theorem 4.5. *Let G and H be two SGs on $\mathcal{V}$. There is a homomorphism from G to H iff $\Psi(\mathcal{V}), \Psi(H) \models \Psi(G)$.*

Ψ has an interesting behavior with respect to homomorphism, but it is not very interesting from a modeling viewpoint. Indeed, it is unusual to consider binary predicates to represent types of entities. An atom such that $Boy(Paul, x_1)$ means that x_1 is a node representing the entity Paul of type Boy. Two sorts of variables must be considered, one sort for the concepts and one sort for the nodes. Consequence of this is that the models with Ψ are less intuitive than the models with Φ, and using Ψ instead of Φ increases the distance from a formal expression to a sentence in natural language.

4.4 Note on the Relationships Between Description Logics and Conceptual Graphs

Description Logics (DLs) and Conceptual Graphs are both rooted in frames and semantic networks. They both remedy two critiques on their ancestors, i.e., the lack of distinction between factual and ontological knowledge, and the lack of precise formal semantics. Indeed, in conceptual graphs, there is a clear distinction between the vocabulary (or support), which represents basic ontological knowledge, and sets of graphs, which basically represent facts. In description logics, a knowledge base is split into a so-called terminological component, the TBox, which can be seen as the ontological part of the KB, and an assertional component, i.e., the ABox, which contains assertions about individuals. Both formalisms are provided with set and first-order logical semantics. Due to these common properties, their relationships have often been questioned.

Let us first outline the main characteristics of description logics (for an in-depth presentation of this family of formalisms, see [BCM$^+$03]). DLs represent knowledge in terms of concepts, which denote sets of objects, and roles, which denote binary relations between concept instances. Starting from atomic concepts (which can be seen as CG concept types) and atomic roles (which can be seen as binary CG relation types) and using a set of constructors, complex concepts and roles can be built. The semantics of the concepts and roles is set-theoretic, which gives a direct translation to FOL. We focus here on the logical translation in order to enhance relationships with conceptual graphs. A concept is translated into a unary predicate and a role into a binary predicate.

Let us consider for instance the description logic $\mathcal{AL}$, which is often considered as the minimal description logic with a practical interest. In this DL, concepts can be built with the following elements. Among the atomic concepts, there are two special elements, $\top$ and $\bot$. The construction rules are (where A is an atomic concept, C and D are concepts and R is a role): $\neg A$ (atomic negation), $C \sqcap D$ (concept intersection), $\forall R.C$ (value restriction) and $\exists R.\top$ (limited existential quantification). $\top$ and $\bot$ are logically translated into valid and unsatisfiable formulas, respectively. Any other concept C can be translated into a logical formula $f_C(x)$ with one free variable x as follows:

$f_A(x) = A(x)$ and $f_{\neg A}(x) = \neg A(x)$;
$f_{C \sqcap D}(x) = f_C(x) \wedge f_D(x)$;
$f_{\forall R.C}(x) = \forall y(R(x,y) \rightarrow f_C(y))$;
$f_{\exists R.\top}(x) = \exists y R(x,y)$.

More expressive DLs are obtained by adding other constructors, e.g., full existential quantification $\exists R.C$, which is translated as follows:

$f_{\exists R.C}(x) = \exists y R(x,y) \wedge f_C(y)$.

For instance, let *Man*, *Course* and *CSCourse* be atomic concepts representing all men, all courses and all computer science courses, respectively, and let *teaches* be an atomic role representing the relationship between a teacher and a course that he/she gives; the following concept represents "men who teach at least one course and teach only computer science courses":

$Man \sqcap \exists teaches.Course \sqcap \forall teaches.CSCourse$

Its logical translation can be:

$Man(x) \wedge \exists y(teaches(x,y) \wedge Course(y)) \wedge \forall y(teaches(x,y) \rightarrow CSCourse(y))$.

A DL knowledge base is composed of a TBox $\mathcal{T}$, which is a finite set of declarations about concepts and roles, and an ABox $\mathcal{A}$, which is a finite set of assertions about individuals. The declarations allowed in the TBox are inclusions of form $C \sqsubseteq D$ or equalities of form $C = D$. An inclusion $C \sqsubseteq D$ can be logically translated into an implication: $\forall x(f_C(x) \rightarrow f_D(x))$. An equality $C = D$ is logically translated into an equivalence: $\forall x(f_C(x) \leftrightarrow f_D(x))$. If, in $C = D$, C is an atomic concept, this equality is called a concept definition: C can be seen as a name for the concept D. For instance, the following equality defines the concept $CSProfessor$:

$CSProfessor = \exists teaches.Course \sqcap \forall teaches.CSCourse$.

Its logical translation is $\forall x(CSProfessor(x) \leftrightarrow (\exists y(teaches(x,y) \wedge Course(y)) \wedge \forall y(teaches(x,y) \rightarrow CSCourse(y))))$.

The basic inference problem related to the TBox is the *subsumption* problem: Given a TBox $\mathcal{T}$ and two concepts C and D, is C subsumed by D (which is generally noted $C \sqsubseteq D$), i.e., can D be considered as more general than C with respect to $\mathcal{T}$? In logical terms, if $f(\mathcal{T})$ denotes the set of formulas translating $\mathcal{T}$ and $f_C(x)$ and $f_D(x)$ denote formulas assigned to C and D, C is subsumed by D if and only if $f(\mathcal{T}) \models \forall x(f_C(x) \rightarrow f_D(x))$.

The ABox is composed of assertions of form $C(a)$ or $R(a,b)$, where C is a concept and R is a role. a and b are logically interpreted as constants. The basic inference problem in an ABox is the *instance checking* problem, which asks whether a given individual a is an instance of a given concept C with respect to $\mathcal{A}$ and $\mathcal{T}$. In logical terms, let $f(\mathcal{A})$ be the conjunction of all assertions in the A-box, then one asks if $f(\mathcal{T}), f(\mathcal{A}) \models C(a)$.

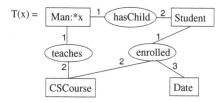

Fig. 4.6 A CG concept type definition

In CGs, similarly to DLs, a concept (or a relation) type can be defined in terms of existing ones. It is not defined via constructors, but rather by a BG, with a distinguished node (or k distinguished nodes for a k-ary relation), called a (k-ary) λ-BG (see Sect. 10.1 for precise definitions related to λ-BGs). For instance, consider the definition $T = (x)G$, where $(x)G$ is the graph of Fig. 4.6. This graph represents a concept type describing all men who teach a computer-science course and have a child who is enrolled in this course (with the distinguished concept node being marked

$*x$). The formula associated to a λ-BG has as many free variables as distinguished nodes. For the λ-BG in Fig. 4.6, one has:

$$\Phi((x)G) = \exists y \exists z \exists t (Man(x) \wedge CSCourse(y) \wedge Student(z) \wedge Date(t) \wedge teaches(x,y)$$
$$\wedge hasChild(x,z) \wedge enrolled(z,y,t))$$

The formula assigned to the definition of a concept type T is $\forall x(T(x) \leftrightarrow \Phi((x)G))$ (and relation type definitions are similarly translated into a logical equivalence). Although type definitions were introduced as early as 1984 in Sowa's book, their processing received little attention. See Sect. 8.5 for a brief presentation of type definitions, and the bibliographic notes in Chap. 8 for further references. Note also that a type definition can be seen as a pair of conceptual graph λ-rules, as defined in Chap. 10.

Since type definitions rely on BGs (or more precisely λ-BGs), as all kinds of knowledge, the BG model can be considered as a core for the comparison of CGs and DLs.

Provided that relations are restricted to binary relations, a BG vocabulary, composed of partially ordered sets of concept types and relation types, can be seen as a set of inclusions between atomic concepts and atomic roles. It can thus be seen as a very basic TBox. On the other hand, the ABox of description logics can be seen as a particular BG without generic concept nodes. If we try to find the "intersection" between DL definitions and CG definitions, we have to considerably restrict the expressiveness of each formalism. For instance, the CG definition of Fig. 4.6 cannot be expressed in DL, because of the ternary relation (which cannot be translated directly into a role), but more fundamentally because it is not possible to translate the cycle, which says that the CSCourse taught by x and the CSCourse in which his child is enrolled are the same entity. On the other hand, the DL definition of a *CSProfessor* given above cannot be expressed by a CG definition due to the value restriction constructor.

A notable work concerning the relationships of CGs and DLs is [BMT99] (which extends the first results of [CF98]). With the aim of characterizing the intersection of CGs and DLs, the authors identified two equivalent fragments. On the CG side, we have rooted BG trees, which are connected graphs with a tree-like structure, with binary relations only and with one distinguished concept node (like in a unary λ-BG)[1]. On the DL side, we have a DL specially tailored for the comparison, called $\mathcal{ELIRO}_1$. The constructors of $\mathcal{ELIRO}_1$ are $\exists R.C$ (existential restriction), $C \sqcap D$ (concept intersection), $R-$ (the inverse of the R role), $R \sqcap R'$ (role intersection, where R and R' are roles) and $\{i\}$ (unary one-of, where i is an individual). The unary one-of constructor allows us to integrate specific individuals in a concept (it corresponds to the possibility of creating an individual concept node). The precise result is that, if concepts contain at most one unary one-of constructor in each concept intersection (which is in accordance with the unique name assumption), then the corresponding

---

[1] By BGs with a tree-like structure, we mean connected BGs that can be transformed into logically equivalent acyclic multigraphs by splitting individual concept nodes. These graphs are exactly those without cycles with more than two distinct concept nodes and passing through generic nodes only (see Sect. 7.2); cycles induced by relations with the same argument lists can be accepted since they would disappear if conjunctive relation types were considered.

syntactic tree can be translated into a rooted BG tree with conjunctive concept types. Conversely, every rooted BG tree with only binary relation types can be translated into an $\mathcal{ELIRO}_1$ concept. This correspondence result allowed us to transfer the tractability result for BG-HOMOMORPHISM for BGs that are trees ([MC92], see also Chap. 7) to $\mathcal{ELIRO}_1$, and computation of subsumption by graph homomorphism was adapted to $\mathcal{ELIRO}_1$ and other description logics to solve other inference problems, namely matching and computing least common subsumers [BK99] [BKM99]. On the other hand, [BMT99] pointed out that tractability can be extended to BGs that are not trees but are logically equivalent to acyclic multigraphs (see above); let us point out however that these BGs can be seen as particular cases of acyclic hypergraphs or, equivalently, of guarded BGs, for which the tractability of homomorphism checking still holds (see Chap. 7).

The preceding correspondence shows that, despite their apparent proximity, DLs and CGs are "orthogonal" formalisms, in the sense that their intersection yields poor fragments of each of them, in which each formalism loses its interesting features: unrestricted cycles of variables and n-ary relations for CGs, and the variety of constructors for DLs. Other results support this claim. On one hand, it is known that even the most expressive DLs cannot express the whole FOL($\wedge$, $\exists$) fragment [Bor96]. On the other hand, BG homomorphism cannot handle negation in a logically complete way, even restricted to atomic negation on primitive types (see Sect. 12.2), thus one cannot add restricted forms of negation "for free" to the BG model. Let us put forward another argument. One reference DL is $\mathcal{ALC}$ which is a fairly general DL in which subsumption remains decidable. It has been pointed out that $\mathcal{ALC}$ can be translated into $\mathcal{L}^2$, the fragment of FOL, which consists of all formulas without functions, but with equality, that can be constructed using two variables only [Bor96]. A result of Kolaitis and Vardi implies that acyclic BGs coincide in expressive power with $\mathcal{L}^2$ [KV00]. Thus, the "intersection" of BGs and DLs is not likely to lead to better results than that cited above.

One could consider more complex kinds of CGs, such as special kinds of rules or of full CGs, with the hope of finding correspondences with some DLs, but once again one would have to give up the natural features. For instance, the transitive closure of a relation cannot be expressed in $\mathcal{ALC}$ whereas it can be handled with a very specific fragment of CG rules (so-called range-restricted rules) for which deduction is decidable (and even NP-complete).

If we turn our attention to extensions of DLs allowing us to query knowledge bases, more relationships between DLs and CGs are to be expected. Indeed, the instance checking problem can only be seen as a very specific querying problem, i.e., asking if a given individual can be considered as belonging to a given concept. Basic queries in databases are conjunctive queries. These queries are also natural queries in conceptual graphs, since they correspond to basic conceptual graphs (see Sect. 5.3). However, extending DLs to cope with conjunctive queries generally has a high computational cost [BCM$^+$03]. The recent DL-Lite approach is interesting from this standpoint [CGL$^+$05][CGL$^+$07]. The leading idea of this approach is to restrict the expressivity of the TBox in order to be able to query the Abox with conjunctive queries, while staying in the same complexity class as the classical conjunctive query answering decision problem. Note that the complexity considered is *data complexity*, a notion imported from databases: It is measured in the size of the ABox only, i.e., the size of the query is considered as a constant because it is assumed to be considerably smaller than the base. Note also that, in these works on the DL-Lite family, the size of the TBox is also ignored in the data complexity. If we take, for instance, the BG-HOMOMORPHISM problem, where the source graph is a query Q and the target graph is a fact base F, then this problem is NP-complete in usual complexity, and polynomial in data complexity, since a brute-force algorithm is in $O(size(F)^{size(Q)})$. The same holds for the classical conjunctive query answering decision problem. The approach developed throughout this book can be seen as motivated by similar objectives as the DL-Lite approach, but starting with a rich assertional component instead of a rich ontological component: The BG/SG fragment can be seen as allowing a rich assertional base, provided with a query mechanism equivalent to conjunctive queries, but with a rudimentary ontological part. We have been trying to extend it by adding constructors, which can be seen as DL constructors, even when they have not been introduced in reference to DLs (e.g., type conjunction, disjointness of types, atomic negation) as well as other kinds of knowledge (e.g., complex constraints or rules), while keeping their essential combinatorial properties.

4.5 Bibliographic Notes

The previous presentation of the model semantic for SGs is mainly issued from Kerdilès [Ker01]. Independently from Kerdilès, the same ideas were developed in [Bag01]. From an historical viewpoint, the denotation operator introduced by Sowa [Sow84] can be considered as the first step towards the definition of a model semantic for SGs, and a definition of the model semantic was given in [MC96].

The FOL semantic Φ was defined in [Sow84] and the soundness of BG homomorphism with respect to logical deduction was shown. The first proof of the completeness result, based on the resolution method, is in [Mug92] (published in [CM92]). Preller pointed out that this first proof was not correct for all graphs and her counter-example led us to define the notion of a normal graph ([CM95] or [MC96]. This notion was independently defined by Ghosh and Wuwongse, who provided another proof of completeness [GW95]. Since then, several other proofs have been provided. In particular, Simonet reformulated the first proof in the context of the Ψ semantic ([Sim98] or [CMS98]).

Concerning Φ, Wermelinger proposed in [Wer95a] another way of interpreting the universal type. If a concept node c is of type $\top$, the associated atom is not $\top(id_c)$ but $(id_c = id_c)$, i.e., an atom true for all interpretations. Then, for any graph G, $f2G(G)$ is equivalent to $\Phi(G)$ (actually the formulas are equal up to the suppression of all atoms $(id_c = id_c)$). Section 4.3.3 can be considered as stating, for a specific case, the relationships between FOL and ordered FOL (cf. [Gal85]).

Chapter 5
BG Homomorphism and Equivalent Notions

Overview

Homomorphism is a key notion for studying algebraic or relational structures, and many combinatorial problems can be reduced to homomorphism problems. In this chapter, it is shown that computing BG homomorphisms between two BGs is "strongly equivalent" to important combinatorial problems in graph theory, algebra, database theory, and constraint satisfaction networks. In Sect. 5.1, basic conceptual hypergraphs (BHs) are introduced, with a BH homomorphism notion, while highlighting relationships between BG homomorphisms and BH homomorphisms. The two notions are so close that they can be simply considered as two different views of the same abstract notion. It is also shown in this section that computing homomorphims for BGs on any vocabulary is equivalent to computing homomorphisms for BGs on specific vocabularies. BGs are kinds of graphs, thus relationships between graph homomorphisms and BG homomorphisms are studied in Sect. 5.2. Section 5.3 concerns relational structures and relational databases. It is especially shown that two fundamental problems dealing with conjunctive, positive and non-recursive queries–the query evaluation problem and the query containment problem–are equivalent to the BG homomorphism problem. Section 5.4 is devoted to the Constraint Satisfaction Problem (CSP), which is the basic problem in the constraint processing domain, and its equivalence to the BG homomorphism problem is stated. This equivalence, that borrows a lot from techniques developed in the CSP framework, will be used in Chap. 6.

Equivalent Problems

Before getting into the heart of the matter, basic notions about "equivalent problems" are briefly reviewed. A *decision problem* is composed of a set of instances and of a yes/no question about these instances. A *yes-instance* is an instance whose answer to the question is *yes*, and a *no-instance* is an instance whose answer to the question is *no*. For example, the decision problem concerning BG homomorphism is:

BG-HOMOMORPHISM

instance: two BGs G and H on the same vocabulary
question: Is there a BG homomorphism from G to H?

If there is a BG homomorphism from G to H, then (G,H) is a yes-instance, and any BG homomorphism from G to H is a *solution*. A yes-instance of a decision problem can have several solutions.

Polynomial reducibility is a fundamental notion in algorithmic complexity. There are different sorts of polynomial reducibility. The common idea is that, if a computational problem P_1 is polynomially reducible to a problem P_2, then any algorithm for solving P_2 can be "simply" transformed into an algorithm for solving P_1. Therefore P_1 can be considered as being no more difficult than P_2. Karp reducibility is considered hereafter.

Definition 5.1 (Reducibility and equivalence of problems).

- A decision problem P_1 is *polynomially reducible* to a decision problem P_2 if there is a transformation τ from the instance set of P_1 to the instance set of P_2 such that I_1 is a yes-instance of P_1 if and only if $\tau(I_1)$ is a yes-instance of P_2 and the transformation τ is computable in a polynomial time (in the size of an instance of P_1). Such a transformation τ is called a *polynomial reduction* from P_1 to P_2.
- Two decision problems P_1 and P_2 are *polynomially equivalent* if each of them is polynomially reducible to the other.
- A *parsimonious* reduction τ from P_1 to P_2 is a polynomial reduction from P_1 to P_2 such that the number of solutions of any yes-instance I_1 of P_1 is equal to the number of solutions of $\tau(I_1)$.
- Two decision problems P_1 and P_2 are *parsimoniously equivalent* if each of them is parsimoniously reducible to the other.

Thus, if two problems are equivalent, algorithms for solving one problem can be polynomially transferred to the other one. Furthermore, if there is a parsimonious reduction, then the (number of) solutions are preserved. Figure 5.1 summarizes the reductions that are stated in this chapter. The rectangles contain the problem names. As they all but one, i.e., CSP, deal with homomorphism, the word "homomorphism" is omitted. A rectangle is labeled by a letter representing the problem (e.g., the rectangle corresponding to the problem concerning the existence of a homomorphism between two (ordinary) graphs is labeled g and the rectangle concerning CSP is labeled c). An arrow from rectangle x to the rectangle y represents a polynomial reduction from problem x to problem y (e.g., $g2b$ is the name of a reduction from the GRAPH-HOMOMORPHISM problem, i.e., the decision problem concerning the existence of a homomorphism between two ordinary graphs, to BG-HOMOMORPHISM). A bold arrow represents a parsimonious reduction. A dotted arrow without name from x to y represents the fact that problem x is a particular case of problem y, so it also represents a parsimonious reduction. The central rectangle, labeled b, gathers the homomorphism problem for BGs and BHs on general, flat or very flat vocabularies. For simplicity, b (resp. h) is used for labeling a reduction to or from any of these three BG homomorphism problems (resp. BH homomorphism problems). The schema can be completed by transitivity. Indeed, the composition of two polynomial reductions is a polynomial reduction.

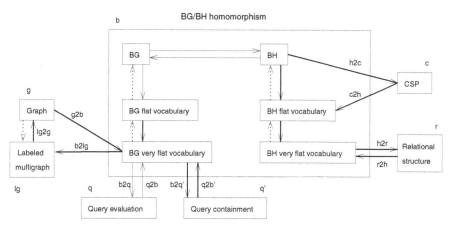

Fig. 5.1 Summary of reductions between BG-HOMOMORPHISM and equivalent problems

5.1 Conceptual Graphs and Conceptual Hypergraphs

5.1.1 Conceptual Hypergraphs

BGs are bipartite graphs. But they could also be seen as ordered hypergraphs, which we call conceptual hypergraphs, using the same vocabularies as BGs.

A (undirected) *hypergraph* is a pair $\mathcal{H} = (X, \mathcal{E})$, such that: X is a set and $\mathcal{E}$ is a family of subsets of X, i.e., the same subset of X can appear several times in $\mathcal{E}$. An element of X is called a vertex of $\mathcal{H}$, and an element of $\mathcal{E}$ is called a hyperedge of $\mathcal{H}$. An *ordered hypergraph* is a pair $\mathcal{H} = (X, \mathcal{E})$, where $\mathcal{E}$ is a family of *tuples* of elements in X, i.e. totally ordered subsets of X.

A conceptual hypergraph is defined as follows:

Definition 5.2 (Conceptual Hypergraphs: BHs). A *conceptual hypergraph* on a vocabulary $\mathcal{V} = (T_C, T_R, \mathcal{I})$ is a triple $\mathcal{H} = (X, \mathcal{E}, l)$, where:

- $\mathcal{H} = (X, \mathcal{E})$ is an ordered hypergraph,
- a vertex x is labeled by a pair *(type(x), marker(x))*, where *type(x)*$\in T_C$ and *marker(x)*$\in \mathcal{I} \cup \{*\}$ (a vertex of a BH is also called a concept node of this BH),
- a hyperedge r is labeled by $l(r) \in T_R$ and the cardinality of hyperedge is equal to the arity of the relation symbols.

The correspondence between conceptual hypergraphs and BGs is a simple extension of a well-known correspondence between hypergraphs and bipartite graphs. Consider a conceptual hypergraph $\mathcal{H} = (X, \mathcal{E}, l)$, then $h2b(H) = (C, R, E, l)$ is defined as follows: C is in bijection with X, R is in bijection with $\mathcal{E}$, C and R are labeled by the labeling function of $\mathcal{H}$ and, if x in X is the i-argument of e in $\mathcal{E}$, then there is an edge labeled i between the concept node associated with x and the relation node associated with e.

The mapping $h2b$ is a bijection from the set of conceptual hypergraphs on a vocabulary $\mathcal{V}$ to the set of BGs on $\mathcal{V}$. The inverse mapping of $h2b$ is denoted $b2h$. Thus, all notions introduced for BGs, e.g., coreference relation, can also be introduced for conceptual hypergraphs. Let us now consider the homomorphism notion between conceptual hypergraphs.

Definition 5.3 (Conceptual Hypergraph Homomorphism). A *BH homomorphism* from the conceptual hypergraph $\mathcal{H} = (X, \mathcal{E}, l)$ to the conceptual hypergraph $\mathcal{H}' = (X', \mathcal{E}', l')$ is a mapping π from X to X', such that:

- for any vertex x in $\mathcal{H}$, its label is greater than or equal to the label of $\pi(x)$, i.e., $l(x) \geq l'(\pi(x))$,
- for any hyperedge $y = (a_1, \ldots, a_k)$ in $\mathcal{E}$, there is a hyperedge $(\pi(a_1), \ldots, \pi(a_k))$ in $\mathcal{E}'$, denoted $\pi(y)$, such that $l(y) \geq l'(\pi(y))$.

It is simple to check that $h2b$ is a polynomial reduction from the decision problem concerning the existence of a homomorphism between conceptual hypergraphs to the decision problem BG-HOMOMORPHISM. But this reduction is not parsimonious.

Indeed, there might be exponentially fewer homomorphisms between conceptual hypergraphs than homomorphisms between the corresponding BGs.

Example. For instance, consider a vocabulary with binary relation symbols $r, r_1, , r_k$, where r is greater than $r_1, \ldots, r_k$, which are pairwise incomparable. Consider the BGs G and H in Fig. 5.2. H is obtained from G by replacing the relation (r) with k relations $(r_1) \ldots (r_k)$ having the same neighbors as (r). There are k homomorphims from G to H, but all of these homomorphisms are the same on concept nodes, thus there is only one homomorphism from $b2h(G)$ to $b2h(H)$.

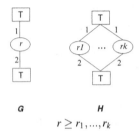

$$r \geq r_1, \ldots, r_k$$

Fig. 5.2 k BG (and only one BH) homomorphisms from G to H

In Fig. 5.3, we consider a BG G' obtained by repeating m times G and another BG H', obtained in the same way by repeating m times H. There are k^m homomorphisms from G' to H', whereas all homomorphisms are the same on concept nodes, and there is only one homomorphism from $b2h(G')$ to $b2h(H')$. Note that G and H do not have twin relation nodes, thus redundancy is irrelevant here.

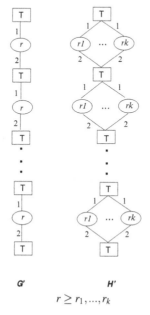

$$r \geq r_1, ..., r_k$$

Fig. 5.3 k^m BG (and only one BH) homomorphisms from G' to H'

Nevertheless, even if the reduction is not parsimonious, it is simple to state the correspondence between the two sets of homomorphisms. The restriction to the node set of any BG homomorphism from G to H is a BH homomorphism from $b2h(G)$ to $b2h(H)$. Conversely, let f be a homomorphism from a conceptual hypergraph $\mathcal{A}$ to a conceptual hypergraph $\mathcal{B}$, then it is simple to build all BG homomorphisms from $h2b(\mathcal{A})$ to $h2b(\mathcal{B})$ having f as a restriction on the concept nodes. Let r be a relation node of $h2b(\mathcal{A})$ with neighbor list $(c_1, \ldots, c_k)$. Any relation node r' in $h2b(\mathcal{B})$, with neighbor list $(f(c_1), \ldots, f(c_k))$ and whose label is less than or equal to the label of r, can be taken as an image for r.

The two notions of conceptual hypergraphs and of BGs are so similar that we will often refer to the first one as the *hypergraph vision* of BGs. But when drawings are considered, graphs are definitely more appropriate than hypergraphs because readable drawings of a hypergraph are generally drawings of its corresponding bipartite graph. Previous results can be theoretically summarized as follows.

Theorem 5.1 (Equivalence between BGs and BHs).

- *The mapping h2b is a bijection from the set of BHs on $\mathcal{V}$ to the set of BGs on $\mathcal{V}$ (b2h denotes the inverse of h2b).*
- *The restriction to the concept node set of any BG homomorphism from G to H is a BH homomorphism from b2h(G) to b2h(H).*
- *Any BH homomorphism f from $\mathcal{A}$ to $\mathcal{B}$ can be extended to a BG homomorphism from h2b($\mathcal{A}$) to h2b($\mathcal{B}$) having f as restriction on the concept nodes.*
- BG-HOMOMORPHISM *and* BH-HOMOMORPHISM *are polynomially equivalent.*

5.1.2 Different Kinds of Vocabularies

The notion of flat vocabulary and the transformation $flat$, which assigns a flat vocabulary to a vocabulary and transforms a BG on $\mathcal{V}$ into a BG on $flat(\mathcal{V})$ while preserving homomorphisms, have been introduced in Chap. 4. They are reviewed here and the notion of a very flat vocabulary is introduced.

Definition 5.4 (Flat and very flat vocabularies). A BG vocabulary $\mathcal{V} = (T_C, T_R, \mathcal{I})$ is called:

- *flat* if T_C is reduced to $\{\top\}$ and the order on T_R is the identity (any relation symbol is only comparable with itself),
- *very flat* if it is flat and if $\mathcal{I}$ is empty.

Let $\mathcal{V} = (T_C, T_R, \mathcal{I})$ be a BG vocabulary, $flat(\mathcal{V}) = (T_C', T_R', \mathcal{I})$ is the $flat$ vocabulary defined as follows:

- $T_C' = \{\top\}$,
- $T_R' = T_R \cup (T_C \setminus \{\top\})$, where the arity of an element of T_C considered as a relation is one, and any two different relation symbols are incomparable (the order relation on T_R' is the identity relation),
- the set of individual markers of $\mathcal{V}$ and of $flat(\mathcal{V})$ are the same.

Let $\mathcal{V} = (T_C, T_R, \mathcal{I})$ be a vocabulary, $vflat(\mathcal{V}) = (T_C', T_R', \emptyset)$ is the very flat vocabulary defined as follows:

- $T_C' = \{\top\}$,
- $T_R' = T_R \cup (T_C \setminus \{\top\}) \cup \mathcal{I}$, where the arity of an element of T_C and of an individual marker considered as a relation symbol is one, and any two different relation symbols are incomparable (the order relation on T_R' is the identity relation),
- the set of individual markers of $vflat(\mathcal{V})$ is empty.

Given a BG G on $\mathcal{V} = (T_C, T_R, \mathcal{I})$, $flat(G)$ is the BG on $flat(\mathcal{V})$ obtained from G as follows: For each concept node x of G of type t, the new type of x is $\top$, and for each concept type t' in T_C greater than or equal to t and distinct from $\top$, one adds a unary relation node labeled t' and linked to x; for each relation node x of type r

in G, for each relation type r' in T_R strictly greater than r, a relation node of type r' with same neighbors as r is added.

Given a BG G on $\mathcal{V} = (T_C, T_R, \mathcal{I})$, $vflat(G)$ is the BG on $vflat(\mathcal{V})$ obtained from $flat(G)$ as follows: For each individual concept node x with marker i, one adds a unary relation node labeled by the relation symbol associated with i and linked to x, and x becomes generic.

Example. An example of the transformations $flat$ and $vflat$ is given in Fig. 5.4.

The transformations $flat$ and $vflat$ are polynomial, and one has:

Property 5.1. The three problems BG-HOMOMORPHISM, BG-HOMOMORPHISM for BGs on a flat vocabulary and BG-HOMOMORPHISM for BGs on a very flat vocabulary are polynomially equivalent.

Proof. Let us prove the result for $vflat$. Let $G = (C_G, R_G, E_G, l_G)$ and $H = (C_H, R_H, E_H, l_H)$ be two BGs on $\mathcal{V}$ and let π be a homomorphism from G to H. Let us define a mapping π' from the node set of $vflat(G)$ to $vflat(H)$ as follows.

1. The concept node set of $vflat(G)$ is equal to the concept node set of G. The restriction of π' to the concept node set is defined as identical to π.
2. Let y be a relation node in $vflat(G)$ with the label $r \in T_R$ and having $x_1, \ldots, x_k$ as neighbors. There is a relation node z in G having $x_1, \ldots, x_k$ as neighbors and the label of z is $r' \leq r$. Let $u = \pi(z)$, i.e., the image of z in H. The label of u is $r'' \leq r'$ and the neighbors of u are $\pi(x_1), \ldots, \pi(x_k)$. In $vflat(H)$, there is a relation node v having $\pi(x_1), \ldots, \pi(x_k)$ for neighbors and with the label r since $r'' \leq r' \leq r$. One takes $\pi'(y) = v$.

   Let us now consider the unary relation nodes of $vflat(G)$ corresponding to concept types or individual markers.
3. Let y be a new unary relation node in $vflat(G)$ linked to x and labeled by a concept type t. t is greater than or equal to the type of x in G. The type of $\pi(x)$ in H is less than or equal to the type of x in G, therefore there is also a new unary relation node y' in $vflat(H)$ linked to $\pi(x)$ and labeled by t. One takes $\pi'(y) = y'$.
4. Let y be a new unary relation node in $vflat(G)$ linked to x and labeled by an individual i. $\pi(x)$ in H is also an individual concept node with marker i and therefore in $vflat(H)$ there is a new unary relation node y' labeled i and linked to $\pi(x)$. One takes $\pi'(y) = y'$.

One can check that π' is a BG homomorphism from $vflat(G)$ to $vflat(H)$.

Conversely, let π be a BG homomorphism from $vflat(G)$ to $vflat(H)$. A BG homomorphism π' from G to H can be defined as follows.

1. For any concept node x in G $\pi'(x) = \pi(x)$.
2. Let y be a relation node in G with the label r and having $x_1, \ldots, x_k$ as neighbors. There is a relation node z in $vflat(G)$ having $x_1, \ldots, x_k$ as neighbors and having the label r. Let $u = \pi(z)$. The label of u is r and its neighbors are $\pi(x_1), \ldots, \pi(x_k)$. There is a relation node v in H with a label $r' \leq r$ and having $\pi(x_1), \ldots, \pi(x_k)$ as neighbors. One takes $\pi'(y) = v$.

One can check that π is a BG homomorphism from G to H. $\quad\square$

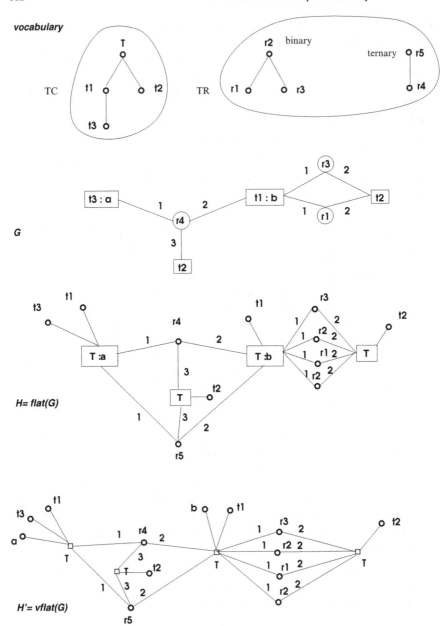

Fig. 5.4 The transformations *flat* and *vflat*

Note that $vflat$ and $flat$ are not parsimonious.

Example. In Fig. 5.5, where the relations are ordered as $r > r'' > r'$, there is only one homomorphism from G to H and there are three homomorphisms from $flat(G)$ to $flat(H)$.

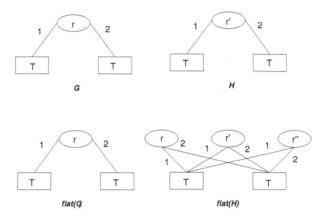

Fig. 5.5 $flat$ is not parsimonious

Note also that the problems BG-HOMOMORPHISM on a flat vocabulary and BG-HOMOMORPHISM on a very flat vocabulary are parsimoniously equivalent.

The same notions can be introduced for conceptual hypergraphs, i.e., one can consider BHs on flat or very flat vocabularies. With a BH homomorphism defined as a mapping only from concept nodes, the transformations $flat$ and $vflat$ applied to BHs are now parsimonious reductions.

Property 5.2. The three problems BH-HOMOMORPHISM, BH-HOMOMORPHISM for BHs on a flat vocabulary and BH-HOMOMORPHISM for BHs on a very flat vocabulary are parsimoniously equivalent.

5.2 Graphs

BG homomorphism is a key notion in this book. A BG is a kind of labeled graph and a BG homomorphism is simply an adaptation of the classical graph homomorphism notion. We show in this section, that the BG-HOMOMORPHISM problem is parsimoniously equivalent to the homomorphism problem on ordinary graphs.

Ordinary graphs can be translated into BGs in such a way that the classical graph homomorphism notion is transformed into BG homomorphism and reciprocally.

5.2.1 From Graphs to BGs

A (ordinary) directed graph G is a pair (X, U), where X is the vertex set of G and $U \subseteq X \times X$ is its arc set. Hereafter we generally consider directed graphs, so we usually omit the term "directed." Nevertheless, when undirected graphs are considered, let us recall (cf. Chap. A) that they can be identified as directed graphs by replacing each edge linking vertices x and y by two symmetrical arcs xy and yx. Let us consider a very simple vocabulary having only two elements, i.e., a concept type $\top$ and a binary relation type $\top_2$ (this vocabulary is very flat). A graph G can be transformed into a BG, denoted $g2b(G)$, on this vocabulary. The concept node set of $g2b(G)$ is (in bijection with) the vertex set of G and the relation node set of $g2b(G)$ is (in bijection with) the arc set of G. More precisely, for each vertex x of G, there is one concept node $g2b(x)$ labeled by $(\top, *)$, and for each arc xy in G, there is one relation node labeled by $\top_2$, with first neighbor $g2b(x)$ and second neighbor $g2b(y)$ (cf. Fig. 5.6).

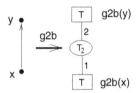

Fig. 5.6 From graphs to BGs

The transformation $g2b$ is polynomial in the size of G. $g2b$ is injective but not surjective and this is due to the existence of twin relation nodes in BGs. Nevertheless, if one considers directed multigraphs, i.e., an arc may appear several times in U, instead of directed graphs, then $g2b$ is bijective.

A *graph homomorphism* h, from a graph $G = (X_G, U_G)$ to a graph $H = (X_H, U_H)$, is a mapping from X_G to X_H which preserves arcs, i.e. if xy is an arc in U_G, then $h(x)h(y)$ is an arc in U_H. The GRAPH-HOMOMORPHISM problem is defined as follows:

GRAPH-HOMOMORPHISM
instance: two (directed) graphs G and H
question: Is there a homomorphism from G to H?
Example. For instance, let G and H be the graphs in Fig. 5.7. There are several homomorphisms from G to H, two of them map the vertex 4 onto the vertex a: $h_1 = \{(4, a), (3, b), (2, c), (1, d)\}$ and $h_2 = \{(4, a), (3, b), (2, c), (1, a)\}$.

Property 5.3. Let G and H be two directed graphs. There is a bijection from the set of homomorphisms from G to H to the set of BG homomorphisms from $g2b(G)$ to $g2b(H)$. Thus, $g2b$ is a parsimonious reduction from GRAPH-HOMOMORPHISM to BG-HOMOMORPHISM.

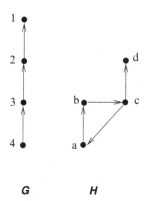

G   **H**

Fig. 5.7 Graph homomorphism

Proof. See the companion Fig. 5.8. Let h be a homomorphism from G to H. From h, a mapping π from $C_{g2b(G)}$ to $C_{g2b(H)}$ can be defined as follows: For all $c \in C_{g2b(G)}$, let $x \in X_G$ such that $c = g2b(x)$, then $\pi(c) = g2b(h(x))$. It is extended to a BG homomorphism from $g2b(G)$ to $g2b(H)$ by assigning, to each relation of $R_{g2b(G)}$ image of an arc xy, i.e. a relation with neighbor list $(g2b(x), g2b(y))$, the relation image of the arc $h(x)h(y)$, i.e. the relation with neighbor list $(\pi(g2b(x)), \pi(g2b(y)))$. Reciprocally, let π be a BG homomorphism from $g2b(G)$ to $g2b(H)$. The restriction of π to concept nodes (mapping from $C_{g2b(G)}$ to $C_{g2b(H)}$) defines a homomorphism h from G to H as follows, for all $x \in X_G$ $h(x) = g2b^{-1}(\pi(g2b(x)))$. Let h and h' be two homomorphisms from G to H and π and π' the associated homomorphisms from $g2b(G)$ to $g2b(H)$. If $h \neq h'$, then $\pi \neq \pi'$. It is also simple to check that this correspondence is surjective. □

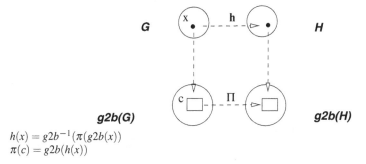

$h(x) = g2b^{-1}(\pi(g2b(x)))$
$\pi(c) = g2b(h(x))$

Fig. 5.8 Graph homomorphism and BG-homomorphism

It is well-known that GRAPH-HOMOMORPHISM is an NP-complete problem. For instance, there is a straightforward reduction from the CLIQUE problem (cf. [GJ79]) to the GRAPH-HOMOMORPHISM problem, since a graph G contains a k-clique as a subgraph if and only if there is a homomorphism from the k-clique to G. Therefore,

Corollary 5.1. BG-HOMOMORPHISM *is an NP-complete problem.*

This corollary is a new proof of Theorem 2.5.

A labeled multigraph is a triple $G = (X, U, l)$, where (X, U) is a multigraph and l is a labeling of X and U. A general definition of a labeled multigraph homomorphism where labels of corresponding vertices or arcs must be compatible is as follows:

Definition 5.5 (Labeled multigraph c-homomorphism). Let $G = (X_G, U_G, l_G)$ and $H = (X_H, U_H, l_H)$ be two labeled multigraphs, and let c be a binary (compatibility) relation between labels. A *labeled multigraph c-homomorphism* from G to H is a mapping π from X_G to X_H and from U_G to U_H, such that:

- for all $x \in X_G$ one has $c(l_H(\pi(x)), l_G(x))$,
- for all $u \in U_G$ having x for origin and y for extremity, $\pi(u)$ has $\pi(x)$ for origin and $\pi(y)$ for extremity and $c(l_H(\pi(u)), l_G(u))$.

A *(labeled multigraph) homomorphism* is a (labeled multigraph) c-homomorphism where c is the equality relation between labels.

Algorithms computing homomorphisms presented in Chaps. 6 and 7 can be transformed into algorithms computing c-homomorphisms by substituting the compatibility relation c to the equality relation between labels.

Example. Let us consider the two labeled multigraphs represented in Fig. 5.9. The application π defined by:

$\pi(x1) = y1$, $\pi(x2) = \pi(x3) = y2$, $\pi(x4) = \pi(x5) = y3$

and naturally extended to the arc sets is a labeled multigraph homomorphism from the multigraph G on the left to the multigraph H on the right.

With a BG being a labeled multigraph with properties on the labels, the definition of a labeled multigraph c-homomorphism can be adapted for BGs as follows:

Definition 5.6 (BG c-homomorphism). Let G and H be two BGs defined on $\mathcal{V}$, and c be a compatibility relation between labels. A *c-homomorphism* from G to H is a mapping π from C_G to C_H and from R_G to R_H, such that:

- $\forall (r, i, c) \in G, (\pi(r), i, \pi(c)) \in H$,
- $\forall e \in C_G \cup R_G$, one has $c(l_H(\pi(e)), l_G(e))$.

A BG homomorphism is a BG c-homomorphism such that the compatibility relation c is the order relation on the labels, i.e., $c(x, y)$ if and only if $x \leq y$.

The mapping $g2b$, from graphs to BGs (on very flat vocabularies), can be easily extended to a mapping $lg2b$ from labeled multigraphs to BGs (on very flat vocabularies) while preserving Property 5.3. Let us consider a very flat BG vocabulary having a set of unary symbols corresponding to a set of vertex labels of multigraphs and a set of binary relation symbols corresponding to a set of arc labels of multigraphs. Then BGs on such a vocabulary and labeled multigraphs are, as mathematical objects, identical.

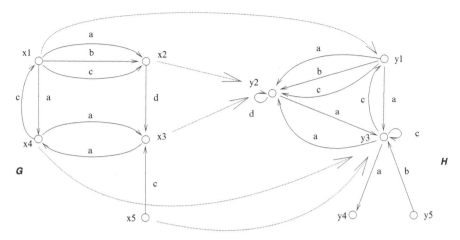

Fig. 5.9 π is a labeled multigraph homomorphism from G to H

Property 5.4. lg2b is a bijection from labeled multigraphs to BGs on very flat vocabularies (associated with multigraph label sets). Let G and H be two labeled multigraphs. There is a bijection from the set of homomorphisms from G to H to the set of BG homomorphisms from $lg2b(G)$ to $lg2b(H)$. Thus $lg2b$ is a parsimonious reduction from LABELED-MULTIGRAPH-HOMOMORPHISM to BG-HOMOMORPHISM.

But, once the vocabulary is enriched, the labeled multigraph homomorphism and BG homomorphism notions diverge. For instance, if an individual marker is introduced, or if two relation symbols are comparable, then the concept label set (or the relation label set) becomes a non-trivially ordered set. Thus, BG homomorphisms and labeled multigraph homomorphisms are no longer equivalent. Nevertheless, it is possible to encode the order on the label sets into the structure of the graph in such a way that BGs can be transformed into unlabeled graphs, while transforming BG homomorphism into (ordinary graph) homomorphism. This is explained hereafter.

5.2.2 From BGs to Graphs

Let us consider BGs on a very flat vocabulary (the transformation from a BG on any vocabulary to a BG on a very flat vocabulary was studied in the previous section). A mapping, denoted *b2lg*, from BGs on a very flat vocabulary to vertex labeled undirected graphs, which preserves homomorphisms, can be defined as follows.

Let $a_1, \ldots, a_n$ be new labels with respect to the set of vertex labels.

1. Let k be the arity of a relation node x labeled r. The edges incident to x are removed, k new nodes, labeled $a_1, \ldots, a_k$ are added, a_i is adjacent to x and to the i-th neighbor of x.
2. The distinction between concept and relation nodes is no longer considered.

The graph $b2lg(G)$ is a (ordinary) undirected graph labeled on the vertices.

Example. In Fig. 5.10 the BG H on a very flat vocabulary (cf. Fig. 5.4) and the labeled graph $b2lg(H)$ are represented.

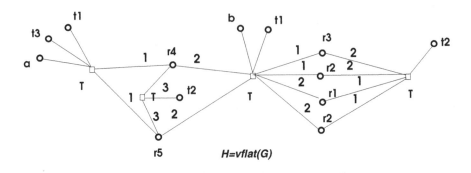

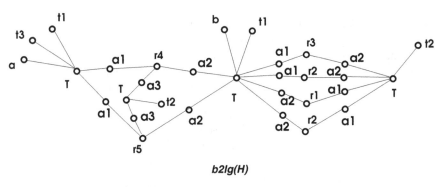

Fig. 5.10 The transformation $b2lg$

The transformation $b2lg(G)$ is polynomial and one has:

Property 5.5. Let G and H be two BGs on a very flat vocabulary $\mathcal{V}$. There is a bijection between the set of BG homomorphisms from G to H and the set of labeled graph homomorphisms from $b2lg(G)$ to $b2lg(H)$. Thus, $b2lg$ is a parsimonious reduction from BG-HOMOMORPHISM to LABELED-GRAPH-HOMOMORPHISM.

The last transformation, denoted $lg2g$, only concerns labeled and unlabeled (ordinary) multigraphs and is classical in graph theory (cf. [HN04]). As this transformation is rather long to define, we only give some hints as to how it works. A set of k graphs, called *replacement* graphs, can be associated with any set of k labels. If a set of replacement graphs fulfills some conditions (e.g., the replacement graphs are pairwise non-comparable or the identity is the only one homomorphism from a replacement graph to itself), then it can be shown that replacing each labeled edge by its associated replacement graph is a parsimonious reduction from LABELED-GRAPH-HOMOMORPHISM to (unlabeled) GRAPH-HOMOMORPHISM.

Property 5.6. *lg2g* is a parsimonious reduction from LABELED-GRAPH-HOMO-MORPHISM to (unlabeled) GRAPH-HOMOMORPHISM.

From Property 5.3, Property 5.4 and Property 5.6, one obtains:

Theorem 5.2 (Equivalence between BG-HOMOMORPHISM and GRAPH-HOMO-MORPHISM). BG-HOMOMORPHISM *and* GRAPH-HOMOMORPHISM *are parsimoniously equivalent.*

It would be rather inefficient to use this latter result to solve a homomorphism problem between BGs, that is to first transform them into ordinary graphs and then to use a graph homomorphism algorithm. A more efficient solution is to build algorithms directly considering BGs (cf. Chap. 6 and Chap. 7).

5.3 Relational Structures and Databases

In Sect. 5.1.1, hypergraphs are considered. A hyperedge is a tuple and as tuples underlie relational structures it is natural to study relational structures and homomorphism of relational structures.

5.3.1 Relational Structures and BGs

A (finite) relational structure C on a finite set R of relation symbols and a universe U is equal to a set $\{r_C \subseteq U^k \mid$ for all relations $r \in R$, k being the arity of $r\}$ with r_C being the finite set of tuples of r in C. Any tuple $(a_1, \ldots, a_k)$ in r_C can be represented by $r(a_1, \ldots, a_k)$ or by the set of k triples $\{(r, i, a_i) | i = 1, \ldots, k\}$. Given two structures C_1 and C_2 on the same R, a homomorphism h from C_1 to C_2 is a mapping from the universe U_{C_1} of C_1 to the universe U_{C_2} of C_2, such that for all relations $r \in R$, if $r(a_1, \ldots, a_k) \in C_1$ then $r(h(a_1), \ldots, h(a_k)) \in C_2$. Equivalently, h is a homomorphism if for all k-ary relations r and for all $i = 1, \ldots, k$ if (r, i, a_i) is in C_1, then $(r, i, h(a_i))$ is in C_2. The RELATION-HOMOMORPHISM problem is as follows:
instance: two (finite) relational structures A and B
question: Is there a homomorphism from A to B?

A relational structure C on R can be transformed, by a mapping denoted $r2b$, into a (normal) BH on the very flat vocabulary with R as relation symbol set as follows. The concept node set of $r2b(C)$ is in bijection with the set of arguments in the tuples of C, i.e., the set $\{a_j \mid \exists r \exists i (r, i, a_j) \in C\}$. Any concept node is generic. For each tuple $r(a_1, \ldots, a_k)$ in C, a hyperedge e labeled by r is built, the i-th element of e is the concept node corresponding to a_i. C and $r2b(C)$ have the same size, thus $r2b$ is a polynomial transformation and one can check that there is a bijection from the set of homomorphisms from C to C' to the set of BH homomorphisms from $r2b(C)$ to $r2b(C')$.

Conversely, let G be a BH without isolated nodes on a very flat vocabulary having R as relation symbol set. G can be transformed into a relational structure $b2r(G)$ as follows. The relation symbol set of $b2r(G)$ is R, the universe of $b2r(G)$ is in bijection with the concept node set of G and, if $e = (c_1,\ldots,c_k)$ is a hyperedge in G labeled r, then $b2r(G)$ contains the tuple $r(a_1,\ldots,a_k)$, where a_i is the element of the universe corresponding to c_i. One can check that if G and H are two BHs without isolated nodes on a very flat vocabulary, then there is a bijection from the set of BH homomorphisms from G to H to the set of relation homomorphisms from $b2r(G)$ to $b2r(H)$.

Theorem 5.3 (Equivalence between RELATION-HOMOMORPHISM and BH-HOMOMORPHISM).

- *r2b is a parsimonious reduction from* RELATION-HOMOMORPHISM *to* BH-HO-MOMORPHISM *on very flat vocabularies.*
- *b2r is a parsimonious reduction from* BH-HOMOMORPHISM *(without isolated nodes) on very flat vocabularies to* RELATION-HOMOMORPHISM.

From a computational viewpoint, conceptual hypergraphs and relational structures are very similar. Nevertheless, there are two kinds of differences:

1. Concerning the structures: Isolated concept nodes and twin (redundant) relation nodes can occur in conceptual hypergraphs but not in relational structures.
2. Concerning the labels: The vertices are labeled in conceptual hypergraphs, and the sets of labels (for the vertices and hyperedges) are ordered.

From a knowledge representation viewpoint, the preceding differences can be important. In BGs, entities are usually considered first; then relationships between entities are taken into account. There can be isolated entities, i.e., entities without any related entity. It may even be possible that there are only entities with no relationships between them. This is for instance the case in information retrieval when documents are represented by a set of keywords and each keyword is represented by a concept node labeled by the keyword. In relational structures, relations are first considered and they necessarily occur.

Relational structure is the basic mathematical notion in relational database theory. Two fundamental problems concerning relational databases, namely the evaluation of a conjunctive query and the query containment problem, are related to BG homomorphism in the next section.

5.3.2 Conjunctive Queries and BGs

Let us begin by reviewing some basic definitions concerning relational databases.

The basic form of a relational database schema $S = (R, dom)$ is composed of a set R of n-ary relation symbols ($n \geq 1$) and of a countably infinite set dom of constants. An instance of such a schema is simply a finite relational structure on R with dom as universe.

A (positive and non-recursive) conjunctive query q on a schema $S = (R, dom)$ is a rule, $q = ans(u) \leftarrow r_1(u_1), \ldots r_n(u_n), n \geq 1$, where:

- $r_1, \ldots, r_n$ belong to R,
- ans is not in R (ans stands for "answer"),
- $u_1, \ldots, u_n$ are tuples of terms (variables or constants of dom), u is a tuple of variables and for any i the length of u_i is equal to the arity of r_i, and $\forall i \neq j$ $r_i(u_i) \neq r_j(u_j)$,
- each variable of u occurs at least once in $u_1, \ldots, u_n$.

Given a query $q = ans(u) \leftarrow r_1(u_1), \ldots r_n(u_n)$ and an instance D of $S = (R, dom)$, $q(D)$ denotes the set of answers to q in D, i.e., $q(D)$ is the set of tuples $\mu(u)$, where μ is a substitution of the variables of q by dom constants such that for any j in $\{1, \ldots, n\}$, $\mu(u_j) \in D(r_j)$, where $D(r_j)$ is the set of r_j-tuples in D. When the arity of ans is 0, i.e, when u is empty, q can be seen as a boolean query whose answer is *yes* or *no*. If there is a substitution μ such that for any j in $\{1, \ldots, n\}$, $\mu(u_j) \in D(r_j)$, then $q(D) = \{()\}$ and the answer is *yes*. Otherwise, $q(D) = \emptyset$ and the answer is *no*.

5.3.2.1 Conjunctive Query Evaluation

The CONJUNCTIVE-QUERY-EVALUATION decision problem is defined as follows:
instance: a database instance D and a conjunctive query q
question: Does D contain an answer to q, i.e. is $q(D)$ not empty?
    A flat vocabulary $\mathcal{V} = (\{\top\}, R, dom)$ can be naturally assigned to a schema $S = (R, dom)$, i.e., the relation symbol set of $\mathcal{V}$ is R and the individual marker set of $\mathcal{V}$ is dom. An instance D of $S = (R, dom)$ can be transformed into a BG $H = q2b(D)$. The concept node set of H is in bijection with the elements of dom appearing in D and the concept node corresponding to a_i is labeled $(\top, a_i)$, thus all concept nodes in H are individual nodes, H has no isolated nodes and H is normal. Each tuple $(a_1, \ldots, a_k)$ of r is represented by a relation node labeled by r and having the concept node associated with a_i for i-th neighbor.
    Let $q = ans(u) \leftarrow r_1(u_1), \ldots, r_n(u_n)$ be a query, it can be transformed into a λ-BG $q2b(q) = (c_1, \ldots, c_k)G$ as follows. G represents the right part of the query q. A generic concept node is associated with each variable occurring in a u_i, an individual node is associated with each constant occurring in a u_i, and one associates with each $r_i(u_i)$ a relation node labeled by r_i having concept nodes associated with u_i as neighbor list. Concept nodes corresponding to variables in $ans(u)$ are the λ-concepts $(c_1, \ldots, c_k)$ of $q2b(q)$.
    Let $q2b(q) = (c_1, \ldots, c_k)G$, $H = q2b(D)$ and π be a BG homomorphism from G to H. Then, the images of the λ-concept nodes constitute an answer to q and, reciprocally, any tuple of $q(D)$ can be obtained from the images of λ-concepts in a BG homomorphism from G to H.
    Conversely, let us consider a normal BG H without generic nodes nodes on a flat vocabulary $\mathcal{V} = (\{\top\}, T_R, \mathcal{I})$. A schema $S = (T_R, \mathcal{I})$ can be associated with $\mathcal{V}$ and an instance of S, denoted $b2q(H)$, can be associated with H.

A conjunctive query $q = b2q(G)$ can be associated with a normal λ-BG $G = (c_1, \ldots, c_k)G'$ on $\mathcal{V}$, with the right part of q corresponding to G' (without isolated nodes) and the left part corresponding to the λ-concepts. Let μ be a substitution that defines an answer to $q = b2q(G)$ in $b2q(H)$. Then μ yields a homomorphism from G to H (since H is normal) and conversely if G has no isolated nodes.

Theorem 5.4 (Equivalence between CONJUNCTIVE-QUERY-EVALUATION and BG-HOMOMORPHISM). *Let D be an instance of a relational schema S and q be a conjunctive query on S.*

- *The answer set $q(D)$ is in bijection with the set of tuples $(\pi(c_1), \ldots, \pi(c_k))$, where π is a BG homomorphism from $q2b(q)$ to $q2b(D)$ and $c_1, \ldots, c_k$ are the λ-concepts of $q2b(q)$.*
- *Let H be a normal BG without generic concept nodes on a flat vocabulary $\mathcal{V}$ and G a λ-BG without isolated nodes on $\mathcal{V}$. The answer set to $q = b2q(G)$ in $b2q(H)$ is in bijection with the set of tuples $(\pi(c_1), \ldots, \pi(c_k))$, where π is a BG homomorphism from G to H, and $c_1, \ldots, c_k$ are the λ-concepts of G.*
- *The problems CONJUNCTIVE-QUERY-EVALUATION and BG-HOMOMORPHISM are polynomially equivalent.*

Note that the number of solutions is not necessarily preserved by these transformations even if conceptual hypergraphs are considered.

Example. Let us consider an instance $D = \{r(a,b), r(a,c)\}$ and a query $q = ans(x) \leftarrow r(x,y)$. $q2b(D)$ and $q2b(q)$ are represented in Fig. 5.11. $q(D) = \{(a)\}$ whereas there are two homomorphisms from $q2b(q)$ to $q2b(D)$.

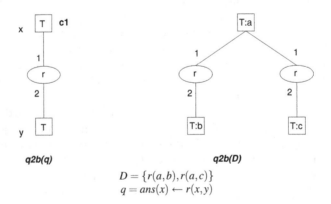

$$D = \{r(a,b), r(a,c)\}$$
$$q = ans(x) \leftarrow r(x,y)$$

Fig. 5.11 The transformation $q2b$

5.3.2.2 Conjunctive Query Containment

A query q is said to contain a query q' ($q' \subseteq q$) if, for any instance D on a schema $S = (R, dom)$, $q'(D) \subseteq q(D)$.

The CONJUNCTIVE-QUERY-CONTAINMENT decision problem is defined as follows:

instance: two queries q and q'
question: Does q contain q'?

It is well-known that this problem can be reformulated as a query homomorphism problem, where a homomorphism between queries is defined as follows:

Definition 5.7 (Query homomorphism). A *query homomorphism* from $q = ans(u) \leftarrow r_1(u_1), \ldots r_n(u_n)$ to $q' = ans'(u') \leftarrow r_1'(u_1'), \ldots r_{n'}'(u_{n'}')$ is a substitution θ of the variables of q by terms (variables or constants of dom) such that $\theta(u) = u'$, and for any $j \in \{1, \ldots, n\}$, there is $i \in \{1, \ldots, n'\}$ with $\theta(r_j(u_j)) = r_i'(u_i')$.

The (query) homomorphism theorem [AHV95] shows that, given two queries q and q', q contains q' if and only if there is a query homomorphism from q to q'.

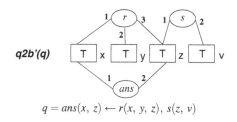

$$q = ans(x, z) \leftarrow r(x, y, z), s(z, v)$$

Fig. 5.12 The transformation $q2b'$

Transformations between CONJUNCTIVE-QUERY-CONTAINMENT and BG-HO-MOMORPHISM, called $q2b'$ and $b2q'$ (cf. Fig. 5.12), are similar to transformations $q2b$ and $b2q$ (and also similar to the transformations $f2G$ and $G2f$ between a logical language of FOL($\exists$, $\wedge$) and a BG language (defined in Chap. 4).

1. Transformation from Query containment to BG homomorphism:
   Let $\mathcal{V} = (\{\top\}, R \cup Ans, dom)$ be a flat vocabulary where $Ans = \{ans_k | k \geq 1\}$, with ans_k being a relation of arity k used to represent the left part of a query having k variables. Then a mapping $q2b'$ from conjunctive queries to normal BGs on $\mathcal{V}$ can be defined as follows (cf. Fig. 5.12 for an example). Let $q = ans(u) \leftarrow r_1(u_1), \ldots, r_k(u_k)$ be a query, then $q2b'(q) = (C_q, R_q, E_q, l_q)$, where,

   - C_q is in bijection with the set of terms occurring in $u, u_1, \ldots, u_n$, and the node of C_q assigned to a term e is generic if e is a variable, and otherwise it is an individual concept with marker e,
   - R_q is a set of (n+1) nodes labeled by $r_1, \ldots, r_n, ans_k$, with k being the number of variables in u,

- if e is the i-th argument of r_j (or *ans*), then the concept node assigned to e is the i-th neighbor of the relation node assigned to r_j (or *ans*).

One can check that, given two queries q and q', any query homomorphism from q to q' induces a BG homomorphism from $q'2b(q)$ to $q'2b(q')$, and reciprocally.

2. Transformation from BG homomorphism to Query containment:
Let us consider BGs on a flat vocabulary $\mathcal{V} = (\{\top\}, T_R, \mathcal{I})$ and let $S = (T_R, \mathcal{I})$ be the relational schema associated with $\mathcal{V}$. A mapping $b2q'$ from normal BGs without isolated concepts on $\mathcal{V}$ to (boolean) queries on S can be defined in a similar way as $b2q$. More precisely, $b2q'(G) = ans() \leftarrow r_1(u_1) \dots r_n(u_n)$, where r_i corresponds to a relation node x in G labeled r_i and where u_i is the neighbor list of x. Given two normal BGs without isolated concepts G and H, it can be checked that there is a query homomorphism from $b2q'(G)$ to $b2q'(H)$ if and only if there is a BG homomorphism from G to H.

Theorem 5.5 (Equivalence between CONJUNCTIVE-QUERY-CONTAINMENT and BG-HOMOMORPHISM).

- *The mapping $q2b'$ is a parsimonious reduction from* CONJUNCTIVE-QUERY-CONTAINMENT *to* BG-HOMOMORPHISM.
- *The mapping $b2q'$ is a polynomial reduction from* BG-HOMOMORPHISM *(without isolated nodes) on a flat vocabulary to* CONJUNCTIVE-QUERY-CONTAINMENT.
- *The* CONJUNCTIVE-QUERY-CONTAINMENT *and* BG-HOMOMORPHISM *problems are polynomially equivalent.*

5.4 Constraint Satisfaction Problem

5.4.1 Definition of CSP

The Constraint Satisfaction Problem is the basic problem of the constraint processing research domain. It has been deeply studied from an algorithm viewpoint. Chapter 6 will borrow a lot from techniques developed in the CSP framework.

The input of a constraint satisfaction problem is composed of a set of variables, with a set of possible values for the variables and a set of constraints on the variables. A constraint on certain variables defines what values these variables can simultaneously take. The question is whether there is an assignment of values to the variables that satisfies the constraints.

We will consider the simplest kind of CSP, where variables have discrete and finite domains. In this case, constraints can be described by the enumeration of all allowed tuples of values. In more general CSPs, variables can have infinite domains (e.g., the set of integers), or constraints can be defined by a language (e.g., algebraic inequalities).

Definition 5.8 (Constraint network). A *constraint network* is a 3-tuple $P = (X, D, C)$ composed of:

- a set of *variables*, $X = \{x_1, \ldots, x_n\}$,
- a set of *values*, $D = D_1 \cup \ldots \cup D_n$, where D_i is the *domain* of the variable x_i,
- a set of *constraints*, $C = \{C_1, \ldots, C_p\}$, where each constraint C_i is composed of a totally ordered subset of variables S_i, and a relation R_i on S_i. $S_i = \{x_{i_1}, \ldots, x_{i_q}\} \subseteq X$ is called the *scope* of C_i. R_i is a subset of the cartesian product $D_{i_1} \times \ldots \times D_{i_q}$ and is called the *definition* of C_i. The *arity* of C_i is q, the size of its scope.

A constraint definition R_i is often represented as a table, where the columns are given by S_i, and the rows are tuples of R_i. Given a constraint C_i and a variable x_j in its scope, we denote $R_i[x_j]$ as the set of values for x_j in R_i. $R_i[x_j]$ is thus obtained from the x_j column in R_i by deleting duplicate values. In relational terms, it is the projection of R_i onto x_j. More generally, if $x_j, \ldots, x_k$ are variables in the scope of C_i, $R_i[x_j, \ldots, x_k]$ denotes the projection of R_i onto $x_j, \ldots, x_k$, i.e., the set of tuples $(a_j, \ldots, a_k)$ obtained by restricting R_i to variables $x_j, \ldots, x_k$.

Note that we allow several constraints with the same scope, and even with the same scope and the same definition. These cases are usually not interesting in the CSP framework, but they bring a simpler correspondence between BGs and constraint networks.

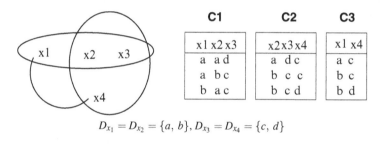

$$D_{x_1} = D_{x_2} = \{a, b\}, D_{x_3} = D_{x_4} = \{c, d\}$$

Fig. 5.13 A constraint network

When all constraints are *binary*, i.e., involve exactly two variables, the pair (X, C) defines an undirected graph, whose vertices are the variables and edges are the scopes of the constraints (as in Fig. 5.14). More precisely, it is a multigraph when several constraints are defined on the same subset of variables. In the general case, (X, C) defines a hypergraph (cf. Fig. 5.13).

Figure 5.14 illustrates how a graph coloring problem can be recast as a CSP. The input of a *graph coloring problem* is composed of an undirected graph, and a set of colors. The question is whether a color can be assigned to each vertex of the graph in such a way that different colors are assigned to vertices connected by an edge. This problem can be seen as an abstraction of various resource allocation problems. For instance, the *map coloring* problem is defined as follows: Given a map of countries and a set of colors, the question is whether the countries can be colored (with the set of colors), in such a way that adjacent countries do not have the same color. The map

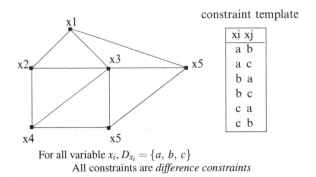

For all variable x_i, $D_{x_i} = \{a, b, c\}$
All constraints are *difference constraints*

Fig. 5.14 A binary constraint network

can be abstracted as an undirected graph, where the vertices represent the countries, with an edge connecting two countries if they are neighbors on the map.

To go from a graph coloring instance to a constraint network, the vertices and edges of the graph become the variables and constraints of the network, respectively. All variables have the same domain, which is the set of colors. Finally, all constraints are "difference" constraints, i.e., their definitions express the fact that the variables have to be assigned different values.

A *solution* to a constraint network P is an assignment of values to the n variables satisfying all constraints, formally:

It is a mapping $S: X \to D$, with $x_i \mapsto a \in D_i$,

and such that for every constraint $C_j = (x_{j_1} \dots x_{j_q})$, $(S(x_{j_1}) \dots S(x_{j_q})) \in R_j$.

A constraint network is (globally) *consistent* if it has at least a solution.

Example. The network in Fig. 5.13 is consistent. It possesses two solutions, $Sol1 = \{(x_1, a), (x_2, a), (x_3, d), (x_4, c)\}$ and $Sol2 = \{(x_1, a), (x_2, b), (x_3, c), (x_4, c)\}$, which correspond to the first two lines of constraint definitions C_1, C_2 and C_3, respectively. The network in Fig. 5.14 is inconsistent (this translates the fact that the graph cannot be colored with three colors).

The CONSTRAINT-SATISFACTION-PROBLEM (CSP) is defined as follows:

instance: a constraint network P

question: Is P consistent, i.e., is there a solution to P?

A lot of combinatorial problems can be expressed as CSP, and BG-HOMOMORPHISM is one of them. It is less obvious that, in turn, CSP can be recast as BG-HOMOMORPHISM. Let us begin with this latter side of the equivalence. We call $c2h$ the transformation from CSP to BG-HOMOMORPHISM, and $h2c$ the transformation in the opposite direction.

5.4.2 From CSP to BGs

Let us first outline the idea of the transformation denoted $c2h$. Consider a constraint network $P = (X, D, C)$. P is transformed into two BGs G and H, with the following idea.

G translates the *macro-structure* of P, i.e., the hypergraph itself. It is built from X and the constraint scopes in C. Each concept node is generic and corresponds to a variable, and each relation corresponds to a constraint, with its arguments corresponding to variables in the constraint scope.

H represents the *micro-structure* of P. It is built from D and the constraint definitions in C. There is an individual concept node for each value of D, and a relation for each tuple of compatible values. Figure 5.15 illustrates this transformation, showing the BGs obtained from the network of Fig. 5.13.

Roughly said, the question is whether there is a mapping from variables (concept nodes of G) to values (concept nodes of H) that satisfies the constraints (maps relations of G to relations of H), i.e., is a homomorphism from G to H.

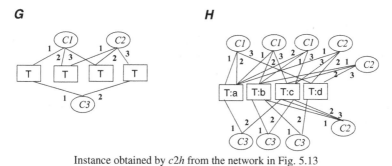

Instance obtained by $c2h$ from the network in Fig. 5.13

Fig. 5.15 The transformation $c2h$

Let us now define $c2h$ more precisely. A flat vocabulary $\mathcal{V} = (\{\top\}, T_R, \mathcal{I})$ is built from P as follows: T_R is in bijection with C, with each relation type having the same arity as the corresponding constraint, and $\mathcal{I}$ is in bijection with D.

G and H are then defined as follows:

- C_G is in bijection with X. Each concept node is labeled by $(\top, *)$.
- R_G is in bijection with C. Each relation is labeled by the corresponding constraint. The i-th argument of a relation is the concept node corresponding to the i-th variable of the constraint scope.
- C_H is in bijection with D. Each concept node is labeled by $(\top, a)$, where a is the individual associated with the element a of D.
- R_H is in bijection with the (disjoint) union of all R_i. In other words, there is one relation for each tuple in a constraint definition; this relation is labeled by the

G   H

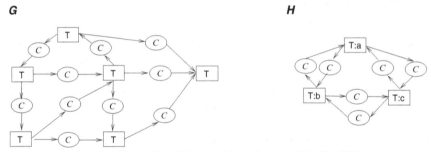

Instance obtained by *c2h* (variant) from the network in Fig. 5.14

Fig. 5.16 The transformation *c2h* (variant)

corresponding constraint, and its i-th argument is the concept node corresponding to the i-th value.

Remark that the mapping *c2h* can be considered as a mapping from a CSP to a BH as well as a mapping from a CSP to a BG.

Note that both graphs G and H are in normal form. It can be simply checked that any solution to the CSP yields a BG homomorphism from G to H, and reciprocally.

Rather than creating one relation type per constraint, a more concise transformation can create one relation type per group of constraints having the same definition (see for instance Fig. 5.16, which shows the graphs obtained from the network in Fig. 5.14: All constraints are *difference constraints*, thus have the same definition, which is translated as a single relation type).

5.4.3 From BGs to CSP

The mapping *h2c* from BGs (or BHs) to CSP is based on the same idea as *c2h*. The source BG G provides the macro-structure of the constraint network, while the target BG H provides the micro-structure. H is supposed to be in normal form. There is one variable for each concept node of G, and one constraint for each relation node of G. The set D of domain values is built from the concept nodes of H: There is one value for each concept node in H. The domain D_i of a variable x_i is composed of the *a priori* possible images for the concept node c_i by a BG homomorphism from G to H. If, for a concept node c_i in G, we note $poss(c_i) = \{c' \in C_H, l_G(c_i) \geq l_H(c')\}$, then D_i is the set of values assigned to the nodes in $poss(c_i)$. The domain of a variable coming from an individual node of G thus contains at most one value. Also note that isolated concept nodes are translated into variables beyond the scope of all constraints. Such variables can be instantiated by any value of their domain.

The definition of a constraint C_i coming from a relation r_i is the set of tuples given by the neighborhood of the relations in H, which might be images for r_i. At

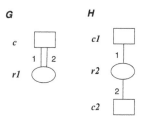

Fig. 5.17 The case of multiedges

this point, we have to pay attention to the fact that the arguments of a relation are not necessarily distinct, that is, in graph terms, there may be several edges between a concept node and a relation node. This case does not occur in a constraint, since its scope is a *set* of variables.

For any relation r of arity q in a BG, let P_r denote the partition on $\{1, ..., q\}$ induced by the equality of arguments of r, i.e., i and j are in the same class of the partition if the ith and the jth arguments of r are the same node. For any c argument of r, $P_r[c]$ denotes the class in P_r corresponding to c, i.e., in graph terms it is the set of numbers labeling the edges between r and c.

Example. In Fig. 5.17, $P_{r_1} = \{\{1,2\}\}$ and $P_{r_2} = \{\{1\},\{2\}\}$. $P_{r_1}[c] = \{1,2\}$, $P_{r_2}[c_1] = \{1\}$, $P_{r_2}[c_2] = \{2\}$.

Given two relations r_1 and r_2 of the same arity, P_{r_2} is thinner than P_{r_1} (notation $P_{r_2} \subseteq P_{r_1}$) if each P_{r_2} class is included in (or equal to) a P_{r_1} class. If a BG homomorphism maps a node r to an node r', then, necessarily, P_r is thinner than $P_{r'}$. Indeed, two nodes can have the same image, whereas a node cannot have two images.

Example. In Fig. 5.17, P_{r_2} is thinner than P_{r_1}, but the opposite is false. Assuming that the conditions on labels are satisfied, there is a homomorphism from H to G but not from G to H.

If P_{r_2} is thinner than P_{r_1}, given any neighbor c of r_2, the *corresponding* neighbor of c in r_1 is the node c' with $P_{r_2}[c] \subseteq P_{r_1}[c']$. If a BG homomorphism maps r_2 to r_1, then necessarily c is mapped to c'.

Each relation r of arity q is transformed into a constraint of arity n ($n \le q$), where n is the number of distinct neighbors of r. Thus, n is not necessarily equal to q, the arity of the *type* of r. n is the number of distinct neighbors of r, which might be strictly smaller than the number q of edges incident to r. There is a bijection between the concept nodes of G and the variables of X, thus the scope of a constraint assigned to a relation node is in bijection with the set of (distinct) neighbors of this relation.

Example. In Fig. 5.18, r is a ternary relation with two distinct neighbors, which leads to a constraint of arity 2 (transformation 1). The definition of this constraint is obtained from $poss(r) = \{r_1, r_3\}$.

Let $(c_{i_1}, \ldots, c_{i_n})$ be the list of distinct neighbors of r ordered in a certain way (for instance by increasing order of their first occurrence in $r(c_1, \ldots, c_q)$, with q being the arity of r). Let $poss(r) = \{r' \in R_H | l_G(r) \ge l_H(r') \text{ and } P_r \subseteq P_{r'}\}$. Let C_i be the

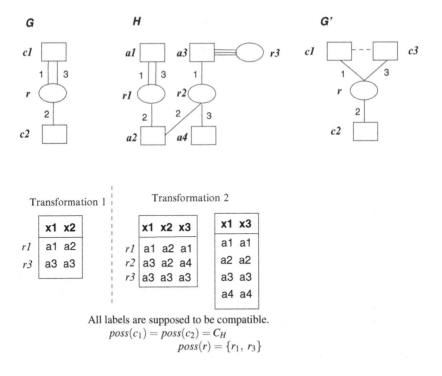

Fig. 5.18 *h2c*: processing multiedges

constraint, with scope $S_i = (x_{i_1}, \ldots, x_{i_n})$, assigned to r. Then, R_i is the set of tuples $(a_{i_1}, \ldots, a_{i_n})$ coming from values assigned to the arguments of relations in $poss(r)$. Thus, two relations with the same label do not necessarily have the same possible images.

The *h2c* transformation is finally defined as follows:

- X is in bijection with C_G (we denote $h2c(C_G) = X$),
- D is in bijection with C_H (we denote $h2c(C_H) = D$); furthermore for each $c \in C_G$, $poss(c)$ is in bijection with $D_{h2c(c)}$, the domain of $h2c(c)$ (we denote $h2c(poss(c)) = D_{h2c(C)}$),
- C is in bijection with R_G (we note $h2c(R_G) = C$),
- the scope of each constraint $h2c(r)$ is in bijection with $\{h2c(c_1), \ldots, h2c(c_n)\}$ ordered in a certain way, where $c_1, \ldots, c_n$ are the *distinct* arguments of r,
- the definition of each constraint $h2c(r)$ is built as follows. Let $poss(r) = \{r' \in R_H | l_G(r) \geq l_H(r')$ and $P_r \subseteq P_{r'}\}$. Let $c_1, \ldots, c_n$ be the list of distinct neighbors of r (listed in the same order as in the scope of $h2c(r)$). Then $R_{h2c(r)}$ is built from $poss(r)$: $R_{h2c(r)} = \{(v_1, \ldots, v_n)|$, $r' \in poss(r)$ and for all $1 \leq i \leq n$, $v_i = h2c(c_i')$, where c_i' is the neighbor of r' corresponding to c_i for $r\}$ (we denote $h2c(poss(r)) = R_{h2c(r)})$. If H does not contain several relations with the same neighbor list, then $poss(r)$ is in bijection with $R_{h2c(r)}$.

It can be simply checked that any BG homomorphism from G to H gives a solution to the CSP, and reciprocally.

$h2c$ is easily extended to deal with *coreference links*. In addition to previous constraints, for each (non-trivial) coreference class of the source BG an *equality constraint* is created, i.e., a constraint expressing that all variables in its scope must be assigned the same value.

Note that, concerning BG homomorphism, the source BG could be processed as if the arguments of any relation were all distinct. Indeed, a concept node linked to a relation node by multiedges can be split into as many nodes as edges, with these nodes being coreferent, while keeping a logically equivalent BG (cf. Fig. 5.18, transformation from G to G'). Thus, each relation could be translated into a constraint with the same arity, plus an equality constraint per coreference class on its neighbors (cf. Fig. 5.18, transformation 2).

We have seen that when BGs are considered as graphs, the number of homomorphisms from one BG to another can be greater than that with the hypergraph vision (cf. Sect. 5.1.1). In this case, by the $h2c$ transformation, a solution to the constraint network can define several BG homomorphisms from G to H. Therefore, BGs have to be considered as hypergraphs so that the $h2c$ transformation will keep the number of solutions.

Example. Let us apply $h2c$ to the BG-HOMOMORPHISM instance (G, H) in Fig. 5.15, which has been obtained by $c2h$ from the P network in Fig. 5.13. The constraint network obtained, say P', is similar to P (in other words $h2c(c2h(P))$ is similar to P), where all X_is come from the concept nodes of G and a, b, c, d from the concept nodes of H. The slight difference between P and P' relies on the domains, since in P' all variables have the same domain, which is equal to D.

The properties of the mappings $c2h$ and $h2c$ considered as mappings concerning BHs instead of BGs lead to the following theorem.

Theorem 5.6 (Equivalence between CSP and BH-HOMOMORPHISM).

- *The $c2h$ mapping is a parsimonious reduction from* CSP *to* BH-HOMOMORPHISM *with flat vocabularies.*
- *The $h2c$ mapping is a parsimonious reduction from* BH-HOMOMORPHISM *to* CSP.

An immediate consequence of the previous result is another proof of NP-completeness for CSP. Furthermore, by the given transformations, the constraint network and the source BG have the same structure, thus, tractable cases based on the structure can be translated from one problem into another. CSP has been extremely well-studied from an algorithm viewpoint. Therefore, importing CSP techniques to BG homomorphism is a natural idea (cf. Chap. 6). However, let us point out that a good algorithm for CSP is not necessarily good for computing BG homomorphisms in practice. Indeed, in the constraint processing field great efforts have been made to design algorithms that perform well on extremely difficult instances, corresponding to the so-called *phase transition*. The networks of these instances are very dense, and generally randomly generated. BGs, at least in a knowledge representation context, have several characteristics, which are usually not observed in constraint networks.

First, they are often written or drawn by a human being, thus have a particular structure: They tend to be of small size, sparse, with a small cyclicity degree (that is with "simple" cycles). Secondly, there is usually dissymmetry between the source BG and the target BG; the source BG (which has the same structure as the constraint network that would be obtained by the *h2c* transformation) is often smaller and less complex than the target BG. Query-answering systems are a typical case, indeed in such a system the source graph is the query and is usually small, and the target graph is (part of) the knowledge base which can be large.

5.5 Bibliographic Notes

It is shown in [Jea98] that many combinatorial problems can be reduced to homomorphism problems. The last transformation in Sect. 5.2.2, allowing us to eliminate the labels, is classical in graph theory (cf.[HN04]). The remark in Sect. 5.1.1 that there might be exponentially fewer homomorphisms from one conceptual hypergraph to another rather than homomorphisms from the corresponding BGs was pointed out by Baget in [Bag01][Bag03]. The hypergraph approach is also favoured in the Darmstadt school viewpoint on CGs (cf. [Wil97], [Pre98b], [Dau03a], and [ABC06], where the equivalence between concept graphs and conceptual graphs is shown).

There are more relationships between BGs and relational databases than the results presented in Sect. 5.3.2. Indeed, it can be proven that the query homomorphism theorem ([AHV95], 6.2.3) is equivalent to the soundness and completeness theorem for BG homomorphism (cf. Chap. 4) [CM92], the minimization theorem ([AHV95], 6.2.6) is equivalent to the theorem concerning irredundant simple graphs (cf.theorem 2.1) [CM92], and the complexity theorem for query decision problems ([AHV95], 6.2.10) is equivalent to the complexity theorem for BG-HOMOMORPHISM (cf.corollary 5.1 [CM92]. See [SCM98] for further details on these equivalences. The translation from the conjunctive query containment problem (CQC) to BG-HOMOMORPHISM was published in [CMS98]. Independently, in [KV98] it was shown that CQC and CSP are essentially the same problem because they can be recast as a relational structure homomorphism problem. Feder and Vardi stated that directed graph homomorphism and CSP are equivalent problems in [FV93]. Since the proof of this result was not given in their paper, we built our own transformations. A first version of the correspondences between CSP and BG-HOMOMORPHISM can be found in [CM95] or in [MC96]. In these previous versions, correspondences were done between classical labeled graph homomorphism and binary CSP. In [Mug00] correspondences between the three problems CQC, CSP and BG-HOMOMORPHISM are presented in a unified framework. For pointers to the constraint processing domain, see the bibliographic notes in Chap. 6.

Part II
Computational Aspects of Basic Conceptual Graphs

Chapter 6
Basic Algorithms for BG Homomorphism

Overview

In this chapter, we study basic algorithms for computing one or all homomorphisms from one BG to another. We first present the basic backtrack scheme, then improve it by consistency-maintaining techniques, essentially adapted from the constraint processing domain (Sect. 6.1). As this book is about conceptual graphs and not constraint networks, these techniques are applied directly on BGs. However, another section is entirely devoted to constraint processing (Sect. 6.2), and it is intended for readers interested in further algorithmic developments. The last section deals with the problem of node label comparisons. It presents and compares data structures and associated algorithms for managing orders on types. General orders are considered, and also orders with a specific structure, e.g., trees or lattices.

6.1 Algorithms for BG Homomorphisms

The fundamental BG-HOMOMORPHISM problem, takes as input two BGs, G and H, and asks whether there is a homomorphism from G to H. However, checking the existence of a homomorphism is often not sufficient. Algorithms able to *exhibit* a homomorphism, or *all* homomorphisms from G to H, are needed. The present chapter is devoted to this issue.

Before getting to the heart of the matter, let us discuss one point that may not be central but cannot be ignored. A first question arising when designing an algorithm exhibiting a homomorphism is about the form of the output, i.e., about the constructed homomorphism. Basically, the output is a mapping from the source graph G to the target graph H. Concerning the *domain* of this mapping, three cases can be considered: (1) The domain is the set of concepts and relations in G; (2) the domain is the set of relations in G; (3) the domain is the set of concepts in G.

Case (1) corresponds to the definition of a BG homomorphism. Case (2) is based on the fact that the image of a relation determines the images of its arguments; if

G has no isolated concept nodes, a homomorphism is uniquely defined by the restriction of its domain to relation nodes. In this case, the algorithms process isolated nodes apart from other nodes. Case (3) considers that the images of the relations are of no importance. It is sufficient to know that a relation is preserved in the image graph; the way it is specialized does not need to be memorized. This view corresponds to the ordered hypergraph viewpoint of conceptual graphs, in which concepts are nodes whereas relations are hyperedges. A homomorphism is then a mapping from the nodes of the source hypergraph to the nodes of the target hypergraph, which preserves hyperedges (cf. Sect. 5.1.1).

We have seen that, if G and H are two BGs and $g2h$ is the mapping translating BGs to hypergraphs, there might be exponentially fewer homomorphisms from $g2h(G)$ to $g2h(H)$ than homomorphisms from G to H (cf. Sect. 5.1.1). Thus, in what follows, we will focus on the hypergraph viewpoint and see the domain of a homomorphism as the set of concept nodes in the BG source. However the algorithms presented in this chapter can be easily adapted for returning images for relations as well, especially since they use propagation techniques that maintain sets of candidate images for concepts *and* relations.

6.1.1 Basic Backtrack Algorithms

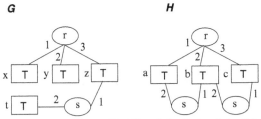

There is one homomorphism from G to H: $\Pi = \{(x,\ a),(y,\ b),(z,\ c),(t,\ b)\}$

Fig. 6.1 Homomorphism

Let us take the classical *backtrack* scheme as the basis. This scheme consists of trying to successively assign an image to each node of the source BG, while keeping the following condition true: At a given step, after a certain number of assignments, the assignments define a *partial* solution, i.e., a partial homomorphism. Let us recall that we only assign images to concept nodes, but the algorithms can be simply extended to assign images to relations as well if desired.

Definition 6.1. A *partial homomorphism* from G to H is a mapping from a subset C' of C_G to C_H, which is a homomorphism from the subBG induced by C' to H (the subBG induced by C' is the subgraph of G defined by C' and relations having all their arguments in C').

Example. In Fig. 6.1, $f = \{(x, a), (y, b), (z, a)\}$ is a mapping from $C' = \{x, y, z\}$ to C_H that is not a partial homomorphism. Indeed, the image of the hyperedge (x, y, z) in G is (a, b, a), which is not a hyperedge in H.

A partial homomorphism with $C' = C_G$ is a homomorphism. Each assignment of a concept node c has to fulfill the condition on labels: If c' is assigned to c, then $l_G(c) \geq l_H(c')$. After each assignment, it is checked whether this assignment is compatible with previous ones, i.e., whether the set of current assignments still defines a partial homomorphism. If this is the case, an unassigned node (if any) is then considered; if all nodes have been successfully assigned, a solution has been found. If the last assignment is not acceptable, another assignment is tried for c. If all candidate assignments for c have been tried without success, a *backtrack* is performed: The algorithm comes back to the node which precedes c in the assignment order, and tries a new assignment for this node, if any exists.

A backtracking search is classically seen as a depth-first traversal of a search tree representing all possible assignments. See Fig. 6.2 for a first picture, and this figure is discussed in detail in the example below.

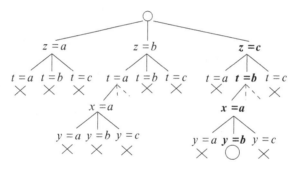

Fig. 6.2 Backtracking search

In this tree, each node except the root corresponds to an assignment. The root has level 0, an assignment occurring at level k is at depth k in the tree. Each path from the root to a node at level k represents a partial mapping of k nodes. The algorithm explores this tree, in a depth-first search manner. A node is said to be visited by the algorithm if the algorithm constructs the partial mapping from the root to this node. The search tree visited by the algorithm is the subtree composed of all visited nodes. Leaves of the visited search tree correspond either to solutions (the path of length $|C_G|$ from the root to this node defines a homomorphism) or to dead-ends (the path to the root to this node yields a mapping which is not a (partial) homomorphism). All internal nodes define (strictly) partial homomorphisms.

Example. Let us consider Fig. 6.2, which pictures a part of the search tree explored by the backtracking algorithm taking the input (G, H) in Fig. 6.1. The concept nodes of G are assumed to be considered in the order $z\ t\ x\ y$, and those of H in the order $a\ b\ c$. Dotted lines indicate parts of the tree explored by the algorithm (but not completely drawn). Crosses below leaves mark dead-ends and circles mark solutions. In this example, there is only one solution; the corresponding assignments are written in bold. The tree is explored in a depth-first manner, from left to right: The first unassigned node considered is z; the first assignment tried is $z = a$, then, for t the values a, b and c are successively tried without success; the algorithm backtracks to z and chooses the assignment $z = b$. The subtree rooted in $z = b$ is explored without success. The algorithm backtracks again to z and tries the only remaining choice for z, i.e., $z = c$. The assignment $t = a$ leads to a failure, thus another assignment for t is tried, i.e., $t = b$. Finally, the single homomorphism is found.

Let us now consider the algorithm GenericBacktrack (Algorithm 1). It returns the first homomorphism found, if any exists, and otherwise a failure. Note that the subalgorithms have access to the local variables and parameters of the main algorithm. All further algorithms will be presented in the same way. The role of the Preprocessing subalgorithm is to prepare the work for the GenericBasic algorithm (Algorithm 2), which does the backtracking search itself. We present this search in a recursive way, which is more "natural," but improvements controlling the way of backtracking (as outlined in Sect. 6.2.1) start mainly from the iterative version. ChooseUnassignedNode returns an unassigned node. ComputeCandidates(c) returns the set of possible images for c. Classical implementations of these subalgorithms are given later (Table 6.1).

If all solutions are required, backtracks are performed until the search tree is exhausted. The generic algorithm is slightly modified and yields GenericBacktrackAllSolutions (see Algorithms 3 and 4). This algorithm returns the set of all homomorphisms from G to H.

Algorithm 1: GenericBacktrack(G,H)

Input: BGs G and H
Output: a homomorphism from G to H if it exists, otherwise Failure
begin
  Preprocessing()
  return GenericBasic($\emptyset$) // see Algorithm 2
end

Several comments can be made concerning these generic algorithms. First, for an efficiency purpose, each connected component in the source BG should be processed by a separate call to the backtrack algorithm. Indeed, each connected component defines an independent subproblem. Imagine, for instance, that G has two connected components C_1 and C_2. If C_2 is processed after C_1 but in the same search, the visited subtree corresponding to C_1 will be repeated as many times as they are successful leaves in the visited subtree of C_2, whereas it would be built only once if C_1 and C_2 were processed separately. It is thus assumed (but this is not mandatory) in all backtrack algorithms that the source BG is connected.

Algorithm 2: GenericBasic(Sol)   *subalgorithm of Algorithm 1*

Input: a partial homomorphism *Sol* from G to H
Output: a homomorphism from G to H extending *Sol* if it exists, otherwise Failure
begin
    if $|Sol| = |C_G|$ **then**
        | **return** *Sol* // `a homomorphism has been found`
    else
        $c \leftarrow$ ChooseUnassignedNode()
        candidates $\leftarrow$ ComputeCandidates(c)
        forall $c' \in$ *candidates* **do**
            $Sol' \leftarrow Sol \cup \{(c, c')\}$
            if *Sol' is a partial homomorphism* **then**
                $S \leftarrow$ GenericBasic(Sol')
                if $S \neq$ *Failure* **then**
                    | **return** S // `S is a homomorphism`

        return *Failure* // `Sol cannot be extended to include c`
end

Algorithm 3: GenericBacktrackAllSolutions(G,H)

Input: BGs G and H
Output: the set of homomorphisms from G to H
begin
    Preprocessing()
    SetOfSols $\leftarrow \emptyset$
    GenericBasicAllSolutions($\emptyset$) // `see Algorithm 4`
    return *SetOfSols*
end

Algorithm 4: GenericBasicAllSolutions(*Sol*)   *subalgorithm of Algorithm 3*

Input: a partial homomorphism *Sol* from G to H
Output: adds to SetOfSols all homomorphisms from G to H extending *Sol*
begin
    if $|Sol| = |C_G|$ **then**
        | add *Sol* to SetOfSols // `a homomorphism has been found`
    else
        $c \leftarrow$ ChooseUnassignedNode()
        candidates $\leftarrow$ ComputeCandidates(c)
        forall $c' \in$ *candidates* **do**
            $Sol' \leftarrow Sol \cup \{(c, c')\}$
            if *Sol' is a partial homomorphism* **then**
                | GenericBasicAllSolutions(Sol')
end

Secondly, the order in which nodes are assigned has an influence on the size of the subtree visited. Some orderings are obviously bad. If one successively assigns t and x, and if t and x do not share any relation (as in Fig. 6.1, assuming that z has not yet been visited), the algorithm will try all assignments on x independently of those done on t. It is better after t to try a neighbor of t (like z in the Figure), since a failure can be detected sooner. That is why nodes are often ordered by a traversal of the BG (depth-first search or a breadth-first search for instance). This is done in the preprocessing phase. Then, the ChooseUnassignedNode algorithm returns the first unassigned node following this order. The backtrack search shown in Fig. 6.2 corresponds to the ordering $(z\ t\ x\ y)$, which can be obtained by a traversal from z.

Thirdly, the preprocessing step can be used to compute the set of *a priori* possible images for each concept node. Basically, a set of candidates, denoted by $poss(c)$ is assigned to each concept node c of C_G, such that for any of these candidates y, (c, y) is a partial homomorphism: $poss(c) = \{y \in C_H | l_G(c) \geq l_H(c)\}$. Then ComputeCandidates(c) returns $poss(c)$. Table 6.1 summarizes the above remarks by presenting classical implementations of the backtrack subalgorithms. Similarly, a kind of preprocessing can be performed for relations, as we will see in the next section.

Preprocessing()
    1. compute $O_G = c_1, \ldots, c_n$ a total ordering on C_G by a traversal
       (depth-first and breadth-first search) of G
    2. for all c_i, compute $poss(c_i) = \{y \in C_H | l_G(c_i) \geq l_H(y)\}$
ComputeCandidates(c)
    return $poss(c)$
ChooseUnassignedNode()
    return $c_{|Sol|+1}$

Table 6.1 Classical implementations of generic algorithms

Let us discuss another point about generic algorithms. After the assignment of a node c, it has to be checked that the mapping is still a partial homomorphism, which requires checking that the newly mapped relations (those which have c as argument and all arguments assigned) are satisfied. More precisely, the order on C_G induces a partial order on R_G. Let us say that a relation has rank i if its greatest argument according to the order on concept nodes is c_i. After the assignment of node c_i, only relations of rank i have to be checked. This is stated in the following property:

Property 6.1. Let π be a partial homomorphism from G to H, with its domain being the subset $C' = \{c_1, \ldots, c_{i-1}\}$ of C_G. Let $c' \in C_H$. $\pi \cup \{(c_i, c')\}$ is a partial homomorphism from G to H if and only if for each relation $r(c_{i_1}, \ldots, c_{i_k})$ of rank i in G, there is a relation $r'(\pi(c_{i_1}), \ldots, \pi(c_{i_k}))$ in H with r' less or equal to r.

Example. Let us consider again Figs. 6.1 and 6.2. With the order on the concept nodes in G being $z\ t\ x\ y$, the rank of the relation $s(z,t)$ is 2 (this relation is thus checked after each assignment of t), and the rank of the relation $r(x,y,z)$ is 4 (this relation is thus checked after the assignment of all concept nodes). Taking relations into account earlier would lead to earlier failure detection. For instance, as soon as a has been assigned to z, it could be detected that the relation $r(x,y,z)$ can no longer be mapped. This issue will be discussed later (see Sect. 6.1.2.2).

In case the target BG is very big, or even cannot be globally accessed, the *poss* sets can be computed *during* the search instead of *before* it. In this case, for $k \geq 1$, $poss(c_k)$ is computed by ComputeCandidates. Instead of C_H, the considered set is computed from the images of c_k's already mapped neighbors: It is basically the set of concepts sharing at least a relation with these images.

If we replace nodes by variables and relations by constraints (see the transformation $h2c$ in Chap. 5), we obtain a basic backtrack algorithm for Constraint Satisfaction Problem (CSP) solving. Improvements of the backtrack scheme have been extensively studied in the CSP community and several approaches for coping with the problem complexity have been developed. We import some of them into the BG framework in the next section.

6.1.2 Backtrack Improvements

The improvements proposed in this section are adapted from algorithmic techniques developed for the CSP. We have selected techniques which are simple and yield good results in practice (see Sect. 6.2 for further details). First, an adaptation of the arc-consistency CSP notion, called *edge-consistency*, is presented. Then this notion is integrated into the backtracking scheme, yielding a method for computing BG homomorphisms similar to the forward-checking method for CSP.

Since this book is about conceptual graphs and not constraint networks, all notions will first be applied directly on BGs. Then, a whole section is devoted to the presentation of their CSP counterpart. This section can be skipped by readers not interested in constraint networks. However, readers interested in efficient algorithms on difficult instances should check out this presentation to find out how CSP techniques can be exploited in the BG framework, as a prelude to a more in depth study of these techniques.

6.1.2.1 Edge-Consistency

Edge-consistency is a necessary, but not sufficient, condition for the existence of a homomorphism. It will be used in the next section to limit the search space in the backtracking procedure. Let G be the source BG and H be the target BG. A concept or relation node x in G can be mapped by a homomorphism to a node x' in H only if $l_G(x) \geq l_H(x')$. Furthermore, a relation r in G can be mapped by a homomorphism to a relation r' in H only if the multi-edges incident to it are also multi-edges incident to r': If there are two edges labeled by i and j between r and a concept node c, then there must be two edges labeled by i and j between r' and a concept node c'. Let us introduce two notations.

Definition 6.2 (Edge partition associated with a relation node). Let r be a relation node of arity q. P_r is the partition on $\{1,\ldots,q\}$ (or equivalently, on the set of edges incident to r) induced by the equality of arguments of r, i.e., i and j are in the same class of P_r if the ith and the jth arguments of r are the same node. For any c neighbor of r, $P_r[c]$ denotes the class in P_r corresponding to c, i.e., the set of numbers labeling edges between r and c.

Definition 6.3 (Corresponding neighbors). Let r and r' be two relation nodes in the same BG or in two different BGs having the same arity and let c be the i-th neighbor of r ($1 \le i \le arity(r)$). The concept node $c' = corresponding(c,r,r')$ is defined as the i-th neighbor of r'.

A relation node r can be mapped to a relation node r' only if P_r is thinner than $P_{r'}$ (notation $P_r \subseteq P_{r'}$), i.e., each class of P_r is included in a class of $P_{r'}$. When $P_r \subseteq P_{r'}$, given any neighbor c of r, $c' = corresponding(c,r,r')$ is the unique neighbor of r' that corresponds to c (i.e. $P_r[c] \subseteq P_{r'}[c']$).

A *filter of* (G,H) maps each concept c of G to a subset of *candidate* concepts for c and each relation r of G to a subset of *candidate* relations for r.

Definition 6.4 (candidates and filters).

A *candidate* for a concept node c is a concept node c' such that $l_G(c) \ge l_H(c')$. A *candidate* for a relation r is a relation r' such that $l_G(r) \ge l_H(r')$ and $P_r \subseteq P_{r'}$. Given two BGs G and H, a *filter* of (G,H) maps each x in $C_G + R_G$ to a *non-empty* subset of $C_H + R_H$ of candidates for x. When (G,H) has at least a filter, the *standard filter* of (G,H) is the maximal filter of (G,H): It maps each x in $C_G + R_G$ to all of its candidates in $C_H + R_H$.

The existence of a filter of (G,H) is a necessary condition for the existence of a homomorphism from G to H. We are now interested in filters restricting the sets of candidates without eliminating homomorphisms. These filters satisfy the following properties. For any concept c in G, for any candidate c' for c, and for any relation neighbor of c, the partial homomorphism $\{(c,c')\}$ can be extended to a partial homomorphism mapping the star graph of r (i.e., r and all its neighbors) to suitable candidates. Furthermore, for any relation r in G, and for any candidate r' for r, the star graph of r can be mapped to the star graph of r', i.e., for each neighbor c of r, $corresponding(c,r,r')$ is a candidate for c. Such filters are called *edge-consistent*. The existence of an edge-consistent filter of (G,H) is a necessary condition for the existence of a homomorphism from G to H. The definition below is based on local conditions on edges.

Definition 6.5 (edge-consistent filter). Given a filter f of (G,H),

- for a concept c in G, $f(c)$ is edge-consistent if for each $c' \in f(c)$, and for each edge (c, i, r) in G, there is $r' \in f(r)$ with an edge (c', i, r') in H;

- for a relation r in G, $f(r)$ is edge-consistent if for each $r' \in f(r)$, and for each edge (c, i, r) in G, there is an edge (c', i, r') in H with $c' \in f(c)$;
- f is edge-consistent if for all x, $f(x)$ is edge-consistent.

Note that if all $f(r)$ are edge-consistent, the $f(c)$ can be restricted to values given by $f(r)$ to obtain an edge-consistent filter. Indeed, edge-consistency does not restrict the candidates of isolated nodes.

Example. Let us consider the BGs in Fig. 6.3, where $f(c_1) = \{d_1\}$, $f(c_2) = \{e_2\}$, $f(r) = \{r_1, r_2\}$. $f(c_1)$ and $f(c_2)$ are edge consistent, but not $f(r)$: $r_1 \in f(r)$ and there is the edge $(r, 2, c_2)$ but no edge $(r_1, 2, e_2)$ in H; the same kind of problem occurs with r_2 since there is no edge $(r_2, 1, d_1)$ in H.

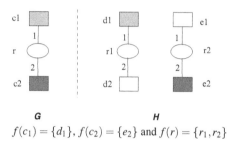

$$G \qquad\qquad\qquad H$$
$$f(c_1) = \{d_1\},\, f(c_2) = \{e_2\} \text{ and } f(r) = \{r_1, r_2\}$$

Fig. 6.3 A non-edge-consistent filter

Property 6.2. Let f be an edge-consistent filter of (G, H). Then, for each concept node c in G, and each concept node c' in $f(c)$, the partial homomorphism (c, c') can be extended to any relation r adjacent to c and its neighbors (i.e., to the star graph of r).

Proof. Let us denote by π the partial homomorphism from G to H defined by: $\pi(c) = c'$. Let (c, i, r) in G. One takes $\pi(r) = r'$, where r' is any node in $f(r)$ such that (c', i, r') is in H (there is at least one such node). Now consider any concept node d adjacent to r. There is a j with (d, j, r) in G (note that d may be equal to c and in this case $i = j$). There is a only one edge incident to r' and labeled by j in H. One takes for $\pi(d)$ the concept extremity of this edge, i.e., $\pi(d) = H_j(r')$ (Fig. 6.4). Furthermore, if $G_j(r) = G_k(r)$, then $H_j(r') = H_k(r')$ since $P_r \subseteq P_{r'}$, and it is straightforward to check that π is a partial homomorphism from the star graph of r to the star graph of r'. $\square$

Given a filter of (G, H), making it edge-consistent can be seen as a closure operation (on the dual inclusion order between filters).

Definition 6.6 (Edge-consistent closure). The *edge-consistent closure* f' of a filter f of (G, H) is the largest edge-consistent filter f' contained in f, more precisely:

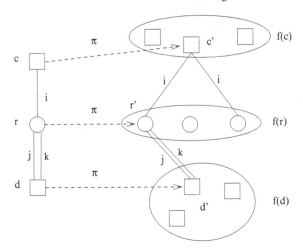

Fig. 6.4 Local extension of a partial homomorphism (proof of Property 6.2)

- f' is an edge-consistent filter of (G,H)
- for all $x \in C_G \cup R_G$, $f'(x) \subseteq f(x)$
- for all $a \in f(x) \setminus f'(x)$, there is no BG homomorphism from G to H mapping x onto a
- there is no other edge-consistent filter f'' of (G,H) that contains f', i.e., such that for all $x \in C_G \cup R_G$, $f'(x) \subseteq f''(x)$

Property 6.3. The algorithm `EdgeConsistency` (Algorithm 5) computes the edge-consistent closure of the standard filter of (G,H), if it exists.

Proof. (Hints) Let us say that a filter f of (G,H) is *locally edge-consistent* at $r \in R_G$ if it is edge-consistent for r and for all the neighbors of r. If a filter is locally edge-consistent at each relation in G, then it is edge-consistent. Given a filter f, `CheckRelation(r)` computes (if it exists) the largest filter included in f and locally edge-consistent at r. The first step consists of removing a relation r' in $f(r)$ if, for a neighbor c of r, the concept $corresponding(c,r,r')$ is not in $f(c)$. The second step consists of removing a concept a in $f(c)$ if c is a neighbor of r, such that there is no r' in $f(r)$ with $a = corresponding(c,r,r')$. The `EdgeConsistency(G,H)` procedure first computes the standard filter of (G,H) then it makes this filter locally edge-consistent for each relation. If for a relation r, CheckRelation(r) restricts the candidate set of one of its neighbors c, all relations neighbors of c have to be checked again. The algorithm terminates because each call to CheckRelation(r), which leads to adding a relation to Q, removes at least one element in a concept candidate set. $\square$

Property 6.4. The time complexity of `EdgeConsistency(G,H)` is in $O(max (|C_G| \times |C_H|, m_G \times |C_H| \times m_H))$, thus $O(m_G \times |C_H| \times m_H)$ if G does not contain isolated concept nodes.

Algorithm 5: EdgeConsistency(G,H)

Input: two normal BGs G and H
Output: the edge-consistent closure of the standard filter of (G,H) if it exists, otherwise
      Failure
begin
    // 1. compute the standard filter
    foreach c in C_G **do**
        $poss(c) = \{c' \in C_H | l_G(c) \geq l_H(c')\}$
    foreach r of R_G **do**
        $poss(r) = \{r' \in R_H | l_G(r) \geq l_H(r')$ and $P_r \subseteq P_{r'}\}$
    // 2. make the filter edge-consistent
    Tocheck $\leftarrow R_G$
    forall c in C_G **do**
        Changed[c] $\leftarrow$ false
    while *ToCheck* $\neq \emptyset$ **do**
        Pick one relation r from ToCheck
        result $\leftarrow$ CheckRelation(r) // see Algorithm 6
        if *result = EmptyPoss* **then**
           return *Failure*
        if *result = Changed* **then**
           forall $r_i \neq r$ *such that there is c common neighbor of r_i and r and*
           Changed[c]= true **do**
               ToCheck $\leftarrow$ ToCheck $\cup \{r_i\}$
    return *poss*
end

Proof. Step 1. Computing the standard filter takes $|C_G| \times |C_H|$ label comparisons for the concept nodes. We assume that label comparison is in constant time. Concerning partitions of the labels of edges incident to a relation, we assume that they are stored in such a way that the concept node corresponding to an element of the partition can be found in constant time (we can use, for instance, an array of size k which, for each index i, refers to the node c extremity of the edge (r,i,c)). Comparison of the partitions of two relations with the same arity k can then be done in $O(k)$, which corresponds to the number of edges incident to each of these relations. For a relation r, at most R_H labels are compared with the label of r, and since the partitions are compared only for relations with the same arity as r, we can bound the time required for partition comparisons by m_H. Thus, we obtain a time complexity for the standard filter computation is in $O(max(|C_G| \times |C_H|, |R_G| \times m_H))$. Step 2. CheckRelation($r$) can be implemented in m_H. Assume that we use vectors of size $|C_H|$ initialized with special values and used to mark the concept nodes of C_H. Different marks can be used for different uses. The first step is done in $arity(r) \times poss(r)$. As it is written, the second step can be very inefficiently carried out. But it can also be performed as follows: For each relation r' in $poss(r)$, for each neighbor a of r', mark a in $poss(c)$ where c is the corresponding neighbor for r. Assume that we memorize the latest mark used to mark the nodes of a $poss(c)$ set: A node of H belongs to $poss(c)$ if it is marked by this mark, otherwise it is not in

---

Algorithm 6: CheckRelation(r)   *subalgorithm of Algorithm 5*

Input: a relation r in G

Output: computes the edge-consistent closure of the star graph of r, if it exists, and marks the concept nodes whose *poss* set has changed; returns EmptyPoss if a *poss*(c) has been emptied, NoChange if no *poss*(c) has been modified, otherwise Changed

begin

    `// 1. remove from poss(r) values not supported by a`
    `   poss(c)`

    foreach *concept c neighbor of r* **do**

        foreach $r' \in poss(r)$ **do**

            let $a = corresponding(c, r, r')$

            if $a \notin poss(c)$ **then**

                remove r' from $poss(r)$

    `// 2. make the poss(c) edge-consistent`

    result ← NoChange

    foreach *concept c neighbor of r* **do**

        foreach $a \in poss(c)$ **do**

            if *there is no* $r' \in poss(r)$ *such that* $a = corresponding(c, r, r')$ **then**

                remove a from $poss(c)$

                if $poss(c) = \emptyset$ **then**

                    return *EmptyPoss*

                else

                    result ← Changed

                    Changed [c] ← true

    return *result*

end

$poss(c)$. The second step is thus done in $O(arity(r) \times |poss(r)|)$, which is bounded by $O(m_H)$. Let us now count the number of total calls to `CheckRelation`. Each time a candidate of a concept node c is removed, the adjacent relations are added to the queue. A relation r can be added to the queue due to each of its neighbors, thus $arity(r) \times |C_H|$ times, where $|C_H|$ is a bound for the $poss(c)$. The total number of calls to `CheckRelation` is thus bounded by $m_G \times |C_H|$.

We obtain a total complexity of step 2 in $O(m_G \times |C_H| \times m_H)$. □

The example in Fig. 6.5 proves that if (G, H) has an edge-consistent filter there is not necessarily a homomorphism from G to H. Indeed, the standard filter is $f(c) = f(d) = f(e) = \{a, b\}$ and $f(p) = f(q) = f(r) = \{u, v\}$ and this filter is edge-consistent but there is no homomorphism from G to H. When the source BG is acyclic, the local consistency defined by edge-consistency implies global consistency, i.e., the existence of a homomorphism. See Chap. 7 devoted to tractable cases for further details.

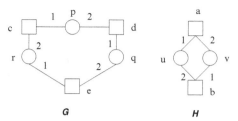

All the concept (resp. relation) nodes have the same label

Fig. 6.5 Edge-consistency does not entail the existence of a homomorphism

6.1.2.2 Forward-Checking

Let us now integrate the edge-consistency check inside the backtrack algorithm. This algorithm can be called in the preprocessing phase to reduce *a priori* the size of the search space. It can also be used to reduce the size of the search tree *dynamically*, i.e., during the search itself. The idea is to propagate the consequences of a new assignment to the remaining non-assigned nodes, thus restricting their candidate sets, and possibly detect a failure sooner. More specifically, a consistency check is performed after each assignment. If the check fails, another assignment has to be tried for the same node, or the algorithm backtracks if all assignments have been tried for this node. The consistency check itself can be more or less strong. It can maintain edge-consistency for all nodes or only for nodes directly connected to the newly assigned node. It can also achieve true edge-consistency, or check each considered star graph only once. A stronger check potentially prunes the search space to a greater extent but is more expensive. The consequences of these choices are investigated more thoroughly in the section about CSP (Sect. 6.2).

The algorithm presented here (Algorithm 7) performs a partial consistency check and is considered as a good tradeoff between the pruning gain and the overhead cost. It is an adaptation to BGs of the popular forward checking method for the CSP (more precisely, the method nFC2 presented in Sect. 6.2).

First, a total ordering $c_1, \ldots, c_n$ is computed on C_G by a traversal of G. Concept nodes are processed according to this ordering. The homomorphisms (if any) are built by extending a partial homomorphism defined for $c_1, \ldots, c_k$ to a partial homomorphism defined for $c_1, \ldots, c_k, c_{k+1}$ by using the following principle:

(BG FC consistency check): *Apply edge-consistency on each relation connecting the current node and at least one non-assigned node, in a single pass.*

Example. Figure 6.6 illustrates the algorithm. It refers to the BGs in Fig. 6.1. Recalling the search tree explored by the classical backtrack (Fig. 6.2), it shows the part explored by the forward checking algorithm (dotted lines). Initially, $poss(z) = \{a, b, c\}$. As soon as a value is assigned to z, the consistency check considers both relations related to z, and restricts all *poss* sets to one element at most.

Property 6.5. The algorithm `AllHomomorphismsByFC(G, H)` (Algorithm 7) computes the set of all homomorphisms from G to H.

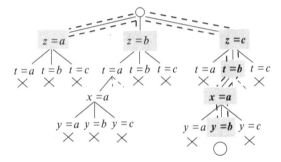

Fig. 6.6 Search tree explored by Forward Checking

Algorithm 7: AllHomomorphismsByFC(G,H)

Input: a source BG G and a target BG H
Output: the set of homomorphisms from G to H
begin

> // preprocessing()
> // 1. total ordering of C_G
> Compute $O_G = c_1, ..., c_n$ a total ordering of C_G by a traversal of G
> // 2. compute the standard filter
> **foreach** c in C_G **do**
> > $poss(c) \leftarrow \{c' \in C_H | l_G(c) \geq l_H(c')\}$
>
> **foreach** r in R_G **do**
> > $poss(r) \leftarrow \{r' \in R_H | l_G(r) \geq l_H(r')$ and $P_r \subseteq P_{r'}\}$
>
> // 3. process relations with a unique neighbor
> **foreach** r in R_G with $|P_r| = 1$ **do**
> > let c be the neighbor of r
> > $poss(c) \leftarrow poss(c) \cap \{c' | c'$ has a neighbor in $poss(r)\}$
>
> **if** there is x in $C_G \cup R_G$ with $poss(x) = \emptyset$ **then**
> > **return** $\emptyset$
>
> // 2 and 3 can be replaced by a call to
> >   EdgeConsistency(G,H): if it fails, return $\emptyset$
>
> $SetOfSols \leftarrow \emptyset$
> **for** level from 1 to n **do**
> > Removed[level] $\leftarrow \emptyset$
>
> RecFC($\emptyset$) // see Algorithm 8
> **return** $SetOfSols$

end

Proof. (Hints) The classical implementations of the subalgorithms of the basic backtrack (Table 6.1) are considered. The preprocessing step of `AllHomomorphismsByFC` involves: first, computing a total ordering on C_G; secondly, computing the standard filter of (G,H); thirdly, taking into account relations with a single neighbor, i.e., either unary relations or relations with multi-edges to a single node. Indeed, the consistency check would never consider them. In this case, the *poss* set of a concept node c that is the single neighbor of relations is restricted to the neighbors of the candidates for these relations.

Algorithm 8: RecFC(*Sol*) *subalgorithm of Algorithm 7*

Input: a partial homomorphism *Sol* from *G* to *H*
Output: puts in *SetOfSols* all homomorphisms from *G* to *H* extending *Sol*
begin

   $k \leftarrow |Sol|$

   if $k = |C_G|$ **then**

      | add *Sol* to SetOfSols // a homomorphism has been found

   else

      $inc(k)$

      $c \leftarrow c_k$ // ChooseUnassignedNode()

      candidates $\leftarrow poss(c)$ // ComputeCandidates(c)

      forall $y \in$ *candidates* **do**

         $Sol' \leftarrow Sol \cup \{(c, y)\}$

         $poss(c) \leftarrow \{y\}$

         consistent $\leftarrow$ true

         forall *relation r neighbor of c with at least an unassigned neighbor and while consistent* **do**

            | consistent $\leftarrow$ checkRelationFC(*r*) // see Algorithm 9

         if *consistent* **then**

            | RecFC(*Sol'*)

         // all poss values suppressed due to the choice (c,y) are restored

         forall (x,y) *in Removed[k]* **do**

            | restore *y* in $poss(x)$

         Removed[k] $\leftarrow \emptyset$

         $poss(c) \leftarrow$ candidates // restore poss(c)

end

The backtrack is implemented in the recursive algorithm RecFC(Sol) (Algorithm 8), where *Sol* is a partial homomorphism defined for $c_1, \ldots, c_k$. When $k = n$, a solution, i.e., a homomorphism, has been computed. The algorithm tries to build an extension of *Sol* to c_{k+1} by searching $poss(c_{k+1})$ using the FC consistency check. If it succeeds, it continues but otherwise it backtracks and restores all $poss(x)$ to the values they had before the extension attempt. For each level k of the tree, corresponding to an assignment of y to c_k, a table *Removed*[k] stores the values deleted from *poss* sets due to this assignment (note that the values deleted from $poss(c_k)$ due to the choice of y are not stored in *Removed*[k], since *candidates* stores the value of $poss(c_k)$ before the assignment). This table is used to restore *poss* sets when the algorithm comes back, and to choose another value for c_k or to backtrack to level $k - 1$; for the node c_k itself, $poss(c_k)$ is restored with *candidates*. The subalgorithm CheckRelationFC() (Algorithm 9) is similar to the subalgorithm CheckRelation() in EdgeConsistency (Algorithms 6 and 5) except that it checks only unassigned concept nodes. Since the consistency check is done in a single step, this subalgorithm does not note which nodes have seen their *poss*

Algorithm 9: CheckRelationFC(r)   *subalgorithm of Algorithm 8*

Input: a relation r
Output: removes from $poss(r)$ values not supported by a $poss(c)$; returns true if no poss has
           been emptied, otherwise false
begin
    foreach *concept c neighbor of r and rank$(c) \geq k$* **do**
        foreach $r' \in poss(r)$ **do**
            let $a = corresponding(c, r, r')$
            if $a \notin poss(c)$ **then**
                remove r' from $poss(r)$
                add (r, r') to Removed[k]

    // make all poss(c) edge-consistent with respect to
    poss(r)
    foreach *concept c neighbor of r and rank$(c) > k$* **do**
        foreach $a \in poss(c)$ **do**
            if *there is no $r' \in poss(r)$ with $a = corresponding(c, r, r')$* **then**
                remove a from $poss(c)$
                add (c, a) to Removed[k]
                if $poss(c) = \emptyset$ **then**
                    return *false*

    return *true*
end

---

sets restricted; on the other hand, it stores suppressed values in *Removed*[k]. More precisely, when calling CheckRelationFC(r), $c_1, \ldots, c_{k-1}$ have been assigned, and one checks if it is possible to extend this partial homomorphism by assigning y to $c = c_k$, and if so the consequences on *poss* are computed. r is a relation node neighbor of c with at least one unassigned neighbor. Two searchs are executed over neighbors of r. During the first search, the relations r' in *poss*(r) that do not satisfy the edge-consistency condition are deleted from *poss*(r). One has to consider the unassigned neighbors of r as well as the newly assigned concept node. For all deleted r', the pair (r, r') is put into *Removed*[k] in order to be restored when backtracking. During the second search, for each unassigned node c neighbor of r, the concepts a in *poss*(c) that do not satisfy the edge-consistency condition are deleted from *poss*(c). For all deleted a, the pair (c, a) is put into *Removed*[k] in order to be restored when backtracking. If a *poss* set is emptied, then the attempted extension is impossible. $\square$

6.2 Constraint Processing

6.2.1 A Panorama of Constraint Processing Techniques

This section presents an overview of main constraint processing techniques. Let us first mention that some of these techniques are at the heart of constraint processing, such as constraint propagation, and have been essentially studied in this framework, while others, such as tree decomposition, are known in various research areas (graph theory, probabilistic reasoning, databases, etc.).

Constraint Propagation

The central concept of constraint processing is *constraint propagation*. This term denotes a set of techniques for propagating the implications of a constraint on one variable to other variables and constraints. Constraint propagation transforms the original network into an equivalent network, i.e., with the same set of solutions, but for which the search space is reduced. Basically, constraint propagation removes hopeless values from variable domains, but it might also add new constraints. Constraint propagation can be seen as a kind of inference.

The basic idea is as follows: If there is a value d in a domain D_i, and a constraint C including x_i, such that d does not appear in the definition of C, then d can be eliminated from D_i since no solution can contain the assignment $x_i = d$. This elimination implies the elimination of all tuples containing d as a value for x_i in other constraints, thus in turn suppressions of values in other variables domains, etc. When no more values can be removed from domains, and if no domain has been emptied, one obtains a so-called *arc-consistent* network: Every assignment of a value to a variable is consistent (relative to the constraints over this variable) and can be extended with an assignment to another variable in a consistent way. Stronger forms of consistency can be defined. Generally speaking, constraint propagation ensures local consistency, which is more or less local depending on the strength of the constraint propagation.

Constraint propagation can be used as a filtering technique in the preprocessing phase. In this case it reduces the size of the *a priori* search space. It can also be used during the search to dynamically reduce the size of the search tree. Once an assignment has been chosen, the idea is to propagate the consequences of this assignment (which can be seen as a restriction of the domain of the assigned variable to exactly one value) in order to detect failures sooner. A crucial point is that the constraint propagation should take significantly less time than exploring the parts of the tree avoided by the check. So a trade-off has to be found between the strength of the consistency check and the complexity of its computation.

Backtrack Improvements

More generally, most strategies for improving the backtrack algorithm can be divided into the so-called *look-ahead* and *look-back* schemes: The first ones are concerned with going forward to extend the current partial solution, while the others are concerned with going back in case of a failure (i.e., when the current partial solution conflicts with every possible value of the next variable).

The look-ahead techniques address the following questions: Which variable should be assigned next, and which value should be tried for this variable? What are the implications of the chosen assignment for the remaining unassigned variables? In the basic backtrack algorithm, variables are assigned according to a static order. But a natural idea is to choose a variable which restricts future searches as much as possible since all variables have to be ultimately assigned. This leads to the "most contrained variable" choice, which consists in choosing a variable with the smallest remaining domain; a companion heuristic to break ties is the "maximum degree variable," which consists in choosing a variable that is involved in the largest number of constraints containing unassigned nodes. The choice of the next value to assign is important if we do not search for all solutions (otherwise we have to consider all values anyway so all orders will do). The idea here is to choose a value that constrains as little as possible future assignments for the still unassigned variables. The "least constraining value" heuristic selects a value that constrains future assignments for the still unassigned variables as less as possible, in other words a value that removes the fewest values for the neighboring variables.

Look-back techniques aim at preventing the construction of the same hopeless partial solution. After a failure, the basic backtrack algorithm comes back to the previous variable following the assignment order so as to try another value for this variable. It is called "chronological" backtracking because it comes back to the most recent assigned variable preceding the dead-end. This way of backtracking may lead to rediscovering the same dead-end. Indeed, let x_i be the variable for which no consistent assignment can be found; the algorithm backtracks to x_{i-1}. Assume x_{i-1} has remaining values to try. Now, if no constraint relates x_i and x_{i-1}, x_{i-1} has no role to play in the failure to instantiate x_i. The same hopeless values will be tried again for x_i, and this for each remaining value for x_{i-1}. The idea of the so-called "backjumping" techniques is to come back to a variable, for which we know that it is not responsible from the failure.

Decomposition of the Network

For many problems, *tree* structures allow for efficient solving. This is the case for CSP: When the constraint network has a tree structure, there is a greedy way of instantiating variables, which allows one to find a solution, if any exists, in polynomial time. Relying on the fact that the tree case is easy to solve, the idea is to decompose the problem structure into a tree.

A simple method consists of "cutting cycles." A subset of variables is extracted so that the remaining network has no cycle. This subset, called a "cycle cutset," defines a subproblem restricted to these variables, their domains and constraints with scope in this cutset (let us consider here that constraints are binary, which allows us to cluster constraints into constraints with scope included in the cutset and constraints with one variable in the cutset at most). Then, if we have a solution to this subproblem, it is easy to check whether this solution can be extended to the whole network, since the remaining network to check is a tree. The idea is thus to enumerate solutions to the subproblem and check each of them. The algorithm is exponential in the size of the cutset, since in the worst case all assignments of this subset have to be tried. Thus if the network is "nearly" a tree, then the cutset will be small, and this method will be efficient. Finding a smallest cutset is an NP-hard problem, but efficient approximation algorithms (thus bringing a cutset not much bigger than a smallest one) are known.

Another family of very general methods used in various research areas is *tree decomposition*. These techniques aim at decomposing a structure (e.g., a constraint network) into a set of substructures as small as possible, whose "intersection" yields a *tree* structure, or more generally a *hypertree*. The precise definition of a tree decomposition will be given in Chap. 7. The algorithm is exponential in the size of the biggest substructure. This size is related to the so-called treewidth of the graph, or the hypertreewidth of the hypergraph. Unfortunately, finding a decomposition with the minimal treewidth or hypertreewidth is NP-hard.

Next Steps

We will first focus on constraint propagation, based on the simplest local consistency, i.e., arc-consistency, which we have translated into *edge-consistency* in the BG framework. As already explained, this technique can be used in the preprocessing phase or during a search to improve the basic backtrack algorithm. Next, we will describe a look-ahead algorithm, called forward-checking, which performs the most limited form of constraint propagation after each assignment: an arc-consistency check in the neighborhood of the new assigned node. Why these choices? The challenge when improving the basic backtrack is to find a trade-off between the reduction of the search tree and the overhead introduced by search tree pruning techniques. For "easy" instances, it is not worthwhile to develop strong consistency techniques. Arc-consistency as a basic technique seems to be appropriate.

Note that most improvements have been tailored for *binary* networks. Indeed, arguing that an n-ary network can always be transformed into an equivalent binary network (equivalent in the sense that it has the same set of solutions modulo the transformation), most research has focused on the binary case. The two prevalent transformations correspond to classical transformation from hypergraphs to graphs, the incidence bipartite graph (or hidden graph), see Sect. 5.1.1 and the dual graph (presented in Sect. 7.3) [BCvBW02]. However, these transformations generate new variables, which make algorithms designed for binary networks inefficient (see for instance the discussion in [BMFL02]). The efficient generalization of techniques developed for binary networks to n-ary networks is an ongoing study in the CSP community. This issue is particularly relevant to BGs and we will keep it in mind throughout the next sections.

We now first present the notions of arc-consistency and forward-checking as introduced for CSP. The translation is not always as direct as one could imagine. It is essential to pay attention to "details," which nevertheless are important for correctness. For instance, as pointed out in the transformation b2c (Sect. 5.4), a variable can occur at most once in a constraint, while a single concept node can occur several times in the argument list of a relation.

6.2.2 Arc-Consistency

We recall in this section the so-called *arc-consistency* technique, in order to clarify how this technique has been translated into the edge-consistency notion studied previously. Arc-consistency, as its name suggests, was first defined for binary networks. The word "arc" refers to a directed edge in the constraint graph. The idea is as follows. Given a binary constraint C on x and y, the arc (x, y) is consistent if for each value in the x domain there is a value in the y domain which is consistent for C. The network is arc-consistent if all of its arcs are consistent. Note that for a constraint C on x and y, both directions (x, y) and (y, x) have to be checked: the subnetwork restricted to C is consistent if and only if (x, y) and (y, x) are both consistent. This view of arc-consistency in terms of domains yields a direct generalization to the n-ary case.

Let us note however that this way of passing from the binary to the n-ary case is not the only possible one (see for instance the "pairwise consistency" presented in Chap. 7).

Definition 6.7 (arc-consistency). Given a variable x_i in the scope of a constraint C_j, a value d of the domain D_i of x_i has a *support* in C_j if d appears in R_j as a possible value for x_i, i.e., there is a tuple t in R_j such that $t[x_j] = d$. t is called a support for the assignment $x_i \leftarrow d$. A domain D_i is *arc-consistent* relative to a constraint C_j having x_i in its scope if (1) it is not empty (2) for all $d \in D_i$, d has a support in C_j. Given a constraint network P, a domain D_i is arc-consistent if it is arc-consistent relative to all constraints in P. P is arc-consistent if all its domains are arc-consistent.

Alternatively, arc-consistency can be defined on constraints: A constraint C_j is arc-consistent if all the domains of all its variables are arc-consistent relative to it. A CSP is arc-consistent if all its domains are non-empty and all constraints are arc-consistent.

It is implicitly assumed that constraints are defined over the current domains: If a value is removed from a domain during the consistency check, in constraint definitions all tuples containing this value have to be removed (actually tuples are physically removed only if constraints are defined in extension, which is our case).

Thus a network is arc-consistent if and only if no domain is empty and for each variable x_i and also for each constraint C_j involving x_i, $D_i = R_j[x_i]$.

C1		
x	y	z
a	a	a
a	b	c
b	a	c
c	c	c

C2		
y	z	u
a	a	b
a	b	c
c	c	c

C3	
u	v
a	a
a	b
c	c

$Dx = Dy = Dz = Du = Dv = \{a,b,c\}$

Fig. 6.7 A non-arc-consistent network

Example. The network in Fig. 6.8 is arc-consistent. The network in Fig. 6.7 is not arc-consistent. None of the constraints is arc-consistent. Consider for instance C_1: $D_z = \{a, b, c\}$ is not arc-consistent since the b value has no support in C_1. Now, if this value is suppressed from D_z, C_1 becomes arc-consistent. Note that this supression leads to the removal of the tuple $(a\ b\ c)$ in C_2.

A network being arc-consistent does not ensure that it is globally consistent (see for instance Fig. 6.8). But it has no solution if it cannot be made arc-consistent.

Before entering into the details of the arc-consistency computation, let us mention that stronger forms of consistency have been defined. Generally speaking, local consistency can be extended to subnetworks on k variables. A partial instantiation (i.e., an assignment of a subset of variables) is *consistent* if it satisfies all constraints whose scope contains no uninstantiated variables. A CSP is said to be k-consistent if any consistent assignment of $(k-1)$ variables can be extended to any other k^{th} variable. 1-consistency is obtained if each variable has a non-empty domain. For binary networks, 2-consistency is arc-consistency. For binary networks, 3-consistency is the consistency called path-consistency.

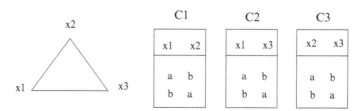

Fig. 6.8 A 2-consistent but not 3-consistent network

For instance the binary network in Fig. 6.8 is 2-consistent but not 3-consistent: e.g., for the assignment $\{(x_1, a), (x_2, b)\}$, which is consistent as it satisfies the constraint C_1 with scope $\{x_1, x_2\}$, there is no consistent value for x_3: the assignment

(x_3, a) satisfies C_3 but not C_2, and conversely for the assignment (x_3, b). A network is strongly k-consistent if it is j-consistent for all $j \leq k$. If the network has n variables and is strongly n-consistent, then it is consistent. Of course there is no miracle: Algorithms to make a network k-consistent are exponential in k in the worst case. Arc-consistency appears to be a good tradeoff between the strength of the consistency check and the computation time of this check.

If the operation of repeatedly suppressing domain values without support in a constraint, until achieving stability, does not lead to a failure, only one arc-consistent network is obtained. More precisely, the *arc-consistent closure* of P is the network P', defined as follows:

- P' has the same set of solutions as P,
- P' is arc-consistent,
- for all domain $D'i$ of P', $D'i \subseteq D_i$,
- there is no network P'' equivalent to P which is arc-consistent and strictly contains P' (with respect to domain containment).

P' is thus the maximum subnetwork of P, obtained by restricting domains, equivalent to P and arc-consistent. A family of algorithms for computing the arc-consistent closure of a network have been designed. They go from AC-1, the brute-force one [Mac77], to AC-7 [BFR95]; all of these algorithms have been designed for binary networks.

The algorithm ArcConsistency (Algorithm 10) is basically AC-3 [Mac77] adapted to n-ary constraints. It maintains a list (ToCheck) of all constraints whose consistency has not been checked yet or must be checked again because the domain of at least one of their variables has been modified. It stops either with a failure (a domain has been emptied), or with the arc-consistent closure of the network (in this case ToCheck is empty). The subalgorithm CheckConstraint (Algorithm 11) takes a constraint C as input and restricts the domains of its variables to make them arc-consistent (if possible). More precisely, it computes the arc-consistent closure of the network restricted to C. Variables whose domain has changed are marked to be taken into account in the main algorithm.

Example. Let us consider the network in Fig. 6.7. Let ToCheck initially contain $\{C_1, C_2, C_3\}$. The following constraint checks are done:

- Check of C_1: there is no support for the assignment $z \leftarrow b$, thus b is removed from D_z; $D_z = \{a, c\}$. C_2 would have been inserted into the queue if not already present.
- Check of C_2: the tuple (a, b, c) no longer exists; $D_y = \{a, c\}$, D_z is unchanged and $D_u = \{b, c\}$. C_1 is inserted into the queue (and C_3 would have been if not already present).
- Check of C_3: the tuples (a, a) and (a, b) no longer exist; $D_u = D_v = \{c\}$. C_2 is inserted into the queue. Now ToCheck contain $[C_1, C_2]$.
- Check of C_1: the tuple (a, b, c) is removed; no domain is modified.
- Check of C_2: the tuple (a, a, b) no longer exists; $D_y = D_z = \{c\}$. C_1 is inserted into the queue.

Algorithm 10: ArcConsistency(P)

Input: a constraint network $P = (X, D, C)$
Output: computes the arc-consistent closure of P if it exists, otherwise returns Failure
begin

> Tocheck $\leftarrow$ C
> **while** *ToCheck* $\neq \emptyset$ **do**
>> pick one constraint C_i from ToCheck
>> **forall** *variable x of* C_i **do**
>>> Changed[x] $\leftarrow$ false
>>
>> result $\leftarrow$ CheckConstraint(C_i)// see Algorithm 11
>> **if** *result = EmptyDomain* **then**
>>> **return** *Failure*
>>
>> **if** *result = Changed* **then**
>>> **forall** $C_j \neq C_i$ *such that there is* $x \in C_i \cap C_j$ *and Changed*[x] = *true* **do**
>>>> ToCheck $\leftarrow$ ToCheck $\cup \{C_j\}$

end

Algorithm 11: CheckConstraint(C_i)   *subalgorithm of Algorithm 10*

Input: a constraint C_i
Output: makes C_i arc-consistent if possible and marks variables whose domain has
          changed; returns EmptyDomain if a domain has been emptied, NoChange if no
          domain has been modified, otherwise Changed
begin

> // remove from C_i values which have been suppressed from
>   domains
> **foreach** *variable x of* C_i **do**
>> remove from C_i every tuple t with $t[x] \notin D(x)$
>
> // make domains arc-consistent with respect to C_i
> result $\leftarrow$ NoChange
> **foreach** *variable x of* C_i **do**
>> **foreach** $v \in D(x)$ **do**
>>> **if** $v \notin C_i[x]$ **then**
>>>> // v has no support in C_i
>>>> remove v from $D(x)$
>>>> **if** $D(x) = \emptyset$ **then**
>>>>> **return** *EmptyDomain*
>>>>
>>>> **else**
>>>>> result $\leftarrow$ Changed
>>>>> Changed[x] $\leftarrow$ true
>
> **return** *result*

end

- Check of C_1: the only remaining tuple is (c, c, c) thus $D_x = \{c\}$. The queue is empty. The network is now arc-consistent with $D_x = D_y = D_z = D_u = D_v = \{c\}$. Note that these values define the only solution to the network.

Property 6.6. The time complexity of this algorithm is in $O(m \times c \times d^{c+2})$ where m is the number of constraints, c the maximal arity of a constraint and d the maximal size of a domain.

Proof. The complexity of `CheckConstraint` (C_i) is $O(d \times t)$, where d is the maximum size of a domain of a variable in the scope of C_i and t is the number of tuples in C_i, since each value in a domain D_x is compared in the worst case to each tuple of C_i. t is bounded by d^c where c is the arity of C_i. Each time a domain of a variable x is modified, the constraints on x are added to the queue. A domain can be modified at most d times (since each time a value is removed). A constraint can be added to the queue due to each of its variables, thus $c \times d$ (actually $min(c \times d, t)$). The number of calls to `CheckConstraint` is thus bounded by $m \times c \times d$. Hence, an upper bound for the whole algorithm is $m \times c \times d^{c+2}$. □

The arc-consistency algorithms from AC-4 to AC-7 yield a better worst-case complexity. However, AC-3 has the advantage of being independent of the specific data structure which would have to be maintained in later algorithms. In addition, AC-3 is almost better than its successor AC-4 in practice (as shown in [Wal93]).

The connection with BG edge-consistency is straightforward. Indeed, $poss(c)$, for a concept c, can be viewed as a variable domain, and $poss(r)$, for a relation r, can be viewed as a constraint definition (if one does not want to assign images to relations, instead of having relations in $poss(r)$, one can have the lists of arguments of these relations, which is close to a constraint definition).

When the original filter is the standard filter, the edge-consistent closure corresponds exactly to the arc-consistent closure of the corresponding constraint network, as expressed by the following property. Recall that $h2c$ is the transformation from the BG-HOMOMORPHISM problem to the CSP, and $c2h$ is the transformation in the converse direction (Chap. 5).

Property 6.7.

- (h2c) Let G and H be normal BGs. Let *poss* be the standard filter of (G, H). If its edge-consistent closure *poss′* exists, then the arc-consistent closure of $h2c(G, H)$ exists, and is obtained by restricting the domains of variables according to *poss′*: for each $x \in X$, $D_x = \{h2c(poss'(c)) \mid x = h2c(c)\}$; otherwise the arc-consistent closure of $h2c(G, H)$ does not exist either.
- (c2h) Let P be a constraint network. If its arc-consistent closure P' exists, then the edge-consistent closure of the standard filter of $c2h(P)$ exists and is exactly the standard filter of $c2h(P')$; otherwise the edge-consistent closure of the standard filter of $c2h(P)$ does not exist either.

6.2.3 Forward Checking

In the so-called look-ahead schemes, a consistency check is performed at each new instantiation, to evaluate the consequences of this assignment on the domains of unassigned variables. Note that in some presentations of look-ahead algorithms this check is considered to be performed just *after* an assignment, while in others it is done just *before* an assignment, in order to guide the choice of a value to assign; anyway, the same actions are done by the algorithms. If we adopt the first formulation, the consistency check can be added to the backtrack algorithm as follows:

(Look-ahead) *After assigning the current variable, apply the consistency check. If successful, continue; otherwise, try another assignment for this variable or backtrack if all assignments have been tried.*

This consistency check can be more or less complete. The so-called *really full-look-ahead* algorithms maintain true arc-consistency on the current network, i.e., the network in which the domain of each instantiated variable is reduced to its instantiation. The popular forward-checking (FC) algorithm performs a partial consistency check considered as a good tradeoff between the pruning gain and the overhead cost. We will consider this algorithm now.

In what follows, the *current variable* is the variable just assigned; a *past variable* is a variable already assigned (the current variable is thus a past variable); a *future variable* is a variable not yet assigned. FC was designed by [HE80] for binary networks. We will denote this algorithm bFC, and nFC its generalization to n-ary networks. The idea of bFC is to remove from the network all values *directly* incompatible with the last assignment. In other words, a *future* variable cannot have its domain inconsistent with a *past* variable. The consistency check performed by FC() is thus as follows:

(bFC consistency check(1)) *Apply arc-consistency on the set of constraints connecting one past variable and one future variable.*

In the binary case, this check can be very efficiently implemented since it suffices to consider the current variable instead of all past variables and the arc-consistency check can be done in a single step (thus considering each constraint only once, in any order). Indeed, let x_i be the current variable. Let x_p be another past variable, $p < i$, and x_f be a future variable, $f > i$. At each instant, the consistency check ensures that the current partial solution can be extended with an assignment of any variable, without conflicting with already assigned variables. Indeed, if a constraint has scope $\{x_i, x_p\}$ or $\{x_p, x_f\}$, previous consistency checks ensure that it is satisfied. Thus only constraints with scope $\{x_i, x_f\}$ have to be checked, possibly reducing D_{x_f}.

(bFC consistency check(2)) *Apply arc-consistency on the set of constraints connecting the current variable and one future variable, in a single step.*

In the n-ary case, the equivalence between the two consistency checks no longer holds, basically because a constraint with a past variable can have more than one future variable. Several alternatives hold, which are exhibited in [BMFL02]. The bFC algorithm can be generalized in different ways depending on the answers to two questions: (1) when is a constraint checked: when it has at most one uninstantiated variable or when at least one of its variables is instantiated? The alternatives are as follows. Constraints checked might be those connecting *at least one past variable* and *at least one future variable*, or connecting *the current variable* and *at least one future variable*, or *the current variable* and *exactly one future variable* (it can be shown that the option *at least one past variable* and *exactly one future variable* is equivalent to the last one). (2) is arc-consistency computed in a *single step* (each constraint is made arc-consistent once) or is true arc-consistency achieved on the subnetwork considered (which might require *multi-steps*)?

Let us say that the propagation is *immediate* if it considers constraints as soon as one of its variables is instantiated, otherwise it is *delayed* (it is considered when all its variables but one has been instantiated). It is *local* if it considers only constraints connecting the current variable, and *global* if it considers all constraints with at least one past variable. Finally, it can be done in *one step* if each constraint is checked once, or in *multi-steps* enforcing true consistency on the considered set of constraints.

Several algorithms can be defined by combining these criteria. The reader is referred [BMFL02] for an experimental comparison of these algorithms. A good tradeoff between the pruning effect and the overhead cost is the so-called nFC2 algorithm, with uses local propagation in a single step:

(nFC2 consistency check): *Apply arc-consistency on each constraint connecting the current variable and at least one future variable, in a single pass.*

C1	C2	C3
x y z	y z u	u v
a a a	a a b	a a
a b c	a b c	a b
b a c	c c c	c c
c c c		

$Dx = Dy = Dz = Du = Dv = \{a, b, c\}$

After assignment (y,a)

	nFC0	nFC2(1)	nFC2(2)	nFC3
Dx	a bc	a b	a b	a
Dy	*a*	*a*	*a*	*a*
Dz	a bc	a	a	a
Du	a bc	b	bc	b
Dv	a bc	a bc	a bc	a bc

nFC0: delayed local propagation
nFC2: immediate local propagation in one step,
checking C_1 before C_2 (nFC2(1)), or in reverse order (nFC2(2))
nFC3: immediate local propagation in multi-steps

Fig. 6.9 n-ary Forward Checking

Figure 6.9 illustrates the behavior of several alternative algorithms. It again considers the network in Fig. 6.7. Note the arc-consistency closure of this network restricts it to the single solution. But let us assume here that the network is not made arc-consistent before launching the forward-checking algorithm. In the figure, the different filterings obtained with three variants are shown when y is instantiated with a. The delayed propagation (nFC0) does not perform any filtering. The immediate local propagation checks arc-consistency of C_1 and C_2. nFC2 does the check in one step and performs a better filtering if C_1 is considered before C_2. nFC3 does the check in multi-steps (enforcing arc-consistency on the network restricted to C_1 and C_2).

The forward-checking algorithm (AllHomByFC(G, H), Algorithm 7) for exhibiting BG homomorphisms corresponds to the nFC2 strategy.

6.3 Label Comparison

A BG is a graph with a specific labeling. Thus, a BG homomorphism is a mapping with two fundamental properties: one dealing with the graph structure, the other concerning the node labeling. One could say that a BG homomorphism is a double homomorphism: It is a (graph) homomorphism for the structure and an (order) homomorphism for the ordered label sets. The fundamental operation for order homomorphisms is the comparison of two labels. In previous algorithms (Sect. 6.1), we considered the label comparison as an operation that is computable in constant time. This section studies its actual complexity, which is related to the data structures implementing the orders on concept and relation types.

Whenever the type sets are small, comparison between labels takes only a short time relative to the total time needed for computing the structural part of a BG homomorphism. When the label sets are large (e.g., when the concept type set corresponds to the concepts of a natural language, it can have hundreds of thousands of items), it is then useful to implement a reasonable data structure to allow quick computing of the comparison between labels.

We will first briefly present classical data structures for representing orders and simple algorithms for comparing two elements. We will also mention related problems on orders. In this section, the classical generic notation $\leq$ is used to denote an order, $<$ denotes its associated strict order and $\prec$ denotes its associated covering relation. Numerous specific orders have been studied: interval orders, bipartite orders, lattices, distributive lattices, etc. [Moh89]. We will retain two classes that are especially important in knowledge representation: trees and lattices. Tree orders are very often used for representing taxonomic knowledge. We will present a specific coding for tree orders, which is both very simple and efficient [Bou71]. When any two types have a least upper bound (lub) and a greater lower bound (glb) then the order is a lattice. Lattices are encountered especially when sets of concept types are built by Formal Concept Analysis techniques [GW99]. Specific techniques have been developed when orders are lattices (e.g., [HHN95], [HNR97], [TV97]).

During the knowledge acquisition phase, especially during construction of the type sets, the ordering is generally given by a set U of comparable type pairs, i.e., by a directed graph on the type set. If this set U is hand-built, then some errors may occur, namely circuits may be constructed. Detecting circuits in a directed graph is a simple problem that can be efficiently solved (e.g., [CLRS01]), and we suppose in the sequel that U has no circuit. U may contain transitivity arcs, i.e., arcs (t, t') such that there is in U a path of length greater than 1 from t to t'. The order relation $\leq$ is equal to the transitive and reflexive closure of U denoted by U^* i.e., $t \leq t'$ if and only if $(t',t) \in U^*$. Thus, when U^* is provided, the comparison problem is reduced to a membership problem. At the other end, the minimum information needed to compare two elements of an ordered set is given by the covering relation of the order. Going from the transitive closure to the transitive reduction, the comparison problem goes from a membership problem in U^* (the maximal representation of the order) to a path existence problem in U^h (the minimal representation of the order). In Fig. 6.10 two directed graphs $G = (X,U)$ and $G' = (X,U')$ are drawn. They have the same transitive reduction (X,U^h) and the same reflexo-transitive closure (X,U^*), thus they define the same order.

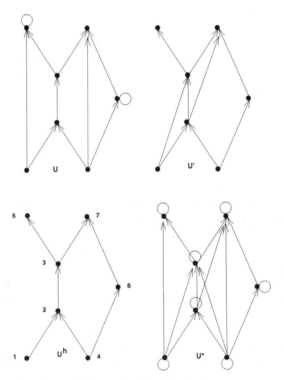

Fig. 6.10 Transitive reduction and reflexo-transitive closure

6.3.1 Basic Data Structures and Algorithms

There are numerous techniques for coding an ordered set. In [Thi01] for instance, a dozen codings are studied, including the three basic techniques that are recalled below, and about twenty other ones are cited. The three basic data structures for representing orders are: the incidence matrix, the lists of descendants (and/or ascendants), the lists of successors (and/or predecessors). In what follows, n denotes the cardinality of the ordered set, i.e. $n = |X|$.

1. **Incidence Matrix**

   Let us suppose that the elements of X are $x_1, \ldots, x_n$. The incidence matrix of $(X, \leq)$ is the $n \times n$ boolean matrix M, such that $M[i,j] = 1$ if and only if $x_i \leq x_j$, and 0 otherwise. This coding has a size of n^2 bits and the comparison between two elements can be done in constant time since it corresponds to an access to an element of the matrix. Whenever a set is represented by its characteristic vector, the membership problem is reduced to an access to an element of this vector. The i-th line of M can be seen as the characteristic vector of the set of descendants of x_i, so the incidence matrix can be seen as the characteristic vector of the set of pairs of comparable elements. It can be proven that under the assumption of a uniform distribution of the orders over a set X the incidence matrix is asymptotically optimal in space [Thi01] and we just saw that this coding is optimal in time for the comparison problem. Thus, whenever the ordered sets are not very large (if $n = 3000$ about one megabyte is needed), and there is no specific information about the orders, and the problems to solve are based on the comparison problem, then the incidence matrix is an efficient coding. The following table shows the incidence matrix representation of the order defined in Fig. 6.10.

	1	2	3	4	5	6	7
1	1	1	1	0	1	0	1
2	0	1	1	0	1	0	1
3	0	0	1	0	1	0	1
4	0	1	1	1	1	1	1
5	0	0	0	0	1	0	0
6	0	0	0	0	0	1	1
7	0	0	0	0	0	0	1

2. **Lists of Descendants**

Instead of representing a set (of couples) by its characteristic vector, it can be represented by the list of its elements. More precisely, to any element $x_i \in X$ is associated the list of its descendants, i.e., the elements x_j such that $x_j > x_i$. If the order has few comparable elements, then this coding is small in size: It is in $O(|U^*| \times log(n))$ bits. But comparing two elements requires searching a list, which can be of length n. This can be improved by sorting the elements of the list because searching an element in a sorted list L is in $O(log(size(L)))$, in our case in $O(log(n))$.

The descendant list representation of the order in Fig. 6.10 is:

1 : 1,2,3,5,7
2 : 2,3,5,7
3 : 3,5,7
4 : 4,2,3,5,6,7
5 : 5
6 : 6,7
7 : 7

3. **Lists of Successors**

To any element $x_i \in X$ is associated the list of its successors, i.e., elements x_j such that $x_j \prec x_i$. This coding is a representation of the covering relation of the order. It is the coding with the smallest size among the three classical codings: Its size is in $O(|U^h| \times log(n))$ bits. But comparing two elements might require a search of the whole structure and thus is in $O(|U^h|)$.

The successor list representation of the order in Fig. 6.10 is:

1 : 2
2 : 3
3 : 5,7
4 : 2,6
5 :
6 : 7
7 :

If the ordered set is very large, then a variant of the lists of successors is probably a good choice. There are a lot of variations using these basic structures. For instance, the time for comparing two elements can be improved by a cache technique for the last pairs of elements compared, and the incidence matrix or the lists of descendants can be used for structuring the cache itself.

6.3.2 Related Problems

Let us mention some related problems dealing with graphs and orders. These problems are defined for general graphs, and we discuss their application to the particular case of orders, when this particular case yields specificities.

1. **Connected Components**

The decomposition of (X, U^*), or equivalently of (X, U^h), in connected components can simplify problems concerning orders since two elements are comparable only if they belong to the same connected component. There are simple and efficient algorithms for computing connected components of a graph (e.g., [CLRS01]). Other decompositions can be useful for studying ordered sets, e.g., the decomposition into modules (e.g., [Pau98]).

2. **Transitive Closure**

   When a graph is represented by its incidence matrix, computing its transitive closure is equivalent to the computation of a product of matrices. More precisely, the transitive closure of a graph of size n can be computed in $O(n^a)$ if and only if the product of two matrices of size $n \times n$ can also be computed in $O(n^a)$. The well-known Roy-Warshall algorithm is in $O(n3)$ (in [CW87] an algorithm with $a = 2.376$ is proposed). With the lists of descendants or of successors, a simple algorithm performs a depth-first search from each vertex keeping track of all vertices encountered. This algorithm is in $O(n \times (n+m))$ where m is the number of arcs of the representation (cf. for example [CLRS01]).

   Transforming the coding using the lists of successors into the two other ones corresponds to computation of the transitive closure of the covering relation. Conversely, transforming the two other codings into the lists of successors corresponds to computation of the transitive reduction.

3. **Transitive Reduction**

   Transitive reduction is the inverse operation of transitive closure, namely the problem is to reduce the number of arcs while maintaining the existence of paths. Computing a minimum transitive reduction of a given graph (i.e., a transitive reduction with the smallest number of arcs) is a difficult problem in the general case. Indeed, a minimal transitive reduction (i.e., which does not strictly contain a transitive reduction) does not necessarily have a minimum number of arcs (cf. K in Fig. 6.11). Determining whether a graph with n vertices has a transitive reduction with n arcs is equivalent to checking the existence of a hamiltonian circuit in this graph, which is an NP-complete problem [GJ79]. The problem becomes easy on graphs without a circuit. Such a graph has exactly one transitive reduction. Thus, any algorithm for computing a minimal transitive reduction gives a minimum transitive reduction for this class of graphs. A naive algorithm is as follows: For any arc (x,y), perform a depth-first from x without using (x,y); if y is reached (y is thus a a descendant of x), then delete (x,y) (it is a transitivity edge). This algorithm is in $O(m \times (n+m)$, thus $O(n^2)$ if $m \in O(n)$. (cf. also [HMR93]). Figure 6.11 shows an example, with H and K being minimal reductions of G.

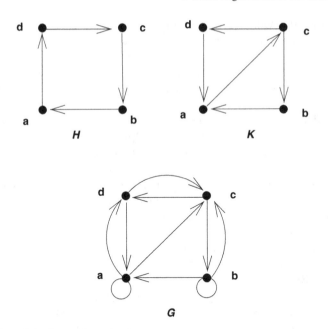

Fig. 6.11 Two minimal transitive reductions

4. Relations with Circuits

Rather than considering that the existence of a circuit of types is an error, one can consider that all types belonging to a circuit are in fact equivalent or, stated otherwise, are aliases of the same type. In this case, the problem is not to find the circuits of the given graph (X,U) but to compute the strongly connected components of (X,U) and the associated quotient graph. Indeed, given two elements x and y, x and y are equivalent if and only if they belong to the same strongly connected component, and x is a (strict) subtype of y if and only if x is less than y in the quotient graph $(X,U)/SCC$, where SCC is the partition of X in strongly connected components. The quotient graph $(X,U)/SCC$ is a graph without a circuit. The strongly connected components and the associated quotient graph of (X,U) can be computed in time $O(|X|+|U|)$ [Tar72].

6.3.3 Tree Orders

Let us call *tree order* an ordered set $(X,\leq)$ such that its covering graph $(X,\prec)$ is a rooted tree. Schubert et al. [SPT83] proposed an elegant and optimal coding for tree orders. Let $n = |X|$. The method consists of numbering the elements by $1,2,\ldots,n$ in such a way that if $num(x)$ is the numbering of x then for any x the set $\{num(y)|x\leq y\}$ is the integer interval $[num(x),num(x)+1,\ldots,M(x)]$. Building such a numbering is easy and can be done by a depth-first search on the rooted tree $(X,\prec)$ from its root.

Indeed, a simple numbering of vertices according to the order in which they are visited satisfies the property. Performing a depth-first search for a tree is in $O(n)$. Figure 6.12 shows a tree order with a depth-first search numbering of the nodes and the associated intervals.

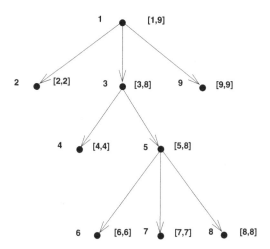

Fig. 6.12 A simple coding for a tree order

Having such a numbering, comparing two elements is reduced to two integer comparisons. Indeed, one has: $x \leq y$ if and only if $num(x) \leq num(y) \leq M(x)$. Having the rooted tree (X, U^h), the method is in time and space $O(n)$ for building the functions num, M, then comparing two elements is in constant time.

6.3.4 Partition in Chains

Let us assume that the ordered set X is partitioned in k chains (i.e., linear or total orders): $C_1, \ldots, C_k$, and $c(x)$ is the subscript of the chain containing x. Let us denote by n_i the cardinality of C_i, then the elements of C_i are numbered by $1, 2, \ldots, n_i$, and $num(x)$ is the number associated with x. For any $x \in X$ and $i \in \{1, 2, \ldots, k\}$ let $t(x, i) = max\{num(y) | y \leq x, y \in C_i\}$ (with $max(\emptyset) = 0$). The tuple of integers $(t(x, 1), t(x, 2), \ldots, t(x, k))$ is associated with $x \in X$. One has: $x \leq y$ if and only if $t(y, c(x)) > num(x)$. Thus, comparing two elements is in constant time. The size of the coding is in $O(k \times log(n))$.

A partition of X in chains has to be built. Dilworth's theorem states that the minimum number of chains which partition an order is equal to its width (i.e., the maximum cardinality of an anti-chain) and such a minimum partition can be computed by a simple algorithm in time $O(n^3)$ by transforming it into a maximum matching problem. Figure 6.13 shows an order, a chain partition of this order in three chains and the associated coding. A node x with the labels (i, j) and $[l, m, n]$ means that x is the j-th node in the chain C_i and $[l, m, n]$ is the coding of x.

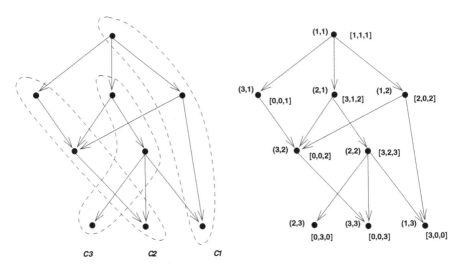

Fig. 6.13 The chain partition coding

6.3.5 Lattices

Besides being a tree another usual property for concept or relation hierarchies is to be a lattice. A classical coding of a lattice consists of using its join-irreducible elements ([DP02]).

Definition 6.8 (Join-irreducible elements). Let $[L, \leq)$ be a finite lattice. An element j of L is *join-irreducible* if j covers exactly one element.

Figure 6.14 gives some examples. The join-irreducible elements are pictured in black.

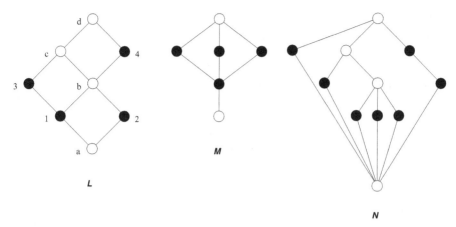

Fig. 6.14 Join-irreducible elements

The covering relation is represented by the lists of predecessors computing the join-irreducible elements of a lattice of n elements is in $O(n)$. The coding of a lattice with the join-irreducible elements consists of representing each element x by the set $J(x)$ of the join-irreducible elements that are greater than or equal to x: $J(x) = \{y$ join-irreducible $\mid y \leq x\}$. One has:

Property 6.8. Let L be a finite lattice. For all x and y in L, $x \leq y$ if and only if $J(x) \subseteq J(y)$.

Let us consider the lattice L in Fig. 6.14. Using a vector of length 4 for representing a set of join-irreducible elements, the coding J of L is as follows:
$J(1) = (1,0,0,0), J(2) = (0,1,0,0), J(3) = (0,0,1,0), J(4) = (0,0,0,1)$
$J(a) = (0,0,0,0), J(b) = (1,1,0,0), J(c) = (1,1,1,0), J(d) = (1,1,1,1)$
The lattice N in Fig. 6.14 is a tree supplemented with a minimum element. Coding it as a lattice is less efficient (for the comparison problem) than coding it as a tree (plus a minimum element).
Dually, a *meet-irreducible* element is an element that is covered by exactly one element, and meet-irreducible elements can be used for coding a lattice.

6.4 Bibliographic Notes

Several algorithms for computing BG homomorphisms have been proposed in the conceptual graph literature. The filtering techniques used in these algorithms can be seen as kinds of local consistency checks, and the notion of edge-consistency filter presented in this chapter seems to generalize most of these proposals. Let us cite in particular [Wil95] with the preprocessing step called polyprojection, [Mug95] with a preprocessing step based on an edge-covering tree of the source graph, and [CC06], where BG homomorphism checking is translated into the problem of finding a clique in a graph representing all possible matchings between relation nodes of both graphs, with the cardinality of this clique being equals to the number of relation nodes of the source graph.

Baget studied the relevancy of CSP algorithms for computing BG homomorphisms (with the hypergraph vision of BGs adopted in this chapter). His evaluating criteria were as follows: (1) The algorithm must be able to find all solutions; (2) the algorithm must be generic, i.e., has to accept any BG as input; (3) the algorithm must add as little overhead cost as possible with respect to the backtrack (because this overcost is offset by a reduction of the explored search tree only on difficult instances, which are unlikely to be encountered in the main uses of conceptual graphs). This study led to an algorithm presented in [Bag01] (in French) and [Bag03]. Let us especially point out the efficient data structure proposed to support the dynamical computation of the candidates for a concept node [Bag03]. This structure was first proposed in [BT01] for constraint networks. It allows efficient implementation of classical constraint processing schemes, i.e., backmarking, forward checking and backjumping.

The subsumption relation on BGs can be used to structure a large set of BGs and reduce the number of homomorphism checks when the base is searched. In this chapter, we did not address retrieval techniques exploiting this structuring. About this aspect, see in particular Ellis and Levinson 's work [LE92] [ELR94].

Pointers to constraint processing Literature

References to specific aspects of constraint processing have been included in the text. Let us end with general introductions to this domain. For a synthetic presentation, the reader is referred to the survey in the book [RN03]. The book [Dec03] provides an in depth presentation, from basic to advanced techniques. The relationships between constraint networks and relational algebra are particularly emphasized. Finally, the recent handbook of constraint programming [RvW06] provides numerous pointers to the state of the art in the domain.

Chapter 7
Tractable Cases

Overview

This chapter presents tractable cases for BG-HOMOMORPHISM. These cases are essentially obtained by restricting the *structure* of the source BG to an acyclic structure, with the definition of acyclicity being more or less restrictive. The simplest acyclicity studied is *multigraph-acyclicity*. When the source BG is multigraph-acyclic, edge-consistency is sufficient to ensure global consistency (i.e., the existence of a solution). Furthermore, it can be computed more efficiently than in the general case, and the number of solutions can be polynomially counted. *Hypergraph-acyclicity* strictly generalizes the previous notion. When the source BG is multigraph-acyclic, global consistency can be polynomially checked using the notion of a join tree issued from database theory, and two kinds of consistency, namely edge-consistency and pairwise-consistency. We give polynomial algorithms to check the global consistency in these two cases; these algorithms compute a filter that allows to enumerate the solutions. We point out the equivalence between hypergraph-acyclic BGs and *guarded* BGs, which correspond to the guarded fragment of existential conjunctive positive first-order logic. We then briefly present generalizations of acyclicity to *bounded treewidth* and *hypertreewidth*, and end with a panorama of expressivity results for all studied definitions of acyclicity. We finally mention the rare tractability results based on labeling properties of the target BG and point out that exhibiting tractable cases combining properties on the structure of BGs and on their labeling is an open challenge.

7.1 Introduction

A *tractable case* of a difficult problem is a case in which the problem can be solved in polynomial time. This chapter is essentially devoted to the presentation of tractable cases for BG-HOMOMORPHISM and associated algorithms. It illustrates one of our claims (see Chap. 1): Expressing reasoning with graph operations allows

us to benefit from important algorithmic results. This chapter also points out the
deep links between particular FOL fragments and graphs with a particular structure.

7.1.1 About Tractability

Tractable cases are classically obtained by restricting the input of the problem to
particular cases, i.e., the target BG or the source BG or both in our case. In this
chapter, we will mainly gather results obtained by restricting the structure of BGs.
Some structures correspond to cases "naturally" occurring in applications and easily
recognizable by a human being. A typical case is that of a tree. This notion is at
the core of this section. We will go from this simple notion to richer structures
decomposable into trees.

Let us, however, keep in mind that we have to distinguish between the theoretical tractability of a problem and efficiency of algorithms solving it in practice. A
problem might be NP-complete but its difficult instances might be rarely found in
practice. Conversely, a problem might be polynomial time solvable but the algorithm solving it might have a complexity involving big constants (for instance, a
complexity in $c_1 \times n^{c_2}$, where n is the size of the input, and c_1 and c_2 are big constants). There are even polynomial problems for which no polynomial algorithm is
known (see for instance [RS04]).

The algorithms given in this chapter that rely on the multigraph-acyclicity
or hypergraph-acyclicity have a low complexity—their worst-case complexity is
bounded by $c \times m_G \times m_H$, where m_G and m_H denote the number of edges of the
source and target BGs, and c is a small constant. For more complex notions of
acyclicity such as bounded treewidth and hypertreewidth, the practical tractability
of the problem is less obvious.

Tractable cases are interesting in practice when they correspond to cases frequently found in a real application. For instance, graphs have a simple structure
when they are manually built: They tend to be trees, and the degree of acyclicity
remains small when they have cycles (this notion will become clearer in next sections).

7.1.2 The Structure of the **Target BG** is of No Help

Structural properties and the complexity of graph homomorphism have been extensively studied in the graph theory domain (see for instance [HT97][HM01] [HN04]).

The complexity of homomorphism checking has been particularly explored in
the framework of the following problem, called the general coloring problem, or H-
COLORING, where H is a fixed graph: Given a graph G, is there a homomorphism
from G to H? The name H-COLORING reflects the fact that classical coloring problems can be seen as special cases of H-COLORING. Given a set of colors, coloring a

graph consists of assigning a color to each of its nodes, such that no adjacent nodes have the same color. A k-coloring is a coloring using at most k colors; formally, it is a mapping c from the node set of G into $\{1 \ldots k\}$, such that for all edges xy, $c(x) \neq c(y)$. The classical k-COLORING problem takes a graph G as input and asks whether there is a k-coloring of G. It can be recast as a H-COLORING problem where H is a clique on k nodes.

GRAPH-HOMOMORPHISM is the problem that takes two graphs G and H as input and asks if there is a homomorphism from G to H (cf. Chap. 5). As in H-COLORING, H is not part of the input of the problem, H-COLORING can be seen as a special case of GRAPH-HOMOMORPHISM where the size of H is bounded by a constant. The H-COLORING complexity study is focused on the structure of H and tries to determine the exact conditions on the structure of H which make the problem polynomial. The answer is known for non-directed graphs: H-COLORING is polynomial if H is bipartite (or if H has a loop or is restricted to one node) and NP-complete in all other cases [HN90]. This result also holds for GRAPH-HOMOMORPHISM. About directed graphs, there are trees H for which H-COLORING is NP-complete [HNZ94]. GRAPH-HOMOMORPHISM, thus BG-HOMOMORPHISM, remains NP-complete even if H has a tree structure.

Corollary 7.1 (from [HNZ94]). BG-HOMOMORPHISM *remains NP-complete in the case where the underlying graph of H is a tree (and the vocabulary is restricted to one concept type and one binary relation).*

What about considering the structure of the *source* graph G rather than that of the target graph H? The question is more relevant to us, since G generally represents a goal or a query (or a part of a rule or constraint, see Chap. 11), while H often encodes the fact base; G then has a small size and a less complex structure than H. The next sections are devoted to this issue.

The tractable cases discussed in the next sections are based on the following property: If G has the required special structure, an ordering on the nodes of G can be computed, such that checking the existence of a solution or computing a solution can be done in a greedy way. In other words, one obtains a "backtrack free" algorithm.

The special structure is related to the acyclicity of the source graph. Let us point out that equivalent results have often been found in several domains (databases, graph theory, CSP, etc.) or transfered from one to the other and enriched. The simplest kind of acyclicity is graph acyclicity (or multi-graph acyclicity since BGs are, strictly speaking, multigraphs and not simply graphs). We shall begin with graph acyclicity before considering generalizations of this notion.

7.2 Tractability Based on the Multigraph-Acyclicity of the *Source* BG

For simplicity, most algorithms in the sequel will be based on the assumption that the source BG is *connected*. Let us however point out that they can be immediately extended to a non-connected source BG. Indeed, if G is a non-connected BG, a homomorphism from G to any BG H can be seen as a set of homomorphisms from each connected component of G to H. In addition, if H is not connected, each homomorphism from a connected component of G to H maps this component to a connected component of H.

7.2.1 Acyclic Multigraphs and Trees

A classical graph is acyclic if it has no cycle. We will say that a BG is *multigraph-acyclic* if it has no cycle except those of length 2, derived from multi-edges between a relation and a concept node. With all definitions of acyclicity that we shall study, a *tree* adds connectivity to acyclicity.

Definition 7.1 ((multi)graph-acyclic). A BG is *multigraph-acyclic* if it has no cycle of form $(c_1\ r_1\ c_2\ ....\ c_n\ r_n\ c_1)$, $n \geq 2$, where all c_i and c_j (resp. r_i and r_j), $i \neq j$, are distinct nodes. It is a *tree* if it is multigraph-acyclic and connected.

The multigraph-acyclic property can be linearly checked, as illustrated by Algorithm 12. Starting from any concept node considered as the root, the graph is traversed, with the multi-edges between two nodes being considered as one edge. Once a (concept or relation) node is reached from a (relation or concept) node x, it is marked by it (x is called its predecessor) and put into a queue, which stores nodes reached but not yet explored. If a node has an already marked neighbor, and this neighbor is not its predecessor, then a cycle has been detected and the algorithm returns false. If all nodes are explored without having detected a cycle, the algorithm returns true and the marks define the rooted trees corresponding to the exploration made by the algorithm. Note that if G is not connected, it is explored connected component by connected component: the algorithm first tries to take a node in the queue (thus an already reached node) and, only if the queue is empty, takes a new node (thus a node from another connected component).

Property 7.1. [MC92] BG-HOMOMORPHISM is polynomial when the source BG is multigraph-acyclic.

Property 7.3 then provides a proof of this property. Let us consider that the source BG is a tree. Moreover, and without loss of generality, we will see it as a *rooted tree*. The root is any concept node whose role is to induce a direction to edges that will be used in algorithms. We will distinguish between two relations induced by this direction: the *parent/child* relation will be reserved for concept nodes; the

---

Algorithm 12: CheckGraphAcyclicProperty(G)

Input: a BG G (not necessarily connected)
Output: true if G is (multi)graph-acyclic, otherwise false
begin

    StillToExplore $\leftarrow |C_G| + |R_G|$ `// number of nodes to explore`
    forall *(concept or relation) node x* **do**
        marked[x] $\leftarrow$ -1
    OK $\leftarrow$ true
    Choose any concept node c as the root
    Let Q be a empty queue`// contains nodes reached but not explored`
    Put c into Q
    marked[c] $\leftarrow$ 0
    while *OK and StillToExplore $\neq 0$* **do**
        if *Q is not empty* **then**
            pick x in Q
        else
            let x be any node with marked[x] = -1
            marked[x] $\leftarrow$ 0
        StillToExplore $\leftarrow$ StillToExplore - 1
        forall *x' neighbor of x, with $x' \neq$ marked[x]* **do**
            if *marked[x'] $\neq -1$* **then**
                OK $\leftarrow$ false
            else
                marked[x'] $\leftarrow x$
                add x' to Q
    return *OK*
end

---

predecessor/successor relation will be used for both concept and relation nodes, according to the direction induced by the root (cf. Fig. 7.1). The successors of a concept node c are the relation nodes that either have c as sole argument, or link c to its children. The successors of a relation node are its arguments, except for the one which is the parent of the others. Given a rooted tree T and any concept node c' of T, the subtree induced by c' is the subtree of T with root c'. A *leaf* is a concept node without child (but it can have successors, which are relations with this concept node as sole argument).

The parent/child relation allows us to focus on concept nodes (as in the hypergraph vision of BGs cf. Chap. 5), whereas the successor relation considers both kinds of nodes.

A homomorphism from a rooted tree T to any BG H can be recursively defined, according to the recursive structure of a tree, as stated in the next property. This property is illustrated by Fig. 7.2. Recall that the standard filter of T with respect to H, denoted by *poss*, assigns to each node of T the list of its candidate images in H according to compatibility conditions based on labels and, for relation nodes, on the partition induced on their neighbors; furthermore, for a neighbor c of a relation r, and for r' a relation in *poss*(r), there is a sole neighbor of r' corresponding to c for

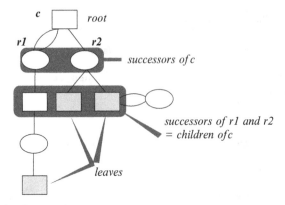

Fig. 7.1 Vocabulary for rooted trees

r: It is the node c' such that for each edge (r, i, c) in T, there is an edge (r', i, c') in H (formally, $P_r[c] \subseteq P_{r'}[c']$). We note $c' = corresponding(c, r, r')$.

Property 7.2. Let *poss* be the standard filter of T with respect to H and let c be the root of T. A mapping π from the nodes of T to the nodes of H is a homomorphism if it satisfies:

1. $\pi(c) \in poss(c)$
2. for all successors r of c, let $r' = \pi(r)$, then $r' \in poss(r)$ and $P_r[c] \subseteq P_{r'}[\pi(c)]$
   ($\pi(c) = corresponding(c, r, r')$), and, for all successor d of r, let d' be the node
   such that $P_r[d] \subseteq P_{r'}[d']$ ($d' = corresponding(d, r, r')$), then (1) $\pi(d) = d'$, and
   (2) the restriction of π to the subtree T_d rooted in d is a homomorphism from T_d
   to H.

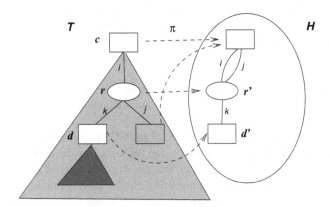

Fig. 7.2 BG Homomorphism from a rooted tree

The choices made for all subtrees T_d are independent since these subtrees do not share any node. This is the key reason why a homomorphism can be found with a backtrack-free search. One can also note in the previous definition that, if the filter *poss* is edge-consistent, a homomorphism can be built by a traversal of the tree from the root c to the leaves: an image for c is first chosen in $poss(c)$; edge-consistency ensures that there are compatible values in $poss(r)$ and $poss(d)$ for all successors r of c and all successors d of r. Each d is itself the root of a subtree, thus the same process is iterated.

Property 7.3. Let G and H be BGs, with G being multigraph-acyclic. There is a homomorphism from G to H if and only if the standard filter of G with respect to H can be made edge-consistent. Furthermore, for any node x and any x' candidate for x in this edge-consistent filter, there is a homomorphism from G to H which maps x onto x'.

Proof. Without loss of generality let us consider that G is connected. Let us take any concept node as the root, say c. The property is proven by induction on the depth of this rooted tree. Let us consider any value c' in $poss(c)$. If G is restricted to c (G is thus of depth 0), $\pi = \{(c, c')\}$ is a homomorphism from G to H (by definition of a standard filter). Let us now assume that G is of depth $n > 0$. Let us consider any relation r successor of c. Since the filter is edge-consistent, there is a relation r' in $poss(r)$ such that the neighbor of r' corresponding to c is c' (i.e. $P_r[c] \subseteq P_{r'}[c']$), and for all successors d of r the neighbor of r' corresponding to d is in $poss(d)$. There is thus a homomorphism from the subgraph induced by r and its neighbors to H, mapping c onto c' (by Property 6.2). Let d be a successor of r, and d' be the value chosen for d. By induction hypothesis, there is a homomorphism from the subtree rooted in d to H. Since G is multigraph-acyclic, the choices of an r' for all successors r of c are independent. By making the union of all homomorphisms, we have a homomorphism from G to H. We have considered any concept node c; now, if we have to start from a relation, say r, we choose any value r' in $poss(r)$ and take a neighbor c of r as root. Since the filter is edge-consistent, there is a concept c' in $poss(c)$ such that c' is the neighbor of r' corresponding to c for r and, with the same arguments as above, we conclude that there is a homomorphism from G to H that maps r onto r'. $\square$

Edge-consistency thus ensures global consistency (i.e., the existence of a solution) for the homomorphism problem when G is acyclic. Moreover, it can be more efficiently computed than in general case, as shown by next algorithms. We will give three versions of algorithms checking edge-consistency (Algorithms 13, 16 and 18). The first algorithm starts from the standard filter, as the algorithm we have seen for the general case (Algorithm 5 in Chap. 6). We then give two variants of this algorithm which are better in practice since they do not precompute the filter. They compute the set of candidates for a node during the search, according to the constraints induced by its already visited neighbors.

The first algorithm starts from the standard filter (step 1); since G is assumed to be connected, isolated concept nodes of H are not considered as soon as G possesses at least a relation. Concept nodes of G are totally ordered by a top-down traversal of the tree, in such a way that a parent precedes its children (step 1). The *level* of a relation is given by the smallest rank in its neighbors. Step 2 considers the concept nodes in reverse order, i.e., by a bottom-up traversal, making each star graph edge-consistent. When a node c of rank i is considered by the algorithm, edge-consistency is checked on all relations of level i, thus all successors of c. After step 2, if Failure has not been returned, it holds that G can be mapped to H. Indeed, for a node c of rank i, for all values of $poss(c)$, for any neighboring relation r of level greater or equal to i, there is a consistent value for r and all its neighbors. However, the filter is not necessarily edge-consistent, since the $poss$ sets of deeper nodes may keep inconsistent values. Actually, at the end of step 2, for each node c, $poss(c)$ contains all possible images for c by a homomorphism from the subtree of G rooted in c to H. Thus, an additional traversal of the tree (step 3), top-down this time, has to be done to clean the $poss$ sets. Note that this second traversal cannot empty a $poss$ set.

Algorithm 14: CheckRelation(r)   *subalgorithm of Algorithm 13*

Input: a relation r in G (and a filter $poss(G,H)$)
Output: EmptyPoss if a $poss(c)$ has been emptied, otherwise OK and make
       the filter edge-consistent on r star graph
```
// This subalgorithm is similar to CheckRelation in
   EdgeConsistency (Algorithm 5) except that the
   Changed[c] booleans need not to be maintained
```

Algorithm 15: Clean(r)   *subalgorithm of Algorithm 13*

---

begin
   Let p be the predecessor of r
```
      // Remove from poss(r) values incompatible with
         poss(p)
```
   forall $r' \in poss(r)$ **do**
      let $a = corresponding(p,r,r')$
      if $a \notin poss(p)$ **then**
         remove r' from $poss(r)$

```
      // Propagate the effects to the successors of r
```
   forall *concept s successor of r* **do**
      forall $a \in poss(s)$ **do**
        if *there is no* $r' \in poss(r)$ *such that* $a = corresponding(s,r,r')$
        then
           remove a from $poss(s)$
end

Property 7.4. The time complexity of the algorithm EdgeConsistencyTreev1(G,H) (Algorithm 13), with G being a tree, is $O(|C_H|)$ if G is restricted to one concept node, otherwise it is $O(m_G \times m_H)$, where m_G and m_H are the number of edges of G and H, respectively.

Algorithm 13: EdgeConsistencyTreev1(G,H)

Input: two BGs G and H, with G being a *tree* (i.e., connected and multigraph-acyclic)
Output: the edge-consistent filter of G with respect to H if it exists, otherwise Failure
begin

1   `// Preprocessing`
      if $R_G \neq \emptyset$ **then**
        let C'_H be obtained from C_H by removing isolated nodes

      `// Computation of the standard filter poss(G,H)`
      forall c *in* C_G **do**
        $poss(c) \leftarrow \{c' \in C'_H | l(c) \geq l(c')\}$
        if $poss(c) = \emptyset$ **then**
          return *Failure*

      forall r *in* R_G **do**
        $poss(r) = \{r' \in R_H | l(r) \geq l(r')$ and $P_r \subseteq P_{r'}\}$
        if $poss(r) = \emptyset$ **then**
          return *Failure*

      `// Top-down traversal of G`
      Choose a concept for the root of G
      Order C_G such that each parent precedes its children

2   `// Bottom-up traversal of G to make the filter`
         `edge-consistent`
      forall c *in* C_G *from rank* $|C_G|$ *to* 1 **do**
        forall r *in* R_G *of level rank*(c) **do**
          result $\leftarrow$ CheckRelation(r) `// see Algorithm 14`
          if *result = EmptyPoss* **then**
            return *Failure*

3   `// Top-down traversal of G to clean the poss sets`
      forall c *in* C_G *from rank* 1 *to* $|C_G|$ **do**
        forall r *in* R_G *of level rank*(c) **do**
          Clean(r) `// see Algorithm 15`

4   **return** *poss*
   end

Proof. First see that $|R_G| \leq m_G$, $|R_H| \leq m_H$ and, since C_G and C'_H contain no isolated nodes, $|C_G| \leq m_G$ and $|C_H| \leq m_H$. The complexity of step 1 is in $O(m_G \times m_H)$: Indeed the ordering of $|C_G|$ is performed by a traversal of G, which is in $O(m_G)$ and the computation of the standard filter is in $O(m_G \times m_H)$. The complexity of CheckRelation(r) is bounded by $arity(r) \times |poss(r)|$. CheckRelation is called once for each relation of G and $|poss(r)|$ is bounded by $|R_H|$. Thus step 2 is in $O(m_G \times |R_H|)$. Clean(r) has same complexity as CheckRelation(r), thus step 3 is in $m_G \times |R_H|$ as well. Hence the global complexity of $O(m_G \times m_H)$. $\square$

Algorithm 16 presents a variant of the previous algorithm, which does not pre-compute the standard filter. The *poss* sets are computed during the search as follows. For all leaves c, $poss(c)$ is computed from all (not isolated) concept nodes of H. For a concept c that is not a leaf, $poss(c)$ is computed from the $poss(r)$ for all relations r that are successors of c. When c is considered by the algorithm, edge-consistency

---

Algorithm 16: EdgeConsistencyTreev2(G, H)

Input: two BGs G and H, with G being a *tree*
Output: the edge-consistent filter of G with respect to H if it exists, otherwise Failure
begin

1   // Preprocessing
     if $R_G \neq \emptyset$ **then**
       $\llcorner$ let C'_H be obtained from C_H by removing isolated nodes
     Choose a concept for the root of G
     // Top-down traversal of G
     Order C_G such that each parent precedes its children

2   // Bottom-up traversal of G to build the poss sets
     forall $c \in C_G$ *from rank* $|C_G|$ *to* 1 **do**
       if c *is a leaf* **then**
         // poss(c) is still empty
         $\llcorner$ $poss(c) = \{c' \in C'_H | l(c) \geq l(c')\}$
       forall $r \in R_G$ *of level rank*(c) **do**
         // this includes relations with c as sole neighbor
         // compute poss(r) and update the poss sets of its
           neighbors
         result $\leftarrow$ ComputeRelation(r) // see Algorithm 17
         if *result = EmptyPoss* **then**
           $\llcorner$ **return** *Failure*

3   // Top-down traversal of G to clean the poss sets
     forall $c \in C_G$ *from rank* 1 *to* $|C_G|$ **do**
       forall $r \in R_G$ *of level rank*(c) **do**
         $\llcorner$ Clean(r) // same as Algorithm 15

4   **return** *poss*

end

---

is checked on all relations with level $rank(c)$, thus all successors of c. The first successor of c considered initializes $poss(c)$. Then, for each other successor r of c, the new value of $poss(c)$ is the intersection of the old value of $poss(c)$ and the set of candidates allowed by $poss(r)$.

Algorithm 17: ComputeRelation(r) *subalgorithm of Algorithm 16*

---

begin

   $poss(r) \leftarrow \emptyset$

   forall *neighbor c of r* **do**

      $poss_r(c) \leftarrow \emptyset$

   forall r' *in* R_H **do**

      `// check that (1), (2) and (3) are all true`

      (1): check if $l(r) \geq l(r')$ and $P_r \subseteq P_{r'}$

      (2): **forall** *successor s of r* **do**

         `// s already has a computed poss set`

         check if $corresponding(s,r,r') \in poss(s)$

      (3): let p be the predecessor of r

      if $poss(p) = \emptyset$ **then**

         `// r is the first sucessor of p considered`
         `   by the algorithm`

         check if $l(p) \geq l(corresponding(p,r,r'))$

      else

         check if $corresponding(p,r,r') \in poss(p)$

      if *(1), (2) and (3) are true* **then**

         add r' to $poss(r)$

      forall *neighbor c of r* **do**

         add $corresponding(c,r,r')$ to $poss_r(c)$

   if $poss(r) = \emptyset$ **then**

      return *EmptyPoss*

   forall *neighbor c of r* **do**

      $poss(c) \leftarrow poss_r(c)$

   return *Done*

end

Algorithm 18 presents a third algorithm which, instead of the leaves, starts from the root of the tree [MC92] (and [Mug95] for a journal version). The set of candidates for the root is computed, then the poss sets of other nodes are computed by a top-down traversal of the tree. This technique has two advantages. First, a source BG often has a natural root: It describes an entity (represented by the root), which has certain properties, is in relationship with other entities, etc. In this case, a computation starting from the root is more likely to rapidly restrict the candidate sets. Secondly, this algorithm is adaptable to very big target BGs. In this case, one wants to precompute as few *poss* sets as possible. With the previous algorithm, these sets have to be precomputed for at least all leaves, whereas here only the *poss* set of the root is precomputed. Moreover, one might select some possible images for the root, possibly not all candidates in the target BG.

This algorithm follows the recursive structure of a rooted tree. The recursive subalgorithm `HomomorphismsRec` takes a concept c as input and returns the set of nodes such that there is a homomorphism from the subtree rooted in c to H. It first

---

Algorithm 18: EdgeConsistencyTreev3(G, H)

Input: two BGs G and H, G being a *tree*
Output: the edge-consistent filter of G with respect to H if it exists; otherwise Failure
begin

> Choose a concept c as the root of G
> **if** $R_G \neq \emptyset$ **then**
>> let C'_H be obtained from C_H by removing isolated nodes
>
> **if** *HomomorphismsRec(c, C'_H)* $= \emptyset$ **then**
>> // see Algorithm 19
>> **return** *Failure*
>
> **else**
>> Clean(c) // see Algorithm 20
>> **return** *poss*

end

---

computes the set of candidates for the root c. There is no homomorphism if this set is empty. Otherwise, if the tree is restricted to the root c, the work is done and this set is returned. If not, the algorithm uses the set of candidates for c to compute the set of candidates for each successor r of c. In turn, the set of candidates for each r induces a set of candidates for each successor s of r. The subalgorithm is recursively called on each s and returns the set of nodes such that there is a homomorphism from the subtree rooted in s to H. Each of these sets is then used to restrict the set of candidates for r and c. The new set of candidates for c is finally returned.

Despite a bit more complex formulation, this algorithm has the same worst-case time complexity as the previous one. We then show how it can be extended to compute the number of homomorphisms from G to H without overhead complexity. A specific notation is introduced. Let c be a concept node of G and c' be a candidate for c. For a successor r of c, $poss(r)_{[c \rightarrow c']}$ denotes the set of candidates for r if c is mapped to c'. Then, for a successor s of r (or, equivalently, for a child s of c), $poss(s)_{[c \rightarrow c']}$ denotes the set of candidates for s if c is mapped to c'. It is assumed in the algorithm that these sets are memorized within the *poss* structure.

Algorithm 19: HomomorphismsRec(c, set)   *subalgorithm of Algorithm 18*

Input: c root of a tree G_c and $set \subseteq C_H$

Output: returns set restricted to $\{c' | \exists \Pi : G_c \rightarrow H, \Pi(c) = c'\}$; fills the *poss* structure for the subtree rooted in c and the computed sets $poss(x)_{[c \rightarrow c']}$ are memorized for all x

begin

1    $poss(c) \leftarrow \{c' \in set | l(c) \geq l(c')\}$

2   **if** $poss(c) = \emptyset$ *or* c *has no successor* **then**
     $\llcorner$ **return** $poss(c)$

3   **forall** *relation* r *successor of* c **do**
       forall *concept* $c' \in poss(c)$ **do**
          $poss(r)_{[c \rightarrow c']} \leftarrow \{r' \in R_H | l(r) \geq l(r')$ and $P_r \subseteq P_{r'}$ and
          $c' = corresponding(c, r, r')\}$
          if $poss(r)_{[c \rightarrow c']} = \emptyset$ **then**
             remove c' from $poss(c)$
             if $poss(c) = \emptyset$ **then**
                $\llcorner$ **return** $\emptyset$

       forall *concept* s *successor of* r **do**
          forall *concept* $c' \in poss(c)$ **do**
             $poss(s)_{[c \rightarrow c']} \leftarrow \{s' | \exists r' \in poss(r)_{[c \rightarrow c']}, s' = $
             $corresponding(s, r, r')\}$
          $poss(s) \leftarrow \cup_{c' \in poss(c)} poss(s)_{[c \rightarrow c']}$
          $poss(s) \leftarrow$ HomomorphismsRec(s, $poss(s)$)
          forall *concept* $c' \in poss(c)$ **do**
             $poss(s)_{[c \rightarrow c']} \leftarrow poss(s)_{[c \rightarrow c']} \cap poss(s)$
             forall *relation* $r' \in poss(r)_{[c \rightarrow c']}$ **do**
                if $corresponding(s, r, r') \notin poss(s)_{[c \rightarrow c']}$ **then**
                   remove r' from $poss(r)_{[c \rightarrow c']}$
                   if $poss(r)_{[c \rightarrow c']} = \emptyset$ **then**
                      remove c' from $poss(c)$
                      exit this "for loop"

4    $\llcorner$ **return** $poss(c)$

end

Algorithm 20: Clean(c)   *subalgorithm of Algorithm 18*

---

begin
    forall r *successor of* c **do**
        $poss(r) \leftarrow \cup_{c' \in poss(c)} poss(r)_{[c \to c']}$
        forall s *successor of* r **do**
            $poss(s) \leftarrow \cup_{c' \in poss(c)} poss(s)_{[c \to c']}$
        Clean(s)
end

Property 7.5. [MC92] The number of homomorphisms from a BG G to a BG H can be computed in polynomial time when G is multigraph-acyclic.

If G is not connected, the number of homomorphisms from G to H is the product of the number of homomorphisms from each connected component of G to H. Let us now assume that G is a tree, and let c be any concept chosen as its root. Given c' any node of H, let $NB(c, c')$ denote the number of homomorphisms from G to H which map c onto c'. The number of homomorphisms from G to H is the sum of the $NB(c, c')$ for all c' in H. $NB(c, c')$ is recursively defined as follows.

$NB(c \in G, c' \in H) =$
    0 if c' is not a candidate for c ($c' \notin poss(c)$)
    1 if c has no successor
    otherwise, $\prod_{r \text{ successor of } c} (\sum_{r' \in poss(r)_{[c \to c']}} nb(r, r'))$
    where $poss(r)_{[c \to c']} = \{r' | r' \in poss(r)$ and $c' = corresponding(c, r, r')\}$
    and $nb(r, r') =$
        1 if r has no successor
        otherwise, $\prod_{s \text{ successor of } r} (NB(s, s') | s' = corresponding(s, r, r'))$

The subalgorithm HomomorphismsRec(c, set) (Algorithm 19) can be adapted to return the set of pairs $(c', NB(c, c'))$ for which $NB(c, c') \neq \emptyset$ rather than just the subset of concept nodes to which c can be mapped.

Note that we compute the number of homomorphisms by considering both concepts and relations of G. If we are only interested in homomorphisms that are *distinct* on the concepts of G, we have to keep only one representative for all relations in $poss(r)_{[c \to c']}$ with the same list of neighbors, as they induce homomorphisms which are identical on the successors of r.

7.2.2 BGs Trivially Logically Equivalent to Acyclic BGs

The previous result can be extended to a class of BGs that can be transformed into logically equivalent acyclic BGs by "cutting cycles", as pointed out in [BMT99]. The idea is that if we split an individual concept node (as in the elementary operation *detach*, see Chap. 2), we obtain a graph logically equivalent to the original one (note however that it can be more general with respect to the $\succeq$ relation). See Fig. 7.3: G is logically equivalent to the tree on the right obtained by splitting the concept node [⊤:a]. This would not be true with a generic concept node—the new graph would be logically deducible from the original one but the opposite would be false.

A BG G can be transformed into a logically equivalent acyclic BG by splitting some individual concept nodes if and only if every cycle of length greater than 2 contains an individual concept node. This property can be checked in polynomial time as follows: Split each individual node into as many nodes as neighboring relations and check whether the obtained BG is acyclic. Consequently, BG-HOMOMORPHISM can then be solved in polynomial time.

Property 7.6. [BMT99] A "true" cycle is a cycle (of length greater than 2) containing no individual concept node. Deduction is polynomial when the source BG has no true cycle.

Instead of actually splitting individual nodes, Algorithm 21 performs a traversal of the graph. The difference with respect to the algorithm for checking multigraph-acyclicity (Algorithm 12) is that it does not check whether individual concept nodes are reached several times and does not explore them. More specifically, when an individual node is reached, it is not marked and it is not added to the queue Q (thus its neighbors are not reached from it). Note that even if the graph is connected, one cannot stop the traversal when the queue is empty, since some paths are broken by individual nodes. This algorithm has linear time complexity in the size of the BG.

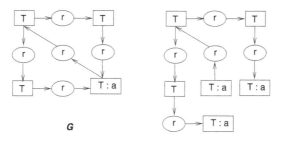

Fig. 7.3 Logically equivalent BGs. G has no true cycle.

Let us now study a generalization of the tree case, which is called "acyclic hypergraph" in graph theory and "guarded BG" when translated from guarded logical formulas.

7.3 Tractability Based on the Hypergraph-Acyclicity of the Source BG

Let us begin with hypergraph-acyclicity. Early works on cyclicity of hypergraphs come from combinatorics, relational databases and also probabilistic networks.

Algorithm 21: CheckExtendedTree(G)

Input: a BG G (not necessarily connected)
Output: true if G has no "true" cycle, otherwise false
begin

    `// let (`C_G`)* be the set of generic concept nodes in` G
    if $|(C_G)^*| = 0$ **then**
      return *true*
    StillToExplore $\leftarrow |(C_G)^*| + |R_G|$ `// number of nodes to explore`
    OK $\leftarrow$ true
    forall *concept or relation node x* **do**
      marked[x] $\leftarrow$ -1
    Choose any generic concept node c as the root
    Let Q be an empty queue
    Put c into Q
    marked[c] $\leftarrow 0$
    while *OK and (StillToExplore $\neq$ 0)* **do**
      if *Q is not empty* **then**
        pick x in Q
      else
        let x be a generic node or relation with marked[x] = -1
        marked[x] $\leftarrow 0$
      StillToExplore $\leftarrow$ StillToExplore - 1
      forall *x' neighbor of x, with x' $\neq$ marked[x]* **do**
        if *x' is a generic node with marked[x'] $\neq -1$* **then**
          OK $\leftarrow$ false
        else
          if *x' is not an individual node* **then**
            marked[x'] $\leftarrow x$
            add x' to Q
    return *OK*
end

Several definitions of acyclicity of a hypergraph have been defined, for which some NP-complete problems on general hypergraphs become polynomial on acyclic instances. All of these definitions are based on associating a graph with the hypergraph; then the notion of acyclicity of the hypergraph is defined by the notion of *acyclicity of the associated graph*. For instance, the *incidence (bipartite) graph* of a hypergraph corresponds to the underlying graph of a BG: hyperedges are transformed into nodes (the relation nodes in a BG) and related to nodes composing the hyperedge (concept nodes in a BG); the only difference in the structure is that the incidence graph does not have multi-edges. Hypergraph-acyclicity based on the acyclicity of the incidence graph is what we call multigraph-acyclicity.

The most general notion of hypergraph-acyclicity among those proposed in the literature can be based on the notion of the *primal graph* of a hypergraph or on that of its *dual graph*, with both notions leading to the same notion of hypergraph acyclicity. When a connected hypergraph is acyclic it can be decomposed into a *join tree*, a name derived from databases [BC81] [BG81] (for several equivalent notions of hypergraph-acyclicity in the context of databases, see [BFMY83]). Hypergraph acyclicity can be polynomially checked. If the test is satisfied, a join tree can be built in polynomial time and is the basis of a polynomial algorithm for homomorphism/homomorphism checking. As we will see, an equivalent notion has been independently defined on BGs, called a *guarded covering* in [Ker01], and inspired from the notion of a guarded first-order logical formula [AvBN96].

7.3.1 Use of a Join Tree

Given a hypergraph $\mathcal{H} = (\mathcal{X}, \mathcal{E})$ the *dual graph* $H = (X, E, \chi)$ of $\mathcal{H}$ is the undirected graph labeled on edges built as follows:

- its nodes are the hyperedges of $\mathcal{H}$: $X = \mathcal{E}$,
- there is an edge between two distinct nodes if they share at least one node of $\mathcal{H}$: $\forall E_1, E_2 \in \mathcal{E}$ with $E_1 \neq E_2$, $E_1 E_2 \in E$ if $E_1 \cap E_2 \neq \emptyset$,
- edges are labeled by the intersection of the hyperedges they connect: $\forall E_1 E_2 \in E$, $\chi(E_1 E_2) = E_1 \cap E_2$.

An edge $E_1 E_2$ of the dual graph is said to be *removable* if there is another path between E_1 and E_2, such that the nodes shared by E_1 and E_2 also belong to all E_i on this path. A partial graph H' of the dual graph H is said to have the *running intersection* property if, whenever in $\mathcal{H}$ the same node x occurs in two hyperedges E_1 and E_2, there is in H' a path between nodes E_1 and E_2 and x occurs in each node on this path. In other words, the subgraph of H' defined by the edges whose label contains x is connected. One can also say that H' is obtained from the dual graph by removing only removable edges. If H' is a tree, it is called a *join tree* of $\mathcal{H}$. Note that if the hypergraphs were not connected, the join tree notion could easily be adapted—it would become a set of join trees.

When a BG is seen as a hypergraph, its nodes are concept nodes and its hyperedges stand for relations and their incident multiedges (see Chap. 5). It is convenient to consider hyperedges as representing star graphs. The dual graph of a BG G can thus be defined as follows: Its nodes are the star graphs of G, there is an edge between two distinct nodes if the corresponding relations share a common concept node in G, and edges are labeled by the sets of common concept nodes of the star graphs they connect. Note however that individual concept nodes can be split without changing the logical meaning of the BG (see Sect. 7.2.2). Thus, edges coming from shared individual concept nodes can be ignored in the construction of the dual graph. Figure 7.4 shows the dual graph resulting from decomposition of the BG G and one of its join trees.

Definition 7.2 (acyclic hypergraph). A connected hypergraph is acyclic if it has a join tree. A hypergraph is acyclic if each connected component has a join tree.

Property 7.7. If a BG is multigraph-acyclic then it is hypergraph-acyclic. But the converse is not true.

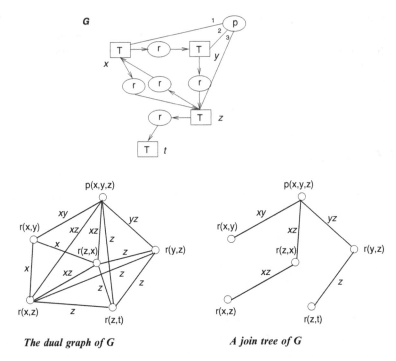

The dual graph of G   **A join tree of G**

Fig. 7.4 Dual Graph and Join Tree

Proof. Let G be a multigraph-acyclic BG. The dual graph of each connected component of G is a join tree. For the converse, see the BG G in Fig. 7.4.   □

Let us come back to homomorphism checking, with G being the source BG and H the target BG. Assume that G has a join tree. Intuitively, and roughly speaking, the edges of this join tree represent constraints on candidates for the concept nodes in G: If an edge is labeled by $c_1 \ldots c_p$, then the nodes it links (representing two star graphs sharing concept nodes $c_1 \ldots c_p$) must have compatible candidates for $c_1 \ldots c_p$. The running intersection property ensures that these constraints are propagated on *all* star graphs sharing nodes. The overall idea is that if the standard filter of G with respect to H can be rendered "consistent" according to the join tree, and then a homomorphism from G to H can be built. More specifically, let us consider Algorithm 22. This algorithm traverses the join tree from its leaves to the root. Edge-consistency is locally achieved for each star graph composed of a relation r_j and its arguments. The result is propagated to the parent node r_k: r_k has to "agree" with r_j on their common arguments, which is another kind of consistency (see pairwise-consistency below). If both kinds of consistency are achieved, the filter built allows us to exhibit a homomorphism from G to H. Otherwise there is no homomorphism from G to H. The algorithm starts from a standard filter, but it could be adapted to compute the *poss* sets during the search as in the algorithm EdgeConsistencyTreev2 (Algorithm 16).

Algorithm 22: ConsistencyCheckwithJoinTree

Input: two BGs G and H, with G being hypergraph-acyclic and connected; T_G a join tree of G

Output: true if there is a homomorphism from G to H and false otherwise; if true is returned, the poss structure satisfies: for any relation r in G, for any $r' \in poss(r)$ $\{(r,r')\}$ can be extended to a homomorphism from the subgraph of G corresponding to the subtree of T_G rooted in r, to H, assigning to each node x an image in $poss(x)$

begin

    Order the relation nodes of G such that each relation appears before its descendant relations in T_G

    Compute the standard filter of G with respect to H

    `// Bottom-up traversal of` T_G

    for j *from* $|R_G|$ *to* 1 **do**

        `// Make the filter edge-consistent for the star graph`
        `   of` r_j

        Failure ← CheckRelation(r_j) `// see Algorithm 5 in Chap. 6`

        if *Failure* **then**

            return *false*

        `// Propagate to the parent relation` r_k
        `// In database terms: compute` $\pi_{r_k}(r_j \bowtie r_k)$

        if $j \neq 1$ **then**

            consider the edge (r_j, r_k) of T_G with $k < j$

            forall r' *in* $poss(r_k)$ **do**

                if *there is no* r'' *in* $poss(r_j)$ *such that for all c common neighbor of* r_j *and* r_k *corresponding*$(c, r_j, r'') = corresponding(c, r_k, r')$ **then**

                    remove r' from $poss(r_k)$

                    if $poss(r_k) = \emptyset$ **then**

                        return *false*

    return *true*

end

The kind of consistency achieved on relations which share arguments is related to a property called "pairwise-consistency" in databases [BFMY83] (see the complementary notes at the end of this chapter).

Definition 7.3 (pairwise-consistency for BGs). A filter of G with respect to H, say *poss*, is *pairwise-consistent* if the following condition is fulfilled for all relations r and s of R_G sharing at least an argument:

- (consistency of s with respect to r) for all $r' \in poss(r)$ there is $s' \in poss(s)$, such that for all c common argument of r and s, $corresponding(c, r, r') = corresponding(c, s, s')$;
- (consistency of r with respect to s) reciprocally, for all $s' \in poss(s)$ there is an $r' \in poss(r)$, such that for all c common arguments of r and s, $corresponding(c, r, r') = corresponding(c, s, s')$.

Pairwise-consistency and edge-consistency are generally not related. Clearly, pairwise consistency does not imply edge-consistency as it does not take the lists

of candidates for concept nodes into account. For instance, if G is restricted to a
star graph, any filter of G with respect to H is pairwise-consistent. The implication
in the other direction does not hold either, as illustrated by Fig. 7.5: Assume that
all concept (or relation) nodes in G and H have the same labels; the standard filter of G with respect to H, which associates $\{c_3, c_4, c_5\}$ to each concept node and
$\{r_3, r_4, r_5\}$ to each relation node, is edge-consistent but not pairwise consistent. For
instance (all nodes play exchangeable roles), r_2 is not pairwise consistent relative to
r_1. Indeed, let us consider $r_3 \in poss(r_1)$. The common arguments of r_1 and r_2 are
$\{c_1, c_2\}$. r_3 leads to map c_1 to c_3 and c_2 to c_4 (i.e., $corresponding(c_1, r_1, r_3) = c_3$ and
$corresponding(c_2, r_1, r_3) = c_4$). But there is no relation in $poss(r_2)$ compatible with
this mapping (i.e. there is no $r' \in poss(r_3)$ such that $corresponding(c_1, r_2, r') = c_3$
and $corresponding(c_2, r_2, r') = c_4$).

Note however, that if any pair of relations in G share at most one node, edge-
consistency implies pairwise-consistency.

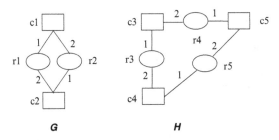

G   **H**

Fig. 7.5 Edge-consistency and pairwise-consistency

Property 7.8. Let G and H be BGs, with G being hypergraph-acyclic. If the standard
filter of G with respect to H can be made edge-consistent and pairwise-consistent,
then it is globally consistent, i.e., there is a homomorphism from G to H.

Proof. If it cannot be made edge-consistent, there is no homomorphism from G to
H. The same holds for pairwise-consistency.

Actually, Algorithm 22 applies *directional* (in *one step*, from the leaves to the
root) edge-consistency and pairwise-consistency; in particular, each r_j is made
pairwise-consistent with respect to its parent relation but not the opposite. If the
algorithm ends with success, a homomorphism from G to H can be constructed by
considering the nodes of T_G from the root to the leaves: an image is taken for the
root relation in its *poss* set, and this choice determines the images of its arguments.
The consequences are propagated to the children relations via the arguments they
share. Directional edge-consistency and pairwise-consistency ensure that no *poss*
set will become empty.   □

Algorithm 22 shows that once a join tree has been constructed, homomorphism
checking can be done in polynomial time. The next section shows that, given a

hypergraph, one can determine in polynomial time (and even in linear time) whether it is acyclic, and if so a join tree can be built. Thus:

Property 7.9. BG-HOMOMORPHISM is polynomial if the source BG is hypergraph-acyclic.

7.3.2 Construction of a Join Tree

Given a join tree of the source BG, homomorphism checking takes polynomial time. But the first problem is to build a join tree. A simple technique is based on the following property:

Property 7.10. [Mai83] If a connected hypergraph is acyclic, then any maximum spanning tree of its dual graph is a join tree.

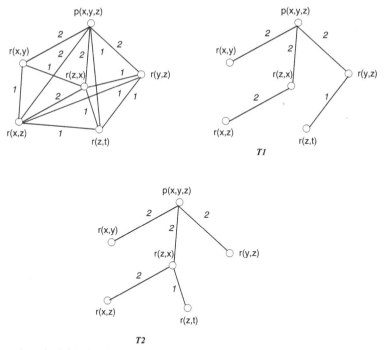

The graph on the left is the dual graph of G in Fig. 7.4, and T_1 is the join tree in the same figure

Fig. 7.6 Maximum spanning trees

This property assumes that the edges of the dual graph are weighted by the number of shared nodes. A spanning tree is a classical notion of graph theory. A spanning subgraph of a graph G is a partial graph of G (thus has the same node set as G). A spanning tree is a spanning subgraph which is a tree. All connected graphs possess spanning trees. Given a (connected) graph with weighted edges, a *maximum* spanning tree is a spanning tree that maximizes the sum of its edge weights (Fig. 7.6). A maximum spanning tree can be built by the following greedy algorithm (Kruskal's algorithm): (1) totally order the edges by the decreasing order of their weights; (2) start from the spanning subgraph without edges; (3) consider each edge according to the order: If this edge produces a cycle, then discard it; otherwise add it to the subgraph. This simple algorithm produces a maximum spanning tree.

The previous property yields a polynomial algorithm to recognize an acyclic hypergraph and build a join tree, if any: First build a maximum spanning tree, then check if it has the running intersection property for any two nodes (which can be done by considering the only path between them and testing if this path satisfies the running intersection property). If the property is not fulfilled, the hypergraph is not acyclic, otherwise one has a join tree.

Another characterization of acyclic hypergraphs relies on properties of the *primal* graph of the hypergraph. As this characterization would lead us to algorithmic techniques beyond the scope of this book, we shall just give pointers to the relevant literature.

The *primal graph* of a hypergraph $\mathcal{H}$ is the undirected graph built as follows (Fig. 7.7):

- it has the same set of nodes as $\mathcal{H}$,
- there is an edge between two distinct nodes if they belong to the same hyperedge of $\mathcal{H}$.

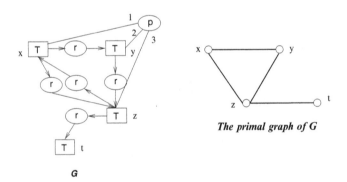

The primal graph of G

Fig. 7.7 Primal graph

Every hyperedge of the hypergraph thus becomes a clique of its primal graph. The characterizing property is as follows:

Property 7.11. [Mai83]: A hypergraph is acyclic if and only if its *primal graph* is chordal and it is conformal.

A *chord* of a cycle is an edge that connects two nodes of the cycle but does not belong to the cycle. A graph is *chordal* if every cycle of length at least 4 has a chord. Many difficult problems on graphs become easy on chordal graphs, such as finding the size of the maximal clique in the graph, which is NP-complete in the general case, or finding all maximal cliques (i.e., not included in another clique) which is NP-hard. A hypergraph is said to be *conformal* if every maximal clique of its primal (chordal) graph form a hyperedge of the hypergraph. [TY84] provides a linear algorithm to check hypergraph-acyclicity and build a join tree based on the previous property (in the context of databases).

Further References

Properties of chordal graphs have been extensively studied in graph theory. In particular, [GHP95] gives a simple algorithm with linear complexity for checking the chordality of a graph and computing a clique-tree if it is chordal; a clique-tree of a chordal graph is a tree where the nodes are the maximal cliques of the graph and satisfying the running intersection property (if the primal graph of the hypergraph is chordal and conformal then the clique tree can be seen as a join tree restricted to maximal hyperedges of the hypergraph). In the context of constraint networks, hypergraph acyclicity is discussed in detail and algorithm schemes are given in [Dec03].

7.3.3 Equivalence with the Existential Conjunctive Guarded Fragment

Other motives have lead to an equivalent notion of acyclicity (up to split of individual concept nodes). They take root in the exploration of relationships between BGs or SGs and a fragment of FOL called the *guarded fragment* introduced in [AvBN96].

In [Ker01], Kerdiles defines the notions of the *crazed form* of a BG and several kinds of covering of this crazed form, namely *acyclic covering*[1] and *guarded covering*. The *crazed form* of a BG G is the SG obtained from the normal form of G with a minimum number of concept splits, such that each concept node is adjacent to at most one relation node. Let us recall that the concept split operation splits a concept node c into two coreferent nodes c_1 and c_2 and attaches each relation that was linked to c to either c_1 or c_2 (see elementary generalization operations in Chap. 3). To build the crazed form of G, each concept node in G attached to n relations ($n \geq 2$) is split n times to obtain n concept nodes attached to one relation. Figure 7.8 shows a BG G and its crazed form.

---

[1] The name used is "tree covering," but as the graphs are not necessarily connected we prefer to use "acyclic" covering.

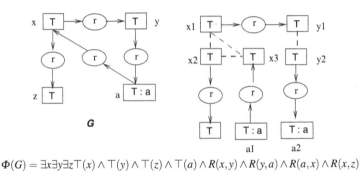

$$\Phi(G) = \exists x \exists y \exists z \top(x) \wedge \top(y) \wedge \top(z) \wedge \top(a) \wedge R(x,y) \wedge R(y,a) \wedge R(a,x) \wedge R(x,z)$$

Fig. 7.8 A BG and its crazed form

This notion thus slightly differs from the antinormal form, where a concept node is incident to at most one edge (cf. Fig. 7.9).

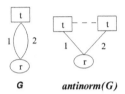

Fig. 7.9 G is in crazed form (but not in antinormal from)

The crazed form of a BG (say G) is unique and has the same logical meaning as G. Let us call *component* of the crazed form of G one of its maximal connected subgraphs without coreferent nodes, i.e., a subgraph composed of one relation node and its neighbors (a star graph), or an isolated concept node (in this case, the node was already isolated in G). In the following, it will be convenient to "materialize" the coreference relation on generic nodes by coreference links (cf. Chap. 3) and to call the other edges *true* edges. A *covering* of G is obtained from the crazed form of G by removing some coreference links (possibly none) in such a way that the coreference relation is kept unchanged: If two generic nodes are coreferent in G, they must be linked by a path of coreference links in the covering. An *extended path* is a path in the covering taking true edges as well as coreference links. An *acyclic covering* of G is a covering with at most one extended path between two concept nodes belonging to distinct components. See Fig. 7.8: Consider the crazed form pictured on the right; an acyclic covering is obtained by removing one coreference link in the triangle on x_1, x_2 and x_3, for instance the edge $x_1 x_3$. The acyclic covering notion characterizes the graphs without true cycles of [BMT99]:

Property 7.12. [Ker01] A BG is transformable into a logically equivalent acyclic BG by splitting *individual* concept nodes if and only if it has an acyclic covering.

Proof. Let T be an acyclic BG equivalent to the BG (say G) and let us build an acyclic covering of G from T. In the crazed form of T, for each coreference clique coming from the split of a generic node c of G, we choose one of the copies of c and keep only the coreference links between this copy and the other copies. Since T has no cycle, we obtain an acyclic covering of the crazed form. Indeed, if there was a cyclic extended path in this covering, one would obtain a cycle in T by merging copies of the same node in T. Conversely, if we merge all generic coreferent nodes of an acyclic covering, a multigraph-acyclic BG is obtained, which is equivalent to G. □

It is convenient to consider an abstraction of a covering: It is the multigraph whose nodes are components of the covering and such that, for each coreference link in the covering between a node in component C_i and a node in component C_j, there is an edge between C_i and C_j. A covering is an acyclic covering if its abstraction has no cycle (even no multiedge). Figure 7.10 shows the abstractions of the crazed form (on the left) and the acyclic covering (on the right) from Fig. 7.8.

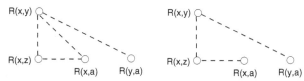

Abstractions of the crazed form and the acyclic covering in Fig. 7.8
Atoms stand for components of the crazed form

Fig. 7.10 Covering abstractions

Let us now generalize the notion of an acyclic covering to that of a *guarded covering*: the abstraction of a guarded covering is an acyclic *multigraph*, i.e., the only cycles authorized are composed of multiedges between two components. More precisely, there is no cycle of form $(C_1 C_2 \ldots C_k C_1)$, $k > 2$, where all C_i, $1 \le i \le k$, are distinct.

Figure 7.11 gives an example of a guarded covering (from left to right: a BG G, a guarded covering of it and the abstraction of this covering). Figure 7.12 shows a non-guarded BG. Actually its crazed form is its sole covering, which is not guarded.

Definition 7.4 (guarded BG). A BG is *guarded* if it has a guarded covering.

The abstraction of the crazed form of a BG has the same structure as the dual graph of this BG (if edges corresponding to shared individual concept nodes are ignored). One goes from the dual graph to a join tree (or a set of join trees if it is not connected) by removing edges while preserving the running intersection property. Equivalently, one goes from the crazed form to a guarded covering by removing coreference links while preserving the coreference relation. Hence the following property.

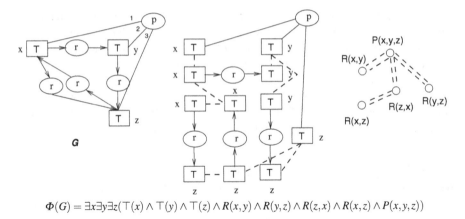

$$\Phi(G) = \exists x \exists y \exists z (\top(x) \land \top(y) \land \top(z) \land R(x,y) \land R(y,z) \land R(z,x) \land R(x,z) \land P(x,y,z))$$

Fig. 7.11 Guarded covering

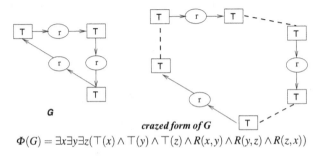

$$\Phi(G) = \exists x \exists y \exists z (\top(x) \land \top(y) \land \top(z) \land R(x,y) \land R(y,z) \land R(z,x))$$

Fig. 7.12 Non-guarded BG

Property 7.13. Given a connected BG G, any join tree of G can be linearly transformed into the abstraction of a guarded covering of G, and reciprocally.

Proof. To go from the join tree of G to the abstraction of a guarded covering of G, transform each edge labeled by a set S into $|S|$ edges with the same extremities. Reciprocally, transform the multiedges (possibly restricted to one edge) between two nodes into a single edge labeled by the set of concept nodes in G yielding these multiedges. □

Corollary 7.2. *A BG is guarded if and only if it is hypergraph-acyclic.*

The notion of a guarded BG is imported from the notion of a guarded FOL formula. The basic idea of the guarded fragment of FOL is to restrict the use of quantifiers: Each quantifier Q is "guarded" by an atom that contains at least all free variables in the subformula generated by Q. Deduction in this fragment is decidable. This fragment turns out to have interesting properties, such as the finite model property (if a guarded formula has a model, it has a finite model). Here we give only the definition of the guarded fragment of FOL($\exists$, $\land$), i.e., the restriction of FOL($\exists$, $\land$)

to guarded formulas. Guarded BGs exactly correspond to closed formulas of the guarded fragment of FOL($\exists$, $\wedge$).

Definition 7.5 (Guarded fragment of FOL($\exists$, $\wedge$)). [Ker01] The guarded fragment of FOL($\exists$, $\wedge$), in short gFOL($\exists$, $\wedge$), is inductively defined as follows:

1. every atomic formula (i.e., every atom) belongs to gFOL($\exists$, $\wedge$),
2. gFOL($\exists$, $\wedge$) is closed under conjunction,
3. –let $free(f)$ denote the set of free variables in a formula f–
   if α is an atom, called a guard, $X \subseteq free(\alpha)$, and f is a formula of gFOL($\exists$, $\wedge$), such that $free(f) \subseteq free(\alpha)$, then $\exists X(\alpha \wedge f)$ is in gFOL($\exists$, $\wedge$).

Any closed formula f of the guarded fragment of FOL($\exists$, $\wedge$) can be polynomially translated into a guarded BG, for instance by the mapping $f2G$ (see Sect. 4.3.3.1). Reciprocally, any guarded BG together with a guarded covering can be polynomially translated into a closed guarded formula: As such, Φ does not produce a guarded formula in the general case, but a slight modification in its definition, arranging quantifiers and atoms in a certain order computed from the guarded covering, produces a trivially equivalent guarded formula. Consider, for instance, the BG G in Fig. 7.11: $\Phi(G)$ is not guarded but the following formula $\Phi'(G)$ obtained by taking the atom $P(x,y,z)$ as a guard is a guarded formula: $\Phi'(G) = \exists x \exists y \exists z (P(x,y,z) \wedge (\top(x) \wedge \top(y) \wedge \top(z) \wedge R(x,y) \wedge R(y,z) \wedge R(z,x) \wedge R(x,z))$. The subformula $f = \top(x) \wedge \top(y) \wedge \top(z) \wedge R(x,y) \wedge R(y,z) \wedge R(z,x) \wedge R(x,z)$ is a guarded formula with free variables x, y and z. Thus $\Phi'(G) = \exists x \exists y \exists z (P(x,y,z) \wedge f)$ is a guarded formula.

More generally, let G be any guarded BG. The closed guarded formula $\Phi'(G)$ assigned to G is inductively defined as follows.

- If G is not a connected BG, then $\Phi'(G)$ is the conjunction of formulas $\Phi'(G_i)$ associated with each connected component G_i of G. It is guarded if each of these subformulas is guarded.
- If G is connected and restricted to an isolated concept, $\Phi'(G) = \Phi(G)$ and $\Phi(G)$ is guarded.
- Otherwise, let us consider one of the guarded coverings of G and choose any of its components as its root. Each component C_i in this rooted covering is the root of a subtree, which is the guarded covering of a subSG G_i of G. If C_i is a leaf, then $\Phi'(G_i) = \exists x_1...x_p(Relation(C_i) \wedge Concepts(C_i))$, where $Relation(C_i)$ is the atom assigned by Φ to the relation of C_i, $x_1...x_p$ are the variables of $Relation(C_i)$ that do not occur in the component parent of C_i and $Concepts(C_i)$ is the conjunction of atoms assigned to the concept nodes of C_i, whose assigned term by Φ is either a constant or one of the variables $x_1...x_p$ (atoms on free variables are not considered in order to avoid duplication of atoms). When G is restricted to a single component G_i, $\Phi'(G_i)$ is thus a closed formula and $Relation(C_i)$ is the guard of $\Phi'(G_i)$. If C_i is not a leaf, let C_{i_1} ... C_{i_k} be its children components. $\Phi'(G_i) = \exists x_1...x_p(Relation(C_i) \wedge Concepts(C_i) \wedge_{j:1...k} \Phi'(G_{i_j}))$. By induction, every $\Phi'(G_{i_j})$ is guarded, and every variable occuring free in $\Phi'(G_{i_j})$ also occurs in $Relation(C_i)$. $\Phi'(G_i)$ is thus guarded. $\Phi'(G_i)$ is closed if C_i is the root (i.e., $G_i = G$).

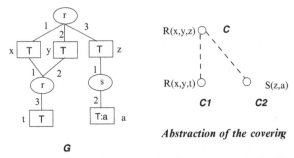

Abstraction of the covering

x, y, z, t, a are the terms assigned to concept nodes by Φ

$$\Phi'(G) = \exists x \exists y \exists z \, (r(x,y,z) \wedge \top(x) \wedge \top(y) \wedge \top(z) \wedge (\exists t(r(x,y,t) \wedge \top(t)) \wedge s(z,a) \wedge \top(a)))$$

Fig. 7.13 From guarded BGs to guarded formulas

Let us consider the example in Fig. 7.13. This figure pictures a BG G and the abstraction of a guarded covering of G, with components C, C_1 and C_2, symbolized by the atom associated to the relation node by Φ. $\Phi'(G_1) = \exists t(r(x,y,t) \wedge \top(t))$. $\Phi'(G_2) = s(z,a) \wedge \top(a)$. One obtains $\Phi'(G) = \exists x \exists y \exists z \, (r(x,y,z) \wedge \top(x) \wedge \top(y) \wedge \top(z) \wedge (\exists t(r(x,y,t) \wedge \top(t)) \wedge s(z,a) \wedge \top(a)))$

Theorem 7.1 (Equivalence of guarded BGs and guarded FOL fragment).
[Ker01] Guarded BGs are equivalent to the (closed) guarded fragment of FOL($\exists, \wedge$).

7.4 Generalizations of Graph-Acyclicity and Hypergraph-Acyclicity

To complete the panorama, let us briefly mention generalizations of graph-acyclicity and hypergraph-acyclicity. We just provide definitions and results, without any illustrations. Readers interested in further developments are referred to the cited papers.

7.4.1 Graphs and Treewidth

The notion of treewitdh was introduced by Robertson and Seymour in the framework of their fundamental research on graph minors (see, e.g., [RS86]). It has proven to be of prime importance in complexity theory. Indeed, many well-known combinatorial NP-hard problems become polynomially solvable when restricted to graphs with treewidth bounded by a constant (see, e.g., [Bod93]). This is the case for BG-HOMOMORPHISM as we shall see later.

The treewidth can be seen as a measure of tree-likehood, or degree of graph-acyclicity. Roughly, a (connected) graph has small treewidth if it can be decomposed into a tree structure where each node represents a small subgraph. More precisely, a *tree decomposition* of a graph G is a pair (T, χ), where $T = (X_T, E_T)$ is a tree and χ is a labeling function that associates a set of nodes of G to each node $x \in X_T$ such that the following conditions are satisfied:

1. For each node v in G, there exists $x \in X_T$ such that $v \in \chi(x)$,
2. for each edge uv in G, there exists $x \in X_T$ such that $u \in \chi(x)$ and $v \in \chi(x)$,
3. for each node v in G, the set $\{x \in X_T | v \in \chi(x)\}$ induces a connected subgraph of T (equivalently: If $v \in \chi(x)$ and $v \in \chi(y)$, then, for every node n in the path between x and y, $v \in \chi(n)$).

Conditions 1 and 2 ensure that all nodes and edges of the graph appear in the decomposition. Condition 3 is the "running intersection property" that we have seen for join trees. A tree decomposition has width k if the largest subgraph occurring in it has $k + 1$ nodes. The *treewitdh* of a graph is the minimum width in all of its tree decompositions. In particular, a (connected) graph is acyclic if and only if it has treewidth 1. A procedure to solve a problem on a graph with bounded treewidth k generally involves two steps: First, a tree decomposition of width k is computed; then, an algorithm based on the tree decomposition is applied.

The treewidth of a graph is NP-hard to compute. More precisely, given a graph G and an integer k, determining whether G has a treewidth less or equal to k is an NP-complete problem, as shown in [ACP87]. However, the same paper shows that if k is fixed (i.e., k no longer belongs to the data of the problem, as is the case for H in the H-COLORING problem, see Sect. 7.1.2), the problem becomes polynomial time solvable; an algorithm is provided that answers the question, and constructs a tree decomposition of width less or equal to k if it exists, using $O(n^{k+2})$ time, where n is the size of the graph. Algorithms with better theoretical complexity were later developed (see [Bod96] and the review paper [Bod98] for more references).

Let now consider the BG-HOMOMORPHISM problem (from G to H). Given a tree decomposition of G of width k, it can be solved in a time exponential in k, and polynomial in the size of G and H. Indeed, there are at most n_H^{k+1} homomorphisms from a subgraph of G with $k + 1$ nodes to H (with n_H being the number of nodes of H). From this observation and the preceding result, we conclude that BG-HOMOMORPHISM is polynomial if k is fixed and G has treewidth bounded by k. The following result is thus obtained with respect to BG-HOMOMORPHISM:

Corollary 7.3. *(from [ACP87]) The following problem "k-tree* BG-HOMOMOR-PHISM*" is polynomial: Given a BG G with treewidth at most k, and a BG H, is there a homomorphism from G to H?*

However, if there are polynomial algorithms computing "good" tree decompositions their practical use is very limited since their complexity contains a large constant factor and has k in the exponent. In other words, the best algorithms from a theoretical viewpoint are exponential in the size of the treewitdh. In [KBvH01], alternatives to these algorithms are studied; they are based on heuristics giving algorithms that are not exponential in the treewidth (but provide no theoretical guarantee on the quality of the result)[2].

7.4.2 Hypergraphs and Hypertreewidth

A measure of hypergraph-acyclicity can be obtained by importing the treewidth notion and applying it to a graph associated with the hypergraph (e.g., its primal graph or incidence graph). In so doing, one obtains classes of hypergraphs whose associated graphs have a treewidth bounded by a certain k, for which HOMOMORPHISM is solvable in polynomial time. However, these notions are incomparable with respect to hypergraph-acyclicity [FFG01] (and [FFG02] for a journal version): There are families of acyclic hypergraphs whose primal graph has unbounded treewidth and, on the other hand, there are families of hypergraphs whose primal graph has bounded treewidth that are not hypergraph-acyclic.

Querywidth [CR97] was an attempt to define a measure of acyclicity on the hypergraph directly, which would play the same role as treewidth for graphs. It generalizes hypergraph-acyclicity (a hypergraph is acyclic if and only if its querywidth is 1) and treewidth (in the sense that the querywidth is always less or equal to the treewidth of the associated primal graph). The negative point is that the problem of determining whether a hypergraph has a querywidth less or equal to k is NP-complete for $k = 4$, as shown by Gottlob, Leone and Scarcello [GLS99][3]. Therefore these authors introduce another notion, i.e., *hypertreewidth*.

Hypertreewidth is defined as follows. A *hypertree* for a hypergraph $\mathcal{H}$ is a triple (T, χ, λ), where $T = (X_T, E_T)$ is a rooted tree and χ and λ are labeling functions associating two sets with each node $x \in X_T$: a set $\chi(x)$ of nodes of $\mathcal{H}$ and a set $\lambda(x)$ of hyperedges of $\mathcal{H}$. A *hypertree decomposition* of $\mathcal{H}$ is a hypertree $\mathcal{HD} = (T, \chi, \lambda)$, such that the following conditions are satisfied:

1. For each hyperedge h of $\mathcal{H}$, there exists $x \in X_T$ such that the nodes of h belong to $\chi(x)$ (it is said that x covers h),
2. for each node v of $\mathcal{H}$, the set $\{x \in X_T | v \in \chi(x)\}$ induces a connected subgraph of T,
3. for each $x \in X_T$, $\chi(x)$ is included in the set of nodes appearing in the hyperedges of $\lambda(x)$,
4. for each $x \in X_T$, let $\chi(T_x)$ be the union of the sets $\chi(y)$ for all y in the subtree T_x rooted in x; then, the intersection between the set of nodes appearing in $\lambda(x)$ and $\chi(T_x)$ is included in $\chi(x)$ (due to (3), the inclusion is in fact an equality).

---

[2] See the web page [Bod] for an up-to-date view of practical algorithms and benchmarks related to tree decompositions.

[3] However, as proven in [CR97], the querywidth of a hypergraph can be approximated by the treewidth of its incidence graph (in particular, it is strictly smaller).

The *width* of a hypertree decomposition is the maximum number of *edges* labeling a node (i.e., $max_{x \in T_x} |\lambda(x)|$). The *hypertreewidth* of a hypergraph is the minimum width throughout all its hypertree decompositions.

Hypertreewidth has nice properties:

- it generalizes hypergraph-acyclicity and querywidth (if a hypergraph has querywidth $\leq k$, it also has hypertreewidth $\leq k$, but the opposite does not hold), thus generalizing the treewidth (as the treewidth of the primal graph is an upper bound of the querywidth),
- it is polynomially recognizable (given a hypergraph, determining whether it has a treewidth less than a constant k can be done in polynomial time),
- (directed) hypergraph homomorphism checking is polynomial on classes of hypergraphs with bounded hypertreewidth.

Corollary 7.4. *(from [GLS99]) The following problem "k-hypertree BG Homomorphism" is polynomial: Given a BG G with hypertreewidth at most k, and a BG H, is there a homomorphism from G to H?*

Note, however, that this beautiful theoretical result does not necessarily lead to practical algorithms (see the discussion on treewidth).

7.4.3 Expressivity Results

Let us end this part with some expressivity results. Correspondences between the treewidth and hypertreewidth notions and fragments of FOL have been established, which give indications on the expressive power of the BG fragment obtained.

Kolaitis and Vardi [KV00] proved that the class of all conjunctive queries (thus BGs) having *treewidth* less than k (i.e., their incidence graphs have treewidth less than k) are equivalent in expressive power to the *k-variable* fragment of FOL($\exists$, $\wedge$), i.e., the class of FOL($\exists$, $\wedge$) formulas using at most k variables.

The class of conjunctive queries (SGs) with *hypertreewidth* at most k coincides in expressive power with the *k-guarded* fragment of FOL($\exists$, $\wedge$) [GLS01]. The k-guarded fragment is a generalization of the guarded fragment, where k atoms can jointly (but not solely) act as guards. More precisely, the condition (3) of definition 7.5 is replaced by:

If $\alpha_1, \ldots \alpha_j$ are atoms, with $j \leq k$, $X \subseteq \bigcup_{i:1 \ldots j} free(\alpha_i)$, and f is a k-guarded formula of FOL($\exists$, $\wedge$), such that $free(f) \subseteq \bigcup_i free(\alpha_i)$, then $\exists X(\alpha_1 \wedge \ldots \wedge \alpha_j \wedge f)$ is a k-guarded formula of FOL($\exists$, $\wedge$). For $k = 1$, one has the guarded fragment of FOL($\exists$, $\wedge$), equivalent to acyclic hypergraphs. Figure 7.14 summarizes the correspondences between classes of BGs and fragments of FOL($\exists$, $\wedge$).

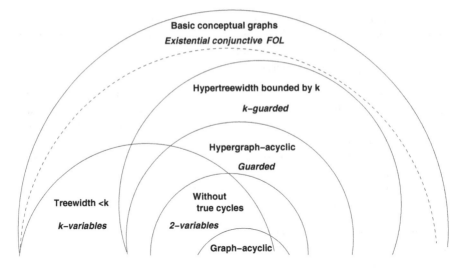

Classes of BGs are in plain text and fragments of FOL($\exists$, $\wedge$) are in italics. The dotted line figures the limit between theoretical tractability and intractability of BG-HOMOMORPHISM.

Fig. 7.14 Classes of SGs and fragments of FOL($\exists$, $\wedge$)

7.5 What About Labels?

All complexity results we have harvested so far were obtained by restricting the *structure* of the graphs/hypergraphs. Now, seen as combinatorial objects, BGs are not only graphs, or hypergraphs; they are *labeled*. Label diversity in the graphs encountered in applications is a major reason why practically efficient algorithms can be built. However, to the best of our knowledge, there are few results about how properties of the labeling function may influence the complexity of labeled graph/hypergraph homomorphism checking.

Let us cite two results, both based on restricting the labeling of the *target* graph. The first one comes from databases. Recall that the CONJUNCTIVE-QUERY-CONTAINMENT (CQC) problem takes two conjunctive queries as input, say q_1 and q_2, and asks if q_1 is contained in q_2 or, equivalently, if there is a query homomorphism from q_2 to q_1 (see Chap. 5). In [SY80], it is shown that this problem becomes polynomial if each atom in q_2 has no more than *two* candidate images in q_1 and an algorithm in $O(n^2)$ time is given, where n is the size of the input [4]. It is also shown

---

[4] The problem actually considered in this paper is "tableau containment" where tableaux represent queries closely related to conjunctive queries.

that the problem remains NP-complete when each atom in q_2 has no more than *three* candidate images in q_1.

It follows from the transformation from CQC to BG-HOMOMORPHISM (detailed in Chap. 5) that BG-HOMOMORPHISM is polynomial when, after a polynomial filtering step (such as computing the standard filter and possibly making it edge-consistent), each relation node in the source BG has at most two candidates. An obvious case for which this property is true if, assuming that the relations in the vocabulary are not ordered (which occurs frequently in conceptual graph applications), each relation label occurs at most twice in the target BG.

Let us roughly outline a polynomial algorithm adapted from the algorithm described in [SY80]. Assume that we start from a standard filter of G with respect to H. For any relation node r in G, let r_1 and r_2 denote its candidates (r possibly has only one candidate). Let us begin with the main idea. The algorithm chooses a relation node r and one of its candidates, say r_1. This defines a partial solution, say $\theta = \{(r, r_1)\}$. The algorithm will try to extend the domain of θ to all relations in G for which an image is *forced* by this choice. Let R be this set. Initially, R contains $\{r\}$ and it grows by traversing G from r. Let s be any relation in R explored at a given step. Each relation s' outside R that shares at least a neighbor with s is considered. If only one of the candidates for s' is compatible with the current θ, the image of s' is forced by the assignment $\{(r, r_1)\}$, thus the corresponding pair is added to θ and s' is added to R. If both candidates for s' are incompatible with the current θ, then there is no solution with the assignment (r, r_1) and the construction of R fails. Assume R has been successfully built. The property that underlines the algorithm is as follows: Let h be any homomorphism from G_R, the subgraph of G induced by the relation nodes outside R, to H; the union of θ and h defines a homomorphism from G to H. Indeed, let v be a relation in R and let w be a relation outside R. $h(w)$ is one of the two candidates of w, and by hypothesis it is compatible with $\theta(v)$, otherwise w would have been either forced or declared incompatible during the construction of R. The algorithm starts from any r and builds a set R for each candidate of r. There is no solution if the construction of both sets fails. If one construction succeeds, then the θ built defines a partial solution. The process is iterated with the remaining unassigned relation nodes in G.

Finally, let us mention the polynomial reductions between CQC and SAT (Is a given conjunction of propositional clauses satisfiable?) described in [Sar95]. More specifically, it is shown that the so-called K-CONTAINMENT problem (in which each atom of q_2 has at most k candidates in q_1) is essentially the same as the K-SAT problem (in which each clause contains at most k literals). The particular case mentioned above thus corresponds to the 2-SAT problem, for which efficient algorithms are known.

The second case relies on a simple observation made in a graph framework applied to machine learning [LB94] [LS98]. An unordered set of labels is considered and used to label nodes of graphs. The fundamental notion is homomorphism on these graphs. First note that checking homomorphism from a graph G to a graph H is polynomial if the nodes of H all have different labels, i.e., the labeling mapping from the nodes of H to the set of labels is injective. Indeed, each node of G can be mapped to at most one node of H (we thus have a special case of the preceding cited case). Such a restriction is too strong to be useful in practice, but let us consider a weaker restriction: Let us say that the labeling mapping is *locally injective* if its restriction to the neighborhood of each node is injective, i.e., all neighbors of a node have distinct labels. A graph with a locally injective labeling mapping is called a *li-graph*. Checking homomorphism from G to H is polynomial if H is a li-graph. Indeed, as soon as a potential image has been assigned to a node z of G, there is at most one possibility for each neighbor of z, thus, by iteration, for each other node of G. Note also that the number of homomorphisms from a li-graph G to a li-graph H is bounded by the number of nodes in H. Thus all solutions can be exhibited in polynomial time. Another interesting property is that a connected li-graph is irredundant. Indeed, since the labels of neighbors of a node are not comparable, there is no folding of the graph (Property 2.9 in Sect. 2.3.2). As for the property of being a tree, the property of being a li-graph corresponds to a case that can be visually checked by a user and can be found in practical applications.

Let us translate this property to BGs. Assume again that the relations in the vocabulary are not ordered. Then enforcing local injectivity of the labeling mapping on concept nodes (i.e., all relations incident to a concept are of different type) in the target BG yields a tractable case for BG-HOMOMORPHISM. The following less restrictive condition on the target BG would also be suitable: For all concept nodes c, if c is the i-th neighbor of two distinct relation nodes r and s, i.e., if there are edges (r,i,c) and (s,i,c) with $r \neq s$, then r and s are of different types.

In conclusion, exhibiting tractable cases combining properties on the structure of BGs and on their labeling is an open challenge.

7.6 Complementary Notes

References to relevant works have been given throughout this chapter. This note adds some explanations about the pairwise consistency notion, as it appears in databases [BFMY83]. A desirable property of a database instance is to be *globally consistent*. Intuitively, this means that it can be represented by a global relation on all attributes, in such a way that each relation is obtained by restriction of this global relation. More specifically, let us consider a database schema $\mathcal{R} = \{R_1, \ldots, R_n\}$, and let $\mathcal{I} = \{I_1, \ldots, I_n\}$ be a database instance on $\mathcal{R}$. $\mathcal{I}$ is globally consistent if there is a universal relation I over the attributes $R_1 \cup \ldots \cup R_n$ such that for all $1 \leq i \leq n$, $I[R_i] = I_i$. If there is such a universal relation, it is obtained by a "full join" $I_1 \bowtie I_2 \ldots \bowtie I_n$; in this case, no tuple of any I_i is lost after this full join. Now, $\mathcal{I}$ is said to be *pairwise consistent* if, for each pair of relations I_i and I_j in $\mathcal{I}$, I_i and I_j are the same on their common attributes, i.e., for all $1 \leq i, j \leq n$, $(I_i \bowtie I_j)[R_i] = I_i$. This ensures that no tuple of I_i is lost after a join with I_j. Checking pairwise-consistency can be done in polynomial time, whereas checking global consistency is NP-complete. A natural question is thus to determine exactly when checking pairwise-consistency is sufficient to garantee global consistency. [BFMY83] provided the answer: If the hypergraph corresponding to the database schema is acyclic then, for every database instance over this schema, pairwise-consistency yields global consistency, and reciprocally.

Chapter 8
Other Specialization/Generalization Operations

Overview

This chapter is about more complex specialization/generalization operations than the elementary specialization/generalization operations studied in Chap. 2. In order to focus on the main ideas, the conceptual graphs considered here are BGs, and not SGs. Moreover, for operations involving compatibility notions (i.e., maximal join and extended join) we consider conjunctive concept types.

In Sect. 8.1, we show how the greatest specialization (or greatest lower bound) and the least generalization (or least upper bound) of two BGs can be computed. Computing the least generalization of two formulas is a fundamental problem in inductive inference. Computing a common specialization of two (or more) BGs is required in various applications. The greatest lower bound is usually not a "good" notion since it does not merge BGs, so several specific common specializations of two graphs were defined. They are often designed for specific applications, and we give here the main ideas that can be used to define operations fitted for a given application. Computing a common specialization of two graphs G and H consists of establishing a correspondence between nodes in G and nodes in H, then merging corresponding nodes. The operations can be distinguished by the kind of correspondence between G and H (e.g., bijection between subsets of concept nodes or general correspondence between a subgraph of G and a subgraph of H), and by a maximality property of the correspondence.

In Sect. 8.2, the notion of a compatible set of concept nodes of a graph is reviewed and compatible sets of relation nodes are introduced. These notions are used for defining maximal join operations. A maximal join operation between two BGs G and H consists of first merging a concept node in G and a concept node in H, and then merging as far as possible neighbors of previously merged nodes. Different neighborhood search strategies lead to different generalized join operations. Usually a generalized join operation stops when it is no longer possible to merge two nodes, so the term "maximal join" is used even though this term represents a set of operations rather than a precisely defined one.

The third section is devoted to the study of compatible partitions of node sets. We define and study compatible partitions of the concept node set of a BG, compatible partitions of the relation node set and compatible partitions of the whole node set of a BG. This last notion is strongly related to surjective homomorphisms. These notions are used for the extended join operation, which generalizes maximal join operations.

The main result in Sect. 8.4 concerns characterization of BGs that can be obtained from a set of given BGs using elementary specialization operations (sometimes known as "canonical" BGs). As a corollary, one obtains an inductive definition of the BGs built on a given vocabulary . This study is done using surjective homomorphisms and the union of a set of BGs.

Finally, in Sect. 8.5, the expansion and contraction operations used when considering defined concept types are presented.

This section is not an in-depth study of problems related to type definitions. The aim is only to provide other examples of conceptual graph operations.

8.1 The Least Generalization and Greatest Specialization of Two BGs

Computing a (or *the*) least generalization of two or more descriptions is a fundamental problem in inductive inference, which occurs particularly in machine learning. This operation can be offered by knowledge-based systems along with classical inference services. Consider, for instance, the tasks of building concept or relation definitions, or schemata typically associated with certain concepts or relations, or rules expressing general properties of certain entities. It may help to start from a set of descriptions assumed to be examples of the same concept (or relation) and consider their least generalization as a working basis. Least generalizations can also be used to organize a large set of descriptions in a hierarchical structure.

If the description language is a BG language, this problem in its basic form takes two BGs as input, say G and H, and asks for a least generalization of G and H, i.e., a BG K such that $K \succeq G$ and $K \succeq H$ and for all BG K', if $K' \succeq G$ and $K' \succeq H$ then $K' \succeq K$. If we restrict ourselves to irredundant BGs, K is unique (up to ismorphism): it is the least upper bound of G and H in the irredundant BG lattice (cf. Sect. 2.3.2). The *categorial product* of two graphs (a graph theoretic notion which can be found in the literature under a variety of other names, e.g., weak product) is used to compute a least generalization of two BGs. Let us review the categorial product of two ordinary graphs before considering BGs.

Definition 8.1 (Categorial Product). The (categorial) product of two (ordinary) graphs $G = (X, U)$ and $H = (Y, V)$ is the graph $G \times H = (Z, W)$ with $Z = \{(x,y) | x \in X \text{ and } y \in Y\}$ and $W = \{((x,y),(z,t)) | (x,z) \in U \text{ and } (y,t) \in V\}$.

Example. Figure 8.1 shows an example of the product of two graphs.

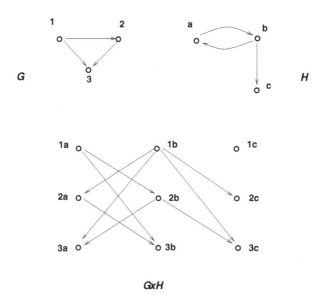

Fig. 8.1 The categorial product of two graphs

Note that $G \times H$ and $H \times G$ are isomorphic. The following properties about homomorphism can be easily observed.

Property 8.1 (Standard properties of the categorial product).

1. There is a homomorphism from $G \times H$ to G and a homomorphism from $G \times H$ to H.
2. If there is a homomorphism from K to G and a homomorphism from K to H, then there is a homomorphism from K to $G \times H$.
3. There is a homomorphism from G to $G \times H$ if and only if there is a homomorphism from G to H.

Proof. 1. Take the homomorphism from $G \times H$ to G (resp. from $G \times H$ to H), which maps each vertex (x,y) to vertex x (resp. y),
2. Take two homomorphisms $f : K \to G$ and $f' : K \to H$. Then the mapping $h : K \to G \times H$, which maps every vertex x to $(f(x), f'(x))$, is a homomorphism,
3. It follows from (1) and (2): If there is a homomorphism from G to $G \times H$, then by (1) there is a homomorphism from G to H; reciprocally, if there is a homomorphism from G to H then, since there is a homomorphism from G to G, by (2) there is a homomorphism from G to $G \times H$. □

Let us extend the definition of the graph product to BGs. Labels have to be taken into account. For ordinary graphs, we had exactly one vertex for each pair of vertices (x,y), with x in G and y in H. Now the number of nodes created from (x,y), where x and y are two concept nodes or two relation nodes in G and H respectively, is the number of minimal upper bounds of x and y labels. Since the concept type set has a greatest element, and individual markers are less than the generic marker, every pair of concept node labels has at least an upper bound. If the concept type set is a sup-semi-lattice (for instance if conjunctive concept types are considered), all pairs have a unique minimal upper bound (called their least upper bound, *lub*); otherwise, some pairs have several minimal upper bounds.

Pairs of relation labels may have any number of minimal upper bounds (including zero). Therefore, if no assumption is made on the structure of the concept and relation type sets, the size of $G \times H$ is no longer bounded by the product of the size of G and H (which is a rough upper bound) but involves the size of the vocabulary, and more precisely the maximal number of minimal upper bounds of two concept or relation types.

For the next definition, we consider the particular case where two concept labels have a least upper bound (i.e., a single least generalization) and two relation labels have either a least upper bound or do not have an upper bound. Then the number of concept nodes in $G \times H$ is $|C_G| \times |C_H|$, the number of relation nodes is bounded by the sum of $(|R_G^i| \times |R_H^i|)$, where R_G^i and R_H^i denote the set of relation nodes with arity i occurring respectively in G and in H, and the number of edges is bounded by the sum of $(|R_G^i| \times |R_H^i| \times i)$.

Definition 8.2 (BG product). Let $\mathcal{V}$ be a vocabulary where the concept type set is a sup-semi-lattice and two relation types have at most one minimal upper bound, and let G and H be two BGs on $\mathcal{V}$. The product of G and H is the BG $K = G \times H$ built as follows:

- $C_K = \{(c,d)|c \in C_G \text{ and } d \in C_H\}$, and the label of (c,d) is the lub of $l_G(c)$ and $l_H(d)$,
- $R_K = \{(r,s)|r \in R_G, s \in R_H \text{ and there is a relation label } t \text{ with } t \geq l_G(r) \text{ and } t \geq l_H(s)\}$, and the label of (r,s) is the lub of $l_G(r)$ and $l_H(s)$,
- $E_K = \{((r,s),i,(c,d))|(r,i,c) \in E_G \text{ and } (s,i,d) \in E_H\}$.

It can be easily checked that $G \times H$ is a BG and that the BG product satisfies the standard properties (cf. property 8.1). This operation can be extended to the product of n BGs.

Property 8.2. The least upper bound, i.e., the least generalization, of two BGs G and H is the irredundant form of $G \times H$.

Example. Consider G and H in Fig. 8.2. These BGs are obtained from the graphs in Fig. 8.1 by transforming the vertices into concept nodes and edges (u,v) into binary relation nodes, with the node assigned to u for first neighbor and the node assigned to v for second neighbor. All pairs of concept node labels have a lub. Concerning relation labels, one assumes that r and s have a lub, as well as r and t, but s and t have no upper bound. Compare the skeleton of $G \times H$ to the product graph in Fig. 8.1: the relation nodes corresponding to the edges $(1b,3a)$ and $(2b,3a)$ do not exist (since s and t have no upper bound). It can be easily checked that $G \times H$ maps to G and to H (assign to each node ij in $G \times H$ the node i in G and the node j in H). One can check that $G \times H$ is not irredundant, the gray part composed of the concept node $1c$,

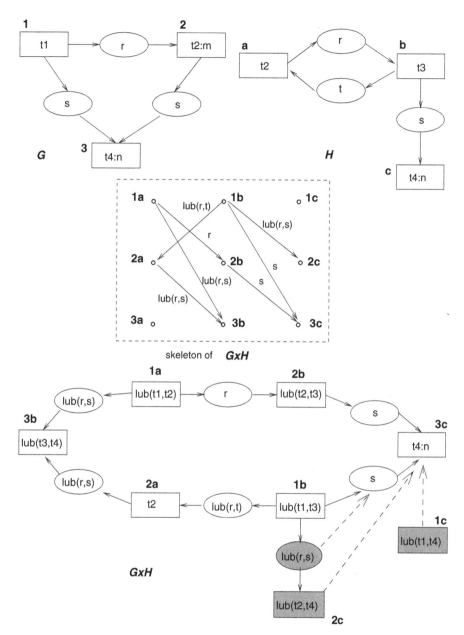

Fig. 8.2 The least generalization of two BGs

the concept node $2c$ and its neighbor relation node can be removed (cf. the dashed arrows in Fig. 8.2). The BG obtained after these deletions is the least generalization of G and H (check that it is irredundant even if some labels might actually be the same in some vocabulary).

With the assumption that the concept type set is a sup-semi-lattice, $G \times H$ is normal if G and H are normal. Indeed, an individual concept node (c, d) with marker m may be created in $G \times H$ only if c and d are both individual nodes with marker m; If G and H are normal, they may each possess at most one node with marker m. As illustrated in Fig. 8.2, $G \times H$ is generally not irredundant, even if G and H are irredundant. As far as we know, the conditions under which $G \times H$ is irredundant have not been characterized.

Computing a common specialization of two (or more) graphs is required in various applications. A *greater specialization* of two BGs is easily built: It suffices to compute their disjoint sum (see Property 2.10). Hence the property:

Property 8.3. The greatest lower bound of two BGs G and H is the irredundant form of their disjoint sum $G + H$.

$G + H$ is generally not normal or irredundant, even though G and H are normal and irredundant. However, note that given G and H irredundant, $G + H$ is irredundant if and only if there is no connected component of one of the two graphs that maps to a connected component of the other. Thus, in the case where G and H are connected irredundant graphs, the greatest lower bound (*glb*) of G and H is G if H maps to G, H if G maps to H, and otherwise it is exactly $G + H$.

The disjoint sum is easily computed but it is generally considered as unsatisfying because it does not "merge" the two graphs. This has led to the introduction of other notions of common specializations which rely on more complex operations than the elementary specialization operations studied in Chap. 2. Among them, the so-called "maximal join" is the most popular. It is studied in the next section.

8.2 Basic Compatibility Notions and Maximal Joins

In this section, we define compatible sets of concept nodes and of relation nodes. A node set can be merged into a single node if and only if it is a compatible set. Compatible sets are then used for defining the maximal join operation.

8.2.1 Compatible Node Set

Let us review the notion of compatible (or mergeable) concept nodes studied in Chap. 3, i.e., when conjunctive concept types are considered.

A set of *compatible concept nodes* (cf. Definition 3.9) is a set of concept nodes such that the set of their labels is compatible, i.e., these labels possess a greatest lower bound. The *label of a compatible concept node set S* is $l(S) = glb(\{l(x) \mid x \in S\})$.

Compatibility is strongly related to homomorphism: If a set of concept nodes is compatible, then these nodes may all be mapped to the same node by a homomorphism, and conversely.

More precisely, let us consider two BGs G and H and a homomorphism π from G to H. If y is a concept node in $\pi(G)$, then $S = \pi^{-1}(y)$ is a compatible concept node set of G. Indeed, for any $x \in S$ one has $l_G(x) \geq l_H(y)$, thus the label set of S has a lower bound (i.e., the conjunction of all types appearing in S is not a banned type and at most one individual marker appears in S), and it has a greatest lower bound.

A set of concept nodes S is compatible if and only the nodes in S can be merged into a single node. Merging these nodes consists of replacing them by a single node labeled $l(S)$ and having for neighbors the union of the neighbors of all nodes in S. In terms of elementary specialization operations: first restricting their type to the conjunction of their types, secondly, if an individual marker m appears in S, restricting their marker to m, and thirdly joining them.

Property 8.4. Let G be a BG and S be a compatible concept node set of G. The graph G/S obtained from G by merging S is a BG and there is a (surjective) homomorphism from G to G/S.

We shall now define the compatibility of relation nodes, which is more complex than the concept node compatibility.

The set of partitions of the integer set $\{1,\ldots,k\}$ is partially ordered by the usual order on partitions. Given two relations r_1 and r_2 of the same arity, P_{r_2} (i.e., the edge partition associated with r_2, cf. Definition 6.2) is thinner than P_{r_1} (notation $P_{r_2} \subseteq P_{r_1}$) if each class of P_{r_2} is included in (or equal to) a class of P_{r_1}. Let π be a BG homomorphism mapping a relation node r in G to a relation node r' in G'. Then, P_r is thinner than $P_{r'}$. Indeed, two concept nodes can have the same image, whereas a node cannot have two images. Moreover, let c' be a neighbor of r'. For any neighbor c of r such that $P_r[c] \subseteq P_{r'}[c']$, one has and $\pi(c) = c'$.

Example. Consider Fig. 8.3, where π is a homomorphism from G to G', $r(a,b,c,d,e,f)$ is in G, $\pi(r) = r'$ and $r'(u,v,u,v,w,z)$ is in G'. $P_r = \{\{1\},\{2\},\ldots,\{6\}\}$ (it is the discrete partition), and $P'_r = \{\{1,3\},\{2,4\},\{5\},\{6\}\}$. As $P_r[a] = \{1\} \subseteq \{1,3\} = P_{r'}[u]$ and $P_r[c] = \{3\} \subseteq \{1,3\} = P_{r'}[u]$, one has $\pi(a) = \pi(c) = u$.

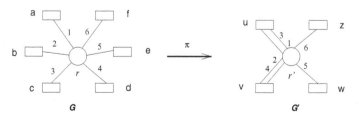

Fig. 8.3 P_r and homomorphism

Similar to concept nodes, a set of relation nodes is said to be compatible if these nodes can be merged, i.e., they may all be mapped to the same node by a homomorphism.

More precisely, let G and H be two BGs, π a homomorphism from G to H and y a relation node in $\pi(G)$. The set of relation nodes $A = \pi^{-1}(y) = \{r_1, \ldots, r_k\}$ satisfies:

1. All r_i have the same arity and furthermore for any $i = 1, \ldots, k$ one has $l_G(r_i) \geq l_H(y)$, therefore the set $\{l_G(r_1), \ldots, l_G(r_k)\}$ has a lower bound.
2. For any $i = 1, \ldots, k$, the partition P_{r_i} is thinner than or equal to P_y, and thus the least upper bound $P(A)$ of all the P_{r_i} is thinner than or equal to P_y.
3. Let $C(A)$ denotes the set of concept nodes which is the union of the neighbors of all nodes in A. For any class X in $P(A)$, let $C(X)$ be the subset of $C(A)$ containing concept nodes linked to a node in A by an edge numbered with an integer in X. Let $\mathcal{C}(A) = \{C(X) \mid X \in P(A)\}$. Every concept node in $C(A)$ belongs to at least a subset $C(X) \in \mathcal{C}(A)$ and possibly to several such subsets. $\mathcal{C}(A)$ is thus a covering of the set $C(A)$. Let $P_C(A)$ denote the thinnest partition of $C(A)$ greater than (or equal to) this covering, $P_C(A)$ is a compatible partition of $C(A)$.

Example. Let us assume that π is the homomorphism from G to H represented in Fig. 8.4, with $\pi(r_1) = \pi(r_2) = r'$. $A = \pi^{-1}(r') = \{r_1, r_2\}$. $P_{r_1} = \{\{1,2\}, \{3\}\}$, $P_{r_2} = \{\{1\}, \{2\}, \{3\}\}$, $C(A) = \{a,b,c,d\}$, $P(A) = \{\{1,2\}, \{3\}\}$, $\mathcal{C}(A) = \{\{a,b,d\}, \{b,c\}\}$, and $P_C(A) = \{\{a,b,c,d\}\}$.

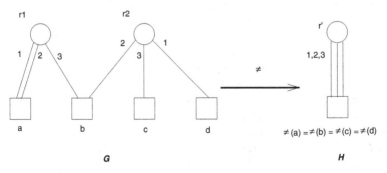

Fig. 8.4 Two relation nodes with the same image

A set of relation nodes satisfying all the previous properties is called a *compatible relation node set*.

Definition 8.3 (Compatible relation nodes). A *compatible relation node set* of a BG G is a set $A = \{r_1, \ldots, r_k\}$ of relation nodes in G satisfying the following constraints:

- All relations in A must have the same arity; moreover, the set of labels of A must have a lower bound (in the vocabulary upon which G is built),
- let $C(A)$ denote the set of concept nodes which are neighbor of at least one relation node in A; let $P(A)$ denote the lub of the partitions P_{r_i}, $i = 1, \ldots, k$; let $P_C(A)$

be the least partition containing the covering $\mathcal{C}(A)$ of $C(A)$ associated with $P(A)$; each class of $P_C(A)$ must be a compatible set of concept nodes.

As for the concept nodes, if a set of relation nodes is compatible, one can define a specialization operation that consists of merging these relation nodes.

Definition 8.4 (Compatible Relation Nodes Merging). Let $A = \{r_1, \ldots, r_k\}$ be a compatible relation node set of a BG G. The graph G/A obtained from G by merging A is defined as follows.

- Compute the partition $P(A)$, i.e., the lub of the partitions P_{r_i}.
- Compute $\mathcal{C}(A) = \{C(X) \mid X \in P(A)\}$ and $P_C(A)$, i.e., the thinnest partition of $C(A)$ greater than or equal to $\mathcal{C}(A)$.
- Merge each class of $P_C(A)$ (which involves a sequence of concept restrict and join).
- Restrict the types of the relation nodes in A to $l(A)$, a maximal lower bound of $\{l(r_1), \ldots, l(r_k)\}$ (now, all relation nodes in A are twin relation nodes).
- Merge A into a single relation node labeled $l(A)$ (i.e. remove all but one of the twin relation nodes in A).

Note that if the relation type set is not an inf-semi-lattice, then a set of types may have several maximal lower bounds and the operation is not deterministic.

Property 8.5. Let A be a compatible relation node set of a BG G. The graph G/A obtained from G by merging A is a BG and there is a (surjective) homomorphism from G to G/A.

Proof. Let $A = \{r_1, \ldots, r_k\}$. Each class of $P_C(A)$ is a compatible concept node set; thus it can be merged (while keeping a BG). Let us check that after step 4 all nodes in A are twin nodes. They have the same label $l(A)$, so it remains to verify that if x and y are i-th neighbors of r_j and r_l, respectively, then they are in the same class of $P_C(A)$ (thus they are merged into a single node in step 3). Indeed, let X be the class of i in $P(A)$; x and y belong to $C(X)$, which is included in a class of $P_C(A)$. Thus step 5 keeps a BG. □

8.2.2 Maximal Join

Maximal join is an important operation in conceptual graph applications. Intuitively, the effect of the maximal join operation is to maximally join, or merge, connected subgraphs of two graphs. It is mainly used to do plausible inference by applying conceptual schemata, typical patterns, etc., to facts.

Although there is agreement on the intuitive purpose of this operation, a lot of variants are considered in practice. In this section, we will thus present the general ideas behind the maximal join, some variations, and give a simple algorithm that illustrates one way of implementing this operation. Section 8.3 will provide the formal foundations required to define the notion of extended join, with maximal join (and its variations) being a particular case of it.

Let us start with the simplest way of joining two graphs, the *external join* operation (by distinction with the elementary join operation already defined, which is also sometimes called *internal* join). The external join consists of merging two concept nodes of two disjoint graphs, say G and H.

Definition 8.5 (External join). Let G and H be two (disjoint) BGs and c and d be two compatible concept nodes in G and H, respectively. The *external join* of c in G and d in H is the BG obtained by first, computing $G + H$, secondly, restricting the labels of c and d to their glb l, then by identifying c and d.

Since an external join can be decomposed into elementary specialization operations, the graph obtained is a common specialization of G and H.

A way of extending an external join of c in G and d in H consists of searching mergeable neighbors (i.e., relation nodes) of c and d, then to check if new concept nodes neighbors of these relations can be merged, and so on. Said otherwise, starting from a pair of compatible concept nodes, the idea is to search, in a greedy way, mergeable neighbors of previously identified mergeable nodes. The algorithm stops when it is impossible to find new mergeable nodes. The result is thus locally "maximal," hence the name *maximal join* for this class of operations.

In order to specify a maximal join operation, one has to define a condition for merging concept nodes and a condition for merging relation nodes (the nodes must be at least compatible, but stronger conditions may be enforced) as well as a strategy for exploring the graphs. Given two mergeable concept nodes as a starting point, there may be several maximal joins, but computing one of them can be done in polynomial time, whereas computing a maximal join with a *maximum* number of nodes is NP-hard (indeed it admits homomorphism or injective homomorphism as a special case, cf. Sect. 5.2).

We now describe a very simple maximal join algorithm. The graph exploration involves extending the initial external join by a breadth-first strategy. One seeks *strongly compatible* pairs of relation nodes defined as follows:

Definition 8.6 (Strongly compatible pair of relation nodes). Two relation nodes $r \in G$ and $s \in H$ are strongly compatible with respect to compatible concept nodes $c \in G$ and $d \in H$ if they fulfill:

- $type(r) = type(s)$,
- there is an integer i with $(r, i, c) \in G$ and $(s, i, d) \in H$,
- $P_r = P_s$ and for each class of P_r (or P_s) if c' (resp. d') is the neighbor of r (resp. s) associated with this class, then $\{c', d'\}$ is a pair of compatible concept nodes.

Starting from the concept node pair (c, d), the algorithm seeks in a greedy way a maximal set of strongly compatible pairs of relations with respect to c and d: a first pair (r, s) is sought, then another pair disjoint from (r, s) is sought and so on until no new pair can be built in the neighborhood of c and d. The found relation pairs yield concept node pairs, which are used as starting points for a new step.

An important point is that the compatibility of concept nodes evolves during the process. For instance, let (c_1, c_3) and (c_2, c_3) be two pairs of initially compatible concept nodes; assume that the process leads to add the pair (c_1, c_3) to f; the label of c_2 may be incompatible with the label of the node that will be obtained by merging c_1 and c_3 (i.e., $l_G(c_2)$ may be incompatible with glb($\{l_G(c_1), l_H(c_3)\}$)). That is why the algorithm maintains two labeling functions memorizing the "current label" of concept nodes: These functions are initially equal to those of the input graphs and, when a pair of concept nodes is added to f, the current label of these nodes becomes the glb of their former current labels. A concept node may appear in several pairs, thus its current label may change several times.

The result of the algorithm is first, the pair (f, g), and secondly, new labeling functions l'_G and l'_H for concept nodes. Effectively computing the maximal join of the input graphs consists of merging the nodes of f and g pair by pair. No computation is necessary to obtain the labels of the new nodes: Merged relation nodes have the same label, and merged concept nodes have the same label by l'_G and l'_H, which is exactly the label of the new node.

If the relation nodes are first merged according to g, there is no need to consider f, since concept merging follows from relation merging. If the concept nodes are first merged according to f, relation nodes of the same pair become twins, and a twin per pair has to be removed (i.e., by a sequence of relation simplify operations).

Hereafter we give the schema of a maximal join algorithm.

The set *ToExplore* contains pairs of concept nodes that have been added to f but whose neighborhood has not yet been explored. The boolean *explored* is used to explore the neighborhood of a given pair (x, y): Initially false, it becomes true when no new pairs of relation neighbors of x and y can be built.

Example. Figure 8.5 presents two BGs G and H. For simplicity, it is assumed that all concept nodes have the same label. Starting with concept nodes 1 and a, two maximal joins that do not have the same number of nodes can be obtained. The first one is defined by the mapping $1 \mapsto a, 2 \mapsto b, 3 \mapsto c$, and the second one by $1 \mapsto a, 2 \mapsto d, 3 \mapsto e, 4 \mapsto f, 5 \mapsto g$.

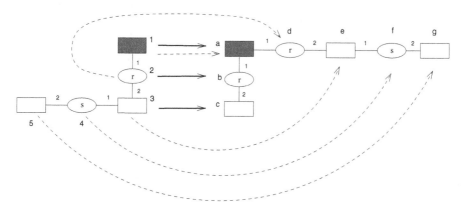

Fig. 8.5 Maximal join

Algorithm 23: Maxjoin(G,H,c,d)

Input: two BGs G and H, two compatible concept nodes $c \in G$ and $d \in H$
Output: a pair (f,g) defining a maximal join of G and H with respect to (c,d), with f being
the set of concept node pairs and g being the set of relation node pairs; l'_G and l'_H
are labeling functions

begin
$\quad$ $l'_G \leftarrow l_G$
$\quad$ $l'_H \leftarrow l_H$
$\quad$ $g \leftarrow \emptyset$
$\quad$ $f \leftarrow \{(c,d)\}$
$\quad$ $ToExplore \leftarrow \{(c,d)\}$
$\quad$ **while** $ToExplore \neq \emptyset$ **do**
$\quad\quad$ Remove (x,y) from $ToExplore$
$\quad\quad$ $explored \leftarrow false$
$\quad\quad$ **while** $not\ explored$ **do**
$\quad\quad\quad$ **if** $there\ is\ a\ pair\ (r,r')\ of\ strongly\ compatible\ relation\ nodes\ with\ respect\ to$
$\quad\quad\quad$ $(x,y),\ l'_G\ and\ l'_H,\ such\ that\ r \notin g,\ r' \notin g$ **then**
$\quad\quad\quad\quad$ $g \leftarrow g \cup \{(r,r')\}$
$\quad\quad\quad\quad$ **foreach** $neighbor\ e\ of\ r,\ with\ e \neq x$ **do**
$\quad\quad\quad\quad\quad$ let $e' = corresponding(e,r,r')$
$\quad\quad\quad\quad\quad$ **if** $(e,e') \notin f$ **then**
$\quad\quad\quad\quad\quad\quad$ $f \leftarrow f \cup \{(e,e')\}$
$\quad\quad\quad\quad\quad\quad$ $ToExplore \leftarrow ToExplore \cup \{(e,e')\}$
$\quad\quad\quad\quad\quad\quad$ $l'_G(e) \leftarrow glb(l'_G(e),l'_H(e'))$
$\quad\quad\quad\quad\quad\quad$ $l'_G(e') \leftarrow glb(l'_G(e),l'_H(e'))$

$\quad\quad\quad$ **else**
$\quad\quad\quad\quad$ $explored \leftarrow true$

$\quad$ **return** $(f,g),\ l'_G,\ l'_H$
end

Figure 8.6 shows that the joined subgraphs are not necessarily isomorphic. Assume that all concept nodes in G and H have the same label. Starting from $(1,a)$,
each relation neighbor of 1 is matched with the relation of the same label neighbor
of a. Thus $g = \{(2,b),(3,c),(4,d)\}$ and $f = \{(1,a),(5,e),(6,e),(7,f),(7,g)\}$. The
concept nodes 5, 6 and e will be merged into a single node, as well as the concept
nodes 7, 8 and f. In this particular example, G and H are completely joined. Thus
they map entirely to the obtained graph.

Let us use this figure to illustrate the role of concept node labels. Assume that
$(5,e)$ and $(6,e)$ are pairs of compatible concept nodes, but that $\{5,6,e\}$ is not a
compatible set. E.g., $l_G(5) = (t,a)$, $l_G(6) = (t,b)$ and $l_H(e) = (t,*)$. If $(2,b)$ is the
first relation pair added to g, then the pair $(5,e)$ is added to f, and the current label
of 5 and e becomes (t,a). Consequently, $(6,e)$ is no longer a compatible pair. Thus
$(3,c)$ is no longer strongly compatible and cannot be added to f. If $(3,c)$ is considered before $(2,b)$, then $(5,e)$ becomes an incompatible pair. Thus, depending on the
order in which the relation neighbors of 1 and a are considered, different maximal
joins are obtained.

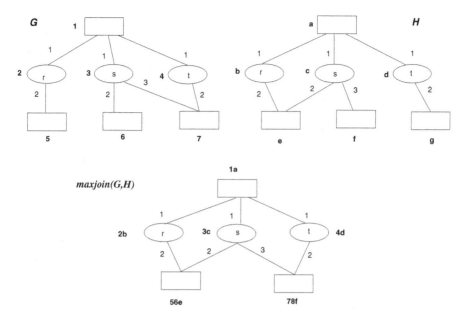

Fig. 8.6 Maximal Join of non isomorphic subgraphs

The algorithm can be seen as defining two homomorphisms from subgraphs of G and H to the same graph. More precisely, let G' (resp. H') be the subgraph of G defined by all nodes which are first (resp. second) components of an ordered pair in f or g. First, let us point out that G' and H' are subBGs of G and H respectively. Indeed, let r be a relation node in G'. There is a single relation node s in H' such that (r, s) is in g and all neighbors of r (resp. s) are first (resp. second) components of f. Thus, all neighbors of a relation node in G' are also in G'. Therefore G' is a BG (the same holds for H'). Let K denote the BG obtained by merging the corresponding nodes in f and g. K is obtained from G' and H' by specialization operations, therefore G' and H' map to K. Furthermore, these homomorphisms are injective on relation nodes, but not necessarily on concept nodes (Indeed, a concept node may appear in several pairs, as illustrated by Fig. 8.6).

Depending on the properties desired for homomorphisms from G' and H' to K, one can adapt the condition for merging relation or concept nodes. By considering stronger conditions for node merging, one obtains homomorphisms with stronger properties, but the number of nodes in G' and H' can be smaller. By considering weaker conditions for node merging, one obtains homomorphisms with weaker properties, but the number of nodes in G' and H' can be greater. Let us give some examples.

Assume we want to make f injective, so that the joined subgraphs are isomorphic. In MAXJOIN, f is initially injective (It is restricted to the ordered pair (c, d) and $c \neq d$). The following condition concerning a pair of relation nodes guarantees that f remains injective after an execution of the while loop.

Let G and H be two BGs, and let f be an injective correspondence between some concept nodes in G and some in H. A pair (r,s) of relation nodes, r in G and s in H, is *strongly compatible with respect to* f if (r,s) is strongly compatible and if, for any neighbor c of r and $c' = corresponding(c,r,s)$, either (c,c') is in f or neither c nor c' appears in f. In MAXJOIN, instead of considering "strongly compatible and disjoint pairs of relation nodes with respect to (x,y)," let us consider "strongly compatible *with respect to* f and disjoint pairs of relation nodes with respect to (x,y)". The correspondence f built by the algorithm is now injective and the restrictions of f and g to nodes in G' and H' are bijective. Thus, there are bijective homomorphisms from G' to K and from H' to K that, furthermore, do not restrict the relation node labels. Note also that the compatibility of concept nodes does not evolve during the algorithm.

Example. Applying this modified algorithm to G and H in Fig. 8.6, assuming that all concept nodes are pairwise compatible, and starting from $(1,a)$ one obtains either K or K' drawn Fig. 8.7. K is obtained with $g = \{(2,b),(4,d)\}$ and $f = \{(1,a),(5,e),(7,g)\}$, and K' with $g = \{(3,c)\}$ and $f = \{(1,a),(6,e),(7,f)\}$.

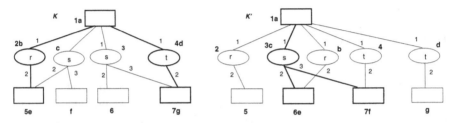

Fig. 8.7 Maximal Joins of isomorphic subgraphs

Instead of reinforcing the condition for merging relation nodes (to obtain bijective homomorphisms keeping relation node labels), it can be relaxed. For instance, one can replace the second point in the definition of a strongly compatible pair of relation nodes by " $type(r)$ and $type(s)$ have a lower bound." In this case (f,g) defines homomorphisms that are still injective on the relation nodes, but the relation node labels can be restricted.

Let us end with an example showing how the maximal join can be used to add plausible information to a BG representing a fact. A schema is a BG associated with a concept type t gathering typical or plausible information commonly accompanying the occurrence of an entity of type t (cf. Chap. 13). In Fig. 8.8, H is a schema for the concept type *Drink*, with h being its privileged concept node. This schema for *Drink* takes meaning in a "baby stories" context, i.e., usually, in a baby stories context, a drink action has for object milk and for instrument a feeding bottle that contains the milk. G is a fact containing a concept node c of type *Drink*. Computing a maximal join between G and H from an external join between c and h adds plausible relevant information to G. In the obtained graph, the soft drink has been specialized into milk, the bottle into a feeding bottle which contains the milk.

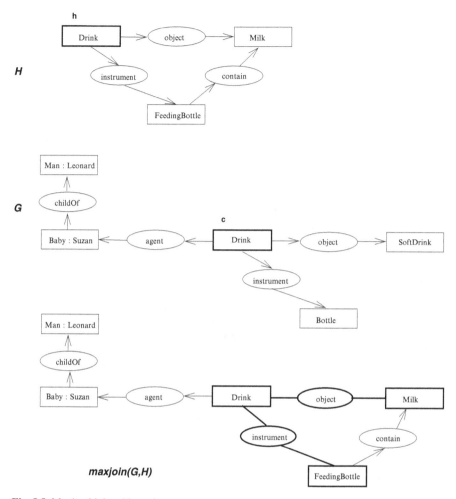

Fig. 8.8 Maximal join with a schema

When several schemas are associated with the same concept type, the situation is more complex. For instance, let us assume that the previous schema H and the schema H' shown Fig. 8.9 are both associated with the concept type *Drink*, with this second schema being related to a context "detective novels." Then maximal join can be used to determine a plausible context of the fact G in Fig. 8.8 as follows. If the number of matched nodes in $maxjoin(G,H)$ is greater than the number of matched nodes in $maxjoin(G,H')$ then "babies stories" is a plausible context of G, otherwise a plausible context of G is "detective novels." As the number of matched nodes in $maxjoin(G,H)$ is 5 whereas the number of matched nodes in $maxjoin(G,H')$ is 1, then the fact G more plausibly concerns a babies story than a detective novel.

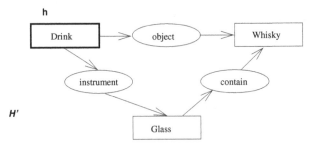

Fig. 8.9 Another schema for the concept type *Drink*

Maximal join operations are particular cases of the extended join that will be defined at the end of the next section.

8.3 Compatible Partitions and Extended Join

In this section, we first study the partition of the node set of a BG induced by a homomorphism. This notion is then used to define the extended join of two BGs. A maximal join of two BGs G and H induces two particular homomorphisms from G and H to the same graph. Similarly, an extended join of G and H induces homomorphisms from G and H to the same graph, but these homomorphisms are more general. Finally, we consider a particular extended join that merges concept node only, and is useful, for instance, in rule processing.

8.3.1 Compatible C-Partition and R-Partition

Let us now consider the partitions of the concept and relation node sets of a BG induced by a homomorphism. Let G and H be two BGs and let π be a homomorphism from G to H. A partition P_π of the node set of G can be defined as follows: Two nodes are in the same class of P_π if they have the same image by π, i.e., two nodes x and y in G are equivalent modulo P_π if $\pi(x) = \pi(y)$. P_π is composed of a partition of the concept node set of G such that any class of this partition is a compatible set of concept nodes and a partition of the relation node set of G such that any class of this partition is a compatible set of relation nodes (cf. Sect. 8.2.1).

Conversely, if P_C (resp. P_R) is a partition of the concept (resp. relation) node set of a BG G, then we give conditions for the quotient graph G/P_C (resp. G/P_R) to be a BG. In this case, there is a (surjective) homomorphism from G to G/P_C (resp. G/P_R).

If P_R is a partition of the relation node set of a BG G, G/P_R is defined like G/P_C (cf. Definition 2.22). Note that if P is a partition of a subset of the node set of a BG, one can still consider G/P by supplementing P with classes having a single node for all nodes that are not in a class of P

Definition 8.7 (Compatible C-partition). A *compatible C-partition* of a BG is a partition P of its concept node set such that any class in P is a compatible concept node set.

Property 8.6. Let G be a (normal) BG and π be a homomorphism from G to H, then π induces a compatible C-partition of G. Conversely, let P_C be a compatible C-partition of G, then the graph G/P_C obtained by merging each class of P_C is a (normal) BG and there is a surjective homomorphism from G to G/P_C.

Proof. The first part comes from the definition of a compatible set of relation nodes. The second part is obtained by recurrence on the number of classes in P_C. Let $P_C = \{S_1, \ldots, S_k\}$. Then G/S_1 satisfies property 8.5. One concludes by noting that $G/\{S_1, \ldots, S_k\} = (G/S_1)/\{S_2, \ldots, S_k\}$. □

Let us now consider the relation nodes.

Definition 8.8 (Compatible R-partition). A *compatible R-partition* of a BG is a partition $P = \{A_1, \ldots, A_k\}$ of its relation node set such that:

- Any class in P is a compatible relation node set,
- the partition $P_C(P_R)$ which is the lub of all the $P_C(A_i)'$-s is a compatible C-partition, where $P_C(A_i)'$ denotes the partition of the whole concept node set obtained by completing $P_C(A_i)$ by a trivial class for each concept node that does not belong to a class of $P_C(A_i)$.

Example. Consider in the BG of Fig. 8.10, $A = \{r_1, r_2\}$. Then $C(A) = \{a, b, c, d\}$, $P_{r_1} = \{\{1\}, \{2, 4\}, \{3, 5\}\}$, $P_{r_2} = \{\{1, 2\}, \{3\}, \{4\}, \{5\}\}$, $P_A = \{\{1, 2, 4\}, \{3, 5\}\}$ and the covering induces by P_A is $\{\{b, c, d\}, \{a, c\}\}$. Thus, $P_C(A)$ is equal to $C(A) = \{a, b, c, d\}$. As $C(A)$ is compatible, A is a compatible set of relation nodes.

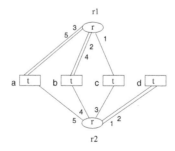

Fig. 8.10 $\{r_1, r_2\}$ is a compatible set of relation nodes

The following property is obtained with a proof similar to that given for Property 8.6:

Property 8.7. Let G be a (normal) BG and let π be a homomorphism from G to H, then π induces a compatible R-partition of G. Conversely, let P_R be a compatible R-partition of G, then the graph G/P_R obtained by merging each class of P_R is a (normal) BG and there is a surjective homomorphism from G to G/P_R.

After considering compatible C-partitions and compatible R-partitions, let us now consider both the concept and relation node sets.

A compatible partition of a BG is composed of a compatible C-partition and a compatible R-partition which are not independent.

Definition 8.9 (Compatible partition of a BG). Let G be a BG, P_C a partition of the concept node set of G and P_R a partition of relation node set of G. $P_C \cup P_R$ is a *compatible partition of G* if:

- P_R is compatible,
- P_C is compatible and $P_C \supseteq P_C(P_R)$.

Let $P = \{P_C, P_R\}$ be a compatible partition of a (normal) BG $G = (C,R,E,l)$. A quotient BG G/P is obtained from G and P as follows. First, the (ordinary) quotient graph G/P is built. Then the node associated with a class of P_C is labeled by the glb of the labels of the class, and the node associated with a class of P_R is labeled by a maximal lower bound of the labels of the class. If there are several such maximal lower bounds, then several G/P can be defined. They all have the same structure but the labels of the relation nodes can be different.

Example. Let G be the BG presented Fig. 8.11. It is assumed that $P_C = \{\{a,b,c,d,f\}, \{e\}\}$ is a compatible C-partition and that $type(r_1) = type(r_2)$ and $type(r_3) = type(r_4)$. Let us check that $P_R = \{\{r_1,r_2\}, \{r_3,r_4\}\}$ is a compatible R-partition. $P_{r_1} = \{\{1\}, \{2,4\}, \{3,5\}\}$, $P_{r_2} = \{\{1,2\}, \{3\}, \{4\}, \{5\}\}$, $P(r_1,r_2) = \{\{1,2,4\}, \{3,5\}\}$. The covering associated with $\{r_1,r_2\}$ is $\{bcd,ac\}$ thus $P_C(\{r_1,r_2\}) = \{abcd\}$.
$P_{r_3} = \{\{1\}, \{2\}\}, P_{r_4} = \{\{1\}, \{2\}\}, P(r_3,r_4) = \{\{1\}, \{2\}\}$. The covering associated with $\{r_3,r_4\}$ is $\{df,e\}$ thus $P_C(\{r_3,r_4\}) = \{df,e\}$. Finally, $P_C(R) = P_C$; therefore P_R is a compatible R-partition and $\{P_C, P_R\}$ is a compatible partition of G.

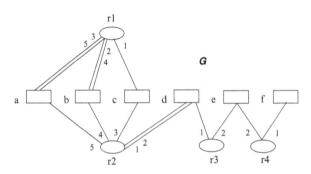

Fig. 8.11 $\{\{r_1,r_2\}, \{r_3,r_4\}\}$ is a compatible partition of the relation node set

The quotient graph G/P is presented Fig. 8.12.

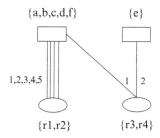

Fig. 8.12 The quotient graph G/P associated with G and P in Fig. 8.11

Theorem 8.1. *1. If P is a compatible partition of a BG G then there is a surjective homomorphism from G to (any) G/P.*

2. Let G and H be two BGs. A mapping π from G to H is a BG homomorphism if and only if the partition of the node set of G induced by the equality of the images by π is a compatible partition of G.

3. If the only compatible partition of a BG is the trivial partition, i.e., each class is restricted a single node, then it is irredundant.

Proof. 1. The mapping which associates to any node of G its class in P is a homomorphism from G to G/P.

2. If there is a homomorphism from G to H, then the partition of the node set of G induced by the equality of the images by π is a compatible partition of G. Indeed, P_C and P_R are compatible partitions by definition of a homomorphism, and P_C is greater than $P_C(P_R)$ by definition of the underlying graph homomorphism. Furthermore, there is a bijective homomorphism from G/P to $\pi(G)$, and for any class $\{c_1, \ldots, c_k\}$ of P_C let y be the concept node in H such that for $i = 1, \ldots, p$, $\pi(c_i) = y$ then $glb(\{l(c_i) \mid i = 1, \ldots, p\} \geq l(y)$. The same applies for any class of P_R.

3. If the only compatible partition of a BG G is the trivial partition, i.e., each class is restricted a single node, then G is irredudant. Indeed, if there is a non-injective homomorphism from G to itself, then there is a non-trivial compatible partition. $\square$

Note that even if a graph is irredundant, it can have non-trivial compatible partitions, as shown in Fig. 8.13 where G is irredundant; nevertheless there is a surjective homomorphism from G to H, and G has a non-trivial compatible partition.

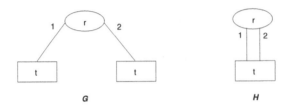

Fig. 8.13 An irredundant graph with a non-trivial compatible partition

8.3.2 Extended Join

We have considered so far two join operations, which build a common specialization of two BGs by merging parts of them. The external join can be considered as a merging of two trivial subgraphs (they are restricted to a single node). In the maximal join algorithm described in Sect. 8.2.2, two connected subgraphs, obtained by maximally extending an external join, can be merged. In this section, we consider the *extended join* operation, which generalizes the previous joins, by allowing subgraphs of any form to be merged.

Definition 8.10 (Fusionnable BGs). Let G_1 and G_2 be two BGs. They are *fusionnable* if there are partitions P_1 and P_2 respectively compatible on G_1 and G_2, such that there is an isomorphism f from G_1/P_1 to G_2/P_2, and f fulfills: For any node x in G_1/P_1, $\{x, f(x)\}$ is compatible.

The fusion of two fusionnable BGs G_1 and G_2 consists of merging any x in G_1/P_1 with $f(x)$. Let H be the BG obtained by such a fusion. H can be obtained from G_1/P_1 by replacing the label of any node x by $glb(l(x), l(f(x)))$. G_1/P_1 and G_2/P_2 are more general or equal to H, and we have:

Property 8.8. If G_1 and G_2 are fusionnable then there is a BG H and two surjective homomorphisms from G_1 to H and from G_2 to H.

Definition 8.11 (Extended join). An *extended join* operation between two BGs consists of fusionning two of their fusionnable subBGs.

The Fig. 8.14 illustrates the extended join between two BGs L and M. f_1 and f_2 are the canonical surjective homomorphisms from G_1 to G_1/P_1 and from G_2 to G_2/P_2, g_1 and g_2 are bijective homomorphisms. $g_1 \circ f1$ and $g_2 \circ f2$ are the surjective homomorphisms of Prop. 8.8.

8.3.3 Join According to a Compatible Pair of C-Partitions

The join defined in this section can be seen as a generalization of the external join, and as a particular extended join. With respect to the external join, instead of only

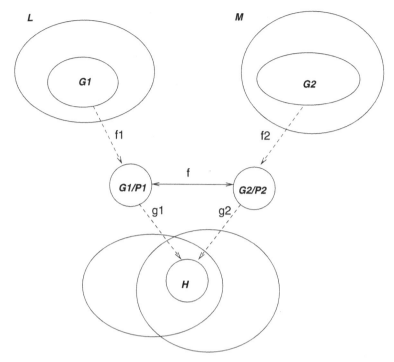

Fig. 8.14 Extended join between the BGs L and M

joining a concept c_1 in G_1 and a concept c_2 in G_2, several compatible sets of concept nodes in G_1 and in G_2 are joined. With respect to the extended join, only concept nodes are joined.

Definition 8.12 (Compatible Pair of C-partitions). Let $P_1 = (p_{1_1}, \ldots, p_{1_k})$ and $P_2 = (p_{2_1}, \ldots, p_{2_k})$ be two ordered partitions with the same cardinality over two subsets of the concept node sets of two BGs G_1 and G_2. $\{P_1, P_2\}$ is a *compatible pair of C-partitions* if for all i, $i = 1, \ldots, k$, the set $p_{1_i} \cup p_{2_i}$ is compatible.

Note that if $\{P_1, P_2\}$ is a compatible pair of C-partitions, then P_1 and P_2 are compatible C-partitions of subsets of the concept node sets of G_1 and G_2.

The join of two graphs with respect to a compatible pair of C-partitions is described in two steps. The first step consists of merging in G_1 (resp. G_2) the classes of P_1 (resp. P_2) using the labels of classes of P_2 (resp. P_1). More precisely,

Definition 8.13 (Specialization of a BG with respect to a compatible pair of C-partitions). Let $P_1 = (p_{1_1}, \ldots, p_{1_k})$ and $P_2 = (p_{2_1}, \ldots, p_{2_k})$ be a compatible pair of C-partitions of G_1 and G_2. The *specialization of G_1 with respect to P_1 and P_2* is the graph $G_1(P_1, P_2)$ built from G_1 as follows:

- the label of any concept node in a class p_{1j} of P_1 is specialized into $glb(p_{1j} \cup p_{2j})$,
- each class of P_1 is merged.

The join of two graphs with respect to a compatible pair of C-partitions can now be defined.

Definition 8.14 (Join of two BGs with respect to a compatible pair of C-partitions). Let P_1 and P_2 be a compatible pair of C-partitions of G_1 and G_2. The *join of G_1 and G_2 with respect to P_1 and P_2* is the graph obtained from G_1 and G_2 as follows. First, build $G_1(P_1, P_2)$ and $G_2(P_2, P_1)$. Then, for all i join c_{1i} and c_{2i}, which are respectively the node of $G_1(P_1, P_2)$ obtained by merging the i-th class of P_1 and the node of $G_2(P_2, P_1)$ obtained by merging the i-th class of P_2.

Said otherwise, the join of G_1 and G_2 with respect to P_1 and P_2 is obtained from $G_1 + G_2$ by merging the nodes of each set $p_{1_i} \cup p_{2_i}$, $1 \le i \le k$.

If the C-partitions are empty, the join is simply a disjoint sum of G_1' and G_2'. If each C-partition is restricted to a single class with a single node, the join is simply an external join. Classes of P_1 and P_2 may contain concept nodes of several connected components. In such cases, the join can stick several connected components.

The join of two graphs with respect to a compatible pair of C-partitions will be used for processing rules in backward chaining (see Chap. 10).

8.4 $\mathcal{G}$-Specializations

Let $\mathcal{G}$ be a set of BGs. The main goal of this section is to give a computational characterization of the set of $\mathcal{G}$-specializations. A $\mathcal{G}$-specialization is simply a BG which can be obtained from $\mathcal{G}$ by using a finite number of elementary specialization operations (cf. Sect. 3.4). Such a set $\mathcal{G}$ is sometimes called a "canonical base" of BGs and, in this case, a $\mathcal{G}$-specialization is called a "canonical BG." BGs built on a given vocabulary can be be considered as canonical BGs generated by star BGs. Thus an inductive definition of BGs is stated. In order to achieve Theorem 8.3, which is the main result of this section, we study surjective homomorphisms and define the union of a set of (non-necessarily disjoint) BGs.

8.4.1 Surjective Homomorphism

In this section, the unary specialization operations and their relationships with homomorphism are only considered. We also state an inductive definition of a BG that differs from the definition given in Chap. 2.

Definition 8.15 (G^α). Let $\alpha = \{s, r, j\}$ be the set of the unary strict specialization operations (s for relation simplify, r for restrict, and j for join) and let G be a BG. G^α is the set of BGs inductively defined by:

- (basis) $G \in G^\alpha$,
- (rules) If $o \in \alpha$ and $H \in G^\alpha$ then $o(H) \in G^\alpha$.

One has:

Property 8.9. Let G and H be two BGs. There is a surjective homomorphism from G to H if and only if H is isomorphic to an element of G^{α}.

Proof. Let H be isomorphic to an element of G^{α}. If $H = o(G)$ with $o \in \alpha$, then there is a surjective homomorphism π from G to H defined as follows.

1. If $o = s$ and r' a twin node of r is deleted then $\pi(r') = r$ and for all node $x \neq r'$ $\pi(x) = x$,
2. if $o = r$ then π is the identity,
3. if $o = j$ and c and c' are identified then $\pi(c') = c$ and for all node $x \neq c'$ $\pi(x) = x$.

The composition of surjective homomorphisms is a surjective homomorphism, so one obtains the sufficient condition. Reciprocally, let π be a surjective homomorphism from G to H. The following sequence of unary specialization operations transforms G into H:

1. Let G_1 be the BG obtained from G by restricting the label of any concept x to the label of its image $\pi(x)$ in H,
2. let G_2 be the BG obtained from G_1 by successively joining all concepts having the same image in H,
3. let G_3 be the BG obtained from G_2 by successively simplifying all relations in G_2 having the same image in H, G_3 is isomorphic to H.
   $\square$

Property 8.10. Let G and H be BGs, then $G \succeq H$ if and only if there is a subBG of H which is isomorphic to an element of G^{α}.

Proof. If $G \succeq H$, let us consider a homomorphism π from G to H. Then there is a surjective homomorphism from G to $\pi(G)$, which is a subBG of H. One concludes with the preceding property for the necessary part. Conversely, let us suppose that K is a subBG of H, with K being isomorphic to $L \in G^{\alpha}$. Therefore, $G \succeq L$, K isomorphic to L, and $K \succeq H$, yield $G \succeq H$. $\square$

In graph theory, a homomorphism is called *complete* if it is faithful and surjective. As a BG homomorphism is faithful, one can replace surjective homomorphism by complete homomorphism.

8.4.2 Union

The union of two (not necessarily disjoint) BGs is a partial operation defined as follows.

Definition 8.16 (Union of two BGs). Let $G=(C_G, R_G, E_G, l_G)$ and $H=(C_H, R_H, E_H, l_H)$ be two BGs such that any node common to G and H has the same label in G and H. The union of G and H is the BG $(C_G \cup C_H, R_G \cup R_H, E_G \cup E_H, l_{G \cup H})$ where $l_{G \cup H}$ is defined by:

- if x is a node in G which is not in H, then $l_{G \cup H}(x) = l_G(x)$,
- if x is a node in H which is not in G, then $l_{G \cup H}(x) = l_H(x)$,
- if x is a node in G and H, then $l_{G \cup H}(x) = l_H(x) = l_G(x)$.

Property 8.11. Let $G = (C, R, E, l)$ and $G' = (C', R', E', l')$ be two BGs such that for any node or edge x in $(C, R, E) \cap (C', R', E')$ $l(x) = l'(x)$. $(C \cup C', R \cup R', E \cup E', l \cup l')$ is a BG if and only if $graph(G) \cap graph(G')$ provided with the labeling function l (or l') is a BG.

Proof. If $(C \cup C', R \cup R', E \cup E', l \cup l')$ is a BG and r is a relation in $graph(G) \cap graph(G')$ all its neighbors must be within G and G'. Otherwise there are two edges (r, i, c) and (r, i, c') with $c \neq c'$; thus r has two i-ith neighbors in the union of G and H. Therefore, the intersection of G and G' is a BG. Reciprocally, let us suppose that $graph(G) \cap graph(G')$ is not a BG, even if for any node or edge x in $graph(G) \cap graph(G')$ $l(x) = l'(x)$. Then there is a relation r in $graph(G) \cap graph(G')$ that does not have all its neighbors in $graph(G) \cap graph(G')$, and the union of G and G' is not a BG because r does not fulfill the condition of BG. $\square$

Whenever the union operation is defined on a set of BGs it is a specialization operation. More precisely,

Property 8.12. If a BG G is equal to the union of a set of BGs $\{G_1, \ldots, G_k\}$, then G is a specialization of the extended disjoint sum $G'_1 + \ldots + G'_k$, such that for all $i \neq j$ G'_i and G'_j are disjoint and G_i and G'_i are isomorphic for every $i = 1, \ldots, k$.

Proof. By recurrence on k. The property is true if $k = 1$. Let us consider a BG G which is the union of $k + 1$ BGs $G_1, \ldots, G_{k+1}$ and let H denote the union of $G_1, \ldots, G_k$. H is a BG and it is a specialization of $G'_1 + \ldots + G'_k$. Let us consider the union of H and G_{k+1}. The intersection K of H and G_{k+1} is a BG K (cf. Property 8.11). Let us consider the disjoint sum $H + G'_{k+1}$. By joining the concepts in H and in G'_{k+1} corresponding to a concept in K, and by deleting twin relations, one obtains the union of H and G_{k+1} which is equal to G. $\square$

Property 8.13. Let T be a specialization tree of G with k leaves labeled by $H_1, \ldots,$ H_k. For all $i = 1, \ldots, k$ there is a homomorphism π_i from H_i to G such that $G = \cup_{1, \ldots, k} \pi_i(H_i)$.

Proof. For each $i = 1, \ldots, k$ H_i is a generalization of G, therefore there is a homomorphism π_i from H_i to G, and $\pi_i(H_i)$ is a subBG of G. Let us prove by recurrence on the number n of nodes of T that the union of these sub-BGs is equal to G. The property holds if $n = 1$. If G has only one predecessor H in T then $G = op(H)$, where op is one of the unary specialization operations. By the recurrence hypothesis, H is equal to $\cup_{1, \ldots, k} \pi'_i(H_i)$. One concludes by considering $\pi_i = \pi \circ \pi'_i$ where π is the surjective homomorphism associated with op as defined in property 8.9. If G is the disjoint sum of two BGs H and K, the set $\{H_1, \ldots, H_k\}$ can be partitioned into $\{K_1, \ldots, K_l\}$ and $\{M_1, \ldots, M_m\}$, and by the recurrence hypothesis, $H = \cup_{1, \ldots, l} \alpha_i(K_i)$ and $K = \cup_{1, \ldots, m} \beta_i(M_i)$. Then G is equal to the union of these two unions, which is precisely $\cup_{1, \ldots, k} \pi_i(H_i)$. $\square$

8.4.3 Inductive Definition of BGs

In this paragraph, BGs that can be obtained with specialization operations from a set of BGs are studied.

Definition 8.17 ($\mathcal{G}$-specialization). Let $\mathcal{G}$ be a set of BGs. A BG G is called a $\mathcal{G}$-specialization if it is obtained by a specialization tree whose leaves are all labeled by BGs in $\mathcal{G}$. Stated otherwise, the set of $\mathcal{G}$-specializations is the set of BGs inductively defined with:

- (basis) the set $\mathcal{G}$,
- (rules) the elementary specialization operations.

In Sect. 2.1.2 a star BG is defined as a BG restricted to a relation node and its neighbors (cf. Definition 2.3). Elementary star BGs are specific star BGs defined as follows:

Definition 8.18 (Elementary Star BG). The *elementary star BG* associated with a relation type r of arity k is a BG restricted to a relation node labeled by r and k neighbors, each labeled by $(\top, *)$.

The BGs on $\mathcal{V}$ can now be inductively defined as follows.

Theorem 8.2. *Let $\mathcal{G}_{\mathcal{V}}$ denote the set of elementary star BGs associated with a vocabulary $\mathcal{V}$ plus a BG, denoted $[\top]$, reduced to a single concept node labeled by $(\top, *)$. The set of $\mathcal{G}_{\mathcal{V}}$-specializations is equal to the set of (non-empty) BGs on the vocabulary $\mathcal{V}$.*

Proof. Any $\mathcal{G}_{\mathcal{V}}$-specialization is a BG on $\mathcal{V}$. Let us prove, by recurrence on the number of relation nodes, that if G is a BG on $\mathcal{V}$ then it is a $\mathcal{G}_{\mathcal{V}}$-specialization. If G has no relation node and k concept nodes x_i labeled by a_i, then G can be obtained from the BG $[\top]$ with $k-1$ disjoint sums, then restricting the labels of k nodes to $a_1, \ldots, a_k$. If a BG G has $k \geq 1$ relation nodes, then let us consider one relation node r. If H is obtained from G by deleting r, then H can be obtained by a $\mathcal{G}_{\mathcal{V}}$-specialization (recurrence hypothesis). Let us consider the subBG K of G induced by r. This BG can be obtained from the star BG associated to the type of r by a sequence of concept restrictions and joins. Finally, one makes the disjoint sum of H and K, and if (r, i, c) is in G, the node c in H is joined to the i-th neighbor of the relation in K. $\square$

Note that if the relation types are equipped with signatures (cf. Sect. 2.1.1), then the elementary star BG associated with the relation type r is the star BG having its relation node labeled r and its i-th neighbor, for $i = 1, \ldots, arity(r)$, labeled by $\sigma(r)$, i.e., the maximal concept type of the i-th neighbor of a relation node labeled r.

8.4.4 *G-Specializations*

G-specializations can occur in design problems, particularly in the synthesis or analysis of complex objects. Let us consider a box with k legs which may represent an abstraction of a simple physical object, e.g., a piece of a jigsaw, an electronic circuit, a mechanical device, an atom (see Fig. 8.15). The box type indicates the type of device, and leg types represent the connections of a box.

Let us assume that boxes can be glued together by means of legs for building complex objects. In Fig. 8.15, two boxes are represented as well as their gluing by identical colors.

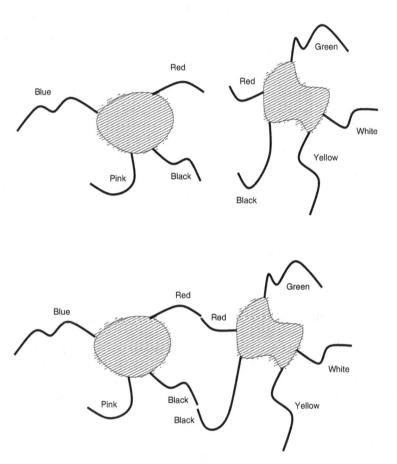

Fig. 8.15 Boxes with legs

A BG representation of boxes with legs in Fig. 8.15 is given Fig. 8.16. In this representation, colors are considered as individuals thus gluing by identical colors corresponds to normalization.

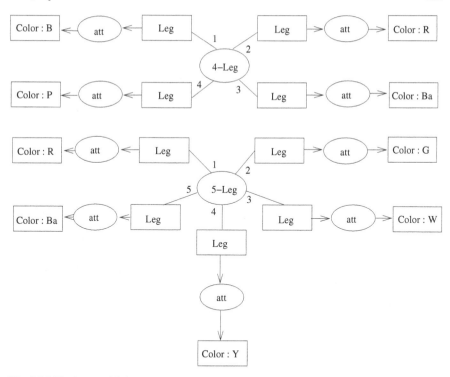

Fig. 8.16 The boxes with legs example represented as BGs

If the rules gathering boxes are elementary specialization rules (or can be defined by composition from specialization rules), then the complex objects obtainable from the boxes correspond to the set of BGs over the vocabulary consisting of the box and leg types (or a part of that BG set). Indeed, such a box with its legs can be represented by a star BG. If the elementary pieces from which more complex objects have to be built are not necessarily boxes but any set G of BGs, then the complex objects are *G*-specializations.

A BG that is a *G*-specialization is sometimes called a *canonical* BG with respect to the set G, which is called a *canon*.

The following property can be used to build an algorithm for recognizing a *G*-specialization, which is polynomial in the complexity of a BG homomorphism algorithm.

Theorem 8.3. *Let G be a set of BGs. A G-cover of a BG G is a set $\{G_1, \ldots, G_p\}$ of elements of G such that: $G = \cup_{i=1,\ldots,p} G_i$. If a BG G has no isolated concepts, then the three following properties are equivalent:*

1. G is a G-specialization
2. there is a G^α-cover of G
3. for all relations r of G there is a homomorphism π from a $B_i \in G$ to G with $r \in \pi(B_i)$

Proof.

- $1 \Rightarrow 2$

  Let us consider a $\mathcal{G}$-specialization G. There is a specialization tree T of G with all its leaves labeled with elements of $\mathcal{G}$. Thus, with property 8.9, $\pi(B_i)$ is an element of B_i^{α}, and one concludes with Property 8.13 in which the union is a $\mathcal{G}^{\alpha}$-cover.

- $2 \Rightarrow 3$

  Let us consider a BG G having a $\mathcal{G}^{\alpha}$-cover $\{G_1, \ldots, G_k\}$. Any relation r of G belongs to some G_i, a subBG of G which is an element of $\mathcal{G}^{\alpha}$. One concludes with Property 8.9.

- $3 \Rightarrow 1$

  Let $B_{r_1}, \ldots, B_{r_k}$ denote the elements of $\mathcal{G}$ associated with relations $r_1, \ldots, r_k$ of G, and π_i a homomorphism such that $r \in \pi_i(B_{r_i})$. Each B_{r_i} is specialized into $\pi_i(B_{r_i})$. Then one makes the union of these BGs, which is equal to G (as there are no isolated concepts in G, a cover of the relations is a cover of the whole BG). One concludes with property 8.12.

  $\square$

The third characterization of a $\mathcal{G}$-specialization in Theorem 8.3 leads to the following CANONICAL algorithm for $\mathcal{G}$-specialization recognition.

Let us assume that there is a function $HOM(B_i, G, r)$, where $B_i \in \mathcal{G}$, G is a BG and r is a relation node in G, which returns $\emptyset$ if there is no homomorphism from B_i to G that maps a relation node in B_i to r, otherwise it returns such a homomorphism π (and one says that "π covers r").

Algorithm 24: CANONICAL (G is a BG, $\mathcal{G}$ is a canonical base)

Input: a BG G not reduced to a single concept node
Output: returns True if G is a $\mathcal{G}$-specialization, otherwise False
begin

    R is the relation node set of G
    $Rel \leftarrow R$
    // relation nodes not yet covered
    while $Rel \neq \emptyset$ **do**
        $covered \leftarrow$ False
        Let $r \in Rel$
        forall $B_i \in \mathcal{G}$ **do**
            $\pi \leftarrow HOM(B_i, G, r)$
            if $\pi \neq \emptyset$ **then**
                $Rel \leftarrow Rel \setminus \{$relation nodes appearing in $\pi\}$
                $covered \leftarrow$ True
                Exit(For loop)
        if $\neg covered$ **then**
            return *False*
    return *True*
end

Initially, *Rel* contains all relation nodes in G. The algorithm iteratively tries to cover a relation node r that has not already been covered (by a graph B_i). Whenever a homomorphism covering r is found, then other relation nodes can be covered by the same homomorphism and all these nodes are removed from *Rel*. If r cannot be covered, the algorithm returns `False`. If *Rel* becomes empty, all relation nodes have been covered and the algorithm returns `True`.

The time complexity of CANONICAL can be roughly bounded by $|R| \times |\mathcal{G}| \times$ *hom*, where *hom* is the maximum complexity of computing a homomorphism from a BG to another BG that covers a specific relation node. For known algorithms, when computing a homomorphism from a BG to another BG the covering condition does not increase the complexity. Since the problem of a homomorphism between two graphs is NP-Complete (cf. Sect. 5.2), CANONICAL is exponential here. Nevertheless, one important point is that the complexity of CANONICAL is polynomially related to the complexity of *hom*. Thus, each time *hom* is polynomial, so is CANONICAL.

8.5 Type Expansion and Contraction

The concept and relation types considered so far are primitive types. They are not explicitly defined, and their meaning is only given by their position in the type hierarchies and their possible occurrences in relation signatures, rules and other sorts of knowledge.

If a type t is less than a type t', this means that every entity of type t is also of type t', but nothing is explicitly said about the properties of entities of type t, nor about what distinguishes a t from any t'. Two kinds of type definitions have been proposed for conceptual graphs: By necessary and sufficient conditions, or with a set of prototypes representing typical instances of the type. With the first approach, an entity is of type t if and only if it fulfills the necessary and sufficient conditions. With the second approach, an entity is of type t if it is sufficiently similar to one of this prototype (We find here one application of the maximal join operations: The similarity can be related to the ratio of nodes in both graphs that are matched by a maximal join). The logical translation of the first kind of definition is simply a logical equivalence, while this is not the case for the second kind of definition.

In this section, we consider type definitions by necessary and sufficient conditions. Note that we shall not conduct an in-depth study of type definitions. Studying how to classify a defined type, i.e., how to find its position in the concept type hierarchy, or extending the specialization/generalization relation between graphs and studying its relationships with FOL semantics, studying recursive type definitions and so on, are beyond the scope of this section. Briefly, in this section we do not extend the conceptual graph model presented in this book with type definitions, we only consider expansion and contraction operations of a defined type.

Furthermore, we describe these operations for concept type definitions only. Similar operations can be provided for relation type definitions (see the bibliographical notes). Finally, let us point out that we consider a classical BG vocabulary, thus without conjunctive types.

Definition 8.19 (Concept type definition). Let $\mathcal{V}$ be a BG vocabulary and t be a symbol which does not belong to $\mathcal{V}$. A *concept type definition* of t is a unary λ-BG $(c)D_t$, where D_t is connected and c is called the *head* of D_t. Such a type definition is noted $t = (c)D_t$.

Example. Assume, for instance, that the vocabulary contains the concept type *Woman* and the relation type *childOf*. One wants to define a new concept type *Mother*, by the λ-graph $(c)G$ in Fig. 8.17. This definition says that an entity m is of type *Mother* if and only if m is of type *Woman* and has a child.

Intuitively, a concept type definition is a necessary and sufficient condition for an entity to belong to this defined type. A definition $t = (c)D_t$ can be considered as an Aristotelician type definition by genus and difference. The genus of t is the type of c; its difference is what completes c in D_t. Logically, the definition $t = (c)D_t$ is naturally translated by the formula $\forall x(t(x) \leftrightarrow \Phi((c)D_t))$, with the term assigned to c in $\Phi((c)D_t)$ being x. For instance, the definition of the concept type Mother (Fig. 8.17) is translated into $\forall x(Mother(x) \leftrightarrow \exists y(Woman(x) \wedge Person(y) \wedge childOf(y,x)))$.

Even if one does not describe the modification of the ordered type set due to the addition of a defined type, let us assume that, in the enriched vocabulary, the defined type t is $\leq$ the type of c (and in general t might be less than other types which are specializations of, or incomparable with, the type of c).

Fig. 8.17 Definition graph of the concept type *Mother*

In the same way, a relation type definition is defined as follows.

Definition 8.20 (Relation type definition). Let $\mathcal{V}$ be a BG vocabulary, r be a symbol that does not belong to $\mathcal{V}$ and k an integer ≥ 1. A *k-ary relation type definition* of r is a k-ary λ-BG $(c_1,\ldots,c_k)D_r$.

For instance, if the vocabulary contains the relation types *parentOf* and *sibling*, then the binary relation type *cousin* can be defined by the λ-graph $(c_1,c_2)G$ in Fig. 8.18. A definition of a relation type of arity k is a necessary and sufficient condition for k entities to be linked by this defined relation type, e.g., the entities represented by c_1 and c_2 are cousins if and only if they have parents who are siblings. It is translated into logics in the same way as a concept type definition: The formula assigned to the definition $r = (c_1,\ldots,c_k)D_r$ is $\forall x_1 \ldots x_k(r(x_1 \ldots x_k) \leftrightarrow \Phi(((c_1,\ldots,c_k)D_r)))$, where the variables assigned to $c_1,\ldots,c_k$ are $x_1,\ldots,x_k$ respectively. For instance, the definition of the concept type cousin (Fig. 8.18) is translated into $\forall x_1 \forall x_2(cousin(x_1,x_2) \leftrightarrow \exists y \exists z(Person(x_1) \wedge Person(x_2) \wedge Person(y) \wedge Person(z) \wedge parentOf(y,x_1) \wedge parentOf(z,x_2) \wedge sibling(y,z)))$.

Note that *cousin* and *sibling* are both symmetrical relations, which can be expressed by BG rules (cf. Chap. 10).

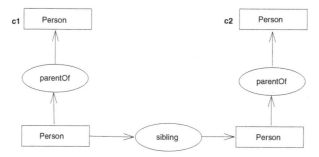

Fig. 8.18 Definition graph of the relation type *cousin*

The expansion of a concept type definition consists of replacing a concept node with a defined type by the graph defining the type. More precisely:

Definition 8.21 (Concept type expansion). Let G be a graph containing a concept node x with a defined type $t = (c)D_t$. The expansion of t at x is the BG $exp(G, x, D_t)$ obtained by merging x and c, the label of the new node being (t', m) where t' is the type of c and m is the marker of x.

An example is given in Fig. 8.19.

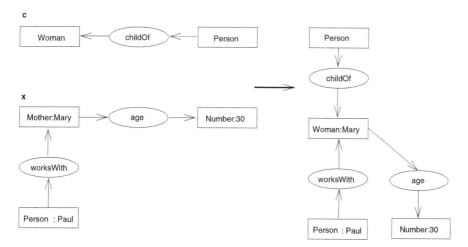

Fig. 8.19 A concept type expansion

Let us point out the relationships with rules. From a logical viewpoint, a concept type definition $t = (c)D_t$ can be considered as equivalent to two specific BG rules (and more precisely, λ-rules, cf. Chap. 10), say R_1 and R_2. The hypothesis of the first rule is a single generic concept node labeled t, its conclusion is D_t, and the node in the hypothesis is in correspondence with the head of D_t; the second rule is the reciprocal rule. See for instance Fig. 8.20, which shows both rules corresponding to the definition of the concept type *Mother*.

At first glance, a type expansion looks like an application of the rule R_1. However, there are two differences. First, R_1 can be applied to concept nodes with a type less than t, whereas an expansion can occur only if the concept node has exactly the type t. Secondly, a rule application cannot generalize the type of an existing node.

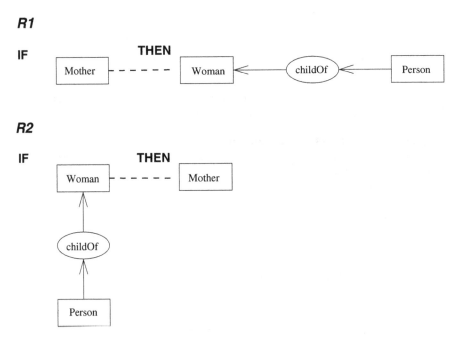

Fig. 8.20 Rules associated with the definition of the concept type Mother

A concept type expansion may add redundant information. This added redundant information can be avoided by a more complex type expansion operation relying on the piece notion that is defined hereafter. This notion will be generalized in Chap. 10, where pieces are used in backward chaining of rules.

Definition 8.22 (Piece). Let G be a BG and let x be a concept node in G. Two nodes u and v in G belong to the same *piece* of x if they are in the connected component of G containing x and if there is a chain between them that does not go through x, i.e., there is a chain $w_0(= u), \ldots, w_i, \ldots, w_k(= v)$ such that $w_i \neq x$ for all $i = 1, \ldots, k-1$.

Note that x belongs to all pieces of x and that x has more than one piece if and only if x is a cut node of G (i.e., the deletion of x strictly increases the number of connected components).

The pieces of x can be built as follows. Let H be the connected component of G containing x, the connected components $H_1,\ldots,H_k$ of the graph $H - x$ obtained from H by deleting x are computed. Then, for $i = 1,\ldots,k$ the piece C_i of x is the BG obtained from H_i by adding x and all edges in H joining x to a node in H_i. If $k = 1$ the single piece is H itself.

The BG $exp(G,x,D_t)$ resulting from a concept type expansion can be reduced, i.e., replaced by an equivalent BG having fewer nodes, by using pieces.

Definition 8.23 (Extended concept type expansion). The *extended concept type expansion* of a defined type $t = (c)D_t$ at x in G is the BG $extexp(G,x,D_t)$ obtained from $exp(G,x,D_t)$, where x still denote the node resulting of the merging of x and c, as follows. Let $C_1,\ldots,C_k$ be the pieces of x in G and let $D_1,\ldots,D_l$ be the pieces of c in D_t. For all $i = 1,\ldots,k$ and $j = 1,\ldots,l$, if D_j maps to C_i with c mapped to x, then D_j is deleted from $exp(G,x,D_t)$.

In Fig. 8.21, G is a BG and D_t is the definition of the type t of x in G, all concept nodes are assumed to be generic, and the type t'' of z in G is less than the type t' of c (i.e., the head of D_t) and t. D_t and G are restricted to a single piece with respect to c and x respectively. G does not map to D_t and D_t does not map to G with c being mapped to x. Thus, $extexp(G,x,D_t)$ is equal to $exp(G,x,D_t)$. One can also check that G and D_t are irredundant. Nevertheless, $extexp(G,x,D_t)$ is redundant. Indeed, by folding z onto xc, $extexp(G,x,D_t)$ is transformed into K, which is equivalent to the subBG H of G.

Note that the reduction used in the extended concept type expansion can also be done after any external join. Figure 8.21 can be used to show (consider that D_t is a BG and not necessarily a type definition, i.e., x and c have the same labels) that after such a reduction the obtained BG can be redundant even though the two joined BGs are irredundant.

Let us now consider the type contraction operation, which can be more or less considered as inverse to the expansion. In order to define this contraction operation, we use the fact that x is a cut node of $exp(G,x,D_t)$, i.e., if x is deleted from $exp(G,x,D_t)$, two disjoint (not necessarily connected) graphs are obtained, which correspond to $D_t - c$ and $G - x$.

Let us assume that a graph G contains a subBG D isomorphic to a concept type definition graph D_t. If D is shrunk into a single node, then all neighbors of the nodes in D that are outside D are linked to the new node resulting from the shrinking of D. This can add irrelevant information, as shown in Fig. 8.22.

In contracting a subgraph corresponding to a type definition, one does not want to add irrelevant information, but one also does not want to lose relevant information. For instance, if one knows that Paul is a child of the woman Mary, i.e., if G in Fig. 8.23 is a fact and if we simply contract the definition of *Mother*, one obtains graph H and the fact that Paul is a child of Mary is lost.

To avoid information loss in type contraction, the parts that strictly specialize the corresponding parts in the type definition are kept. More precisely,

Definition 8.24 (Concept type contraction). Let G be a BG and D_t be a concept type definition, with head c. Let π be a homomorphism from D_t to G, such that c

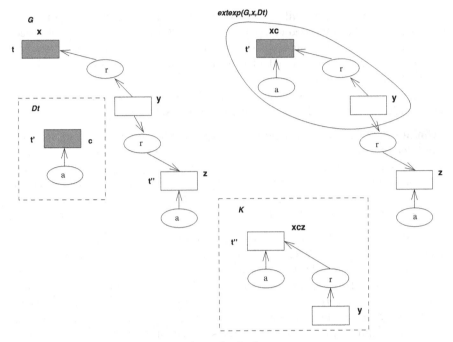

Fig. 8.21 Extended concept type expansion and redundancy

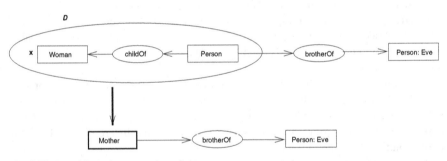

Fig. 8.22 A problematic contraction of a concept type

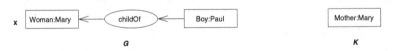

Fig. 8.23 A concept type contraction with information loss

and $\pi(c)$ have the same type. Let $x = \pi(c)$. The BG $contract(G,x,D_t)$ is obtained from G by replacing the type of x with t and, for each piece P of x in G that maps to a piece of c in D_t (without considering the label of x), deleting $P - x$.

An example is given Fig. 8.24, where a *4-Wheel* is a concept type less than the concept type *BigCar*.

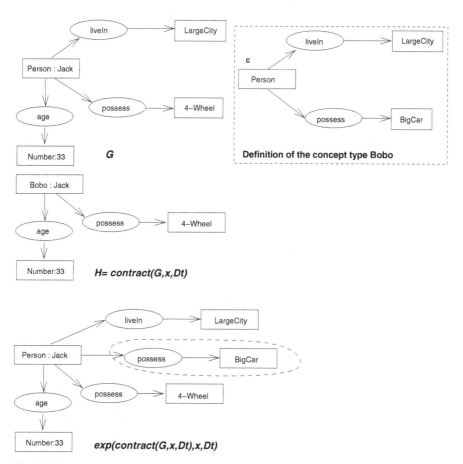

Fig. 8.24 Concept type contraction

Note that $contract(exp(G,x,D_t),x,D_t) = G$. Indeed, the type of x in $exp(G,x,D_t)$ is t and it is unchanged by the contraction. Furthermore, the homomorphism is a BG isomorphism, thus no subBG is added by the contraction. When first computing a contraction then an expansion, i.e., $exp(contract(G,x,D_t),x,D_t)$, the graph obtained is not necessarily equal to G (an example is given in Fig. 8.24 when the type Bobo is expanded in H). Nevertheless, $exp(contract(G,x,D_t),x,D_t)$ is hom-equivalent to G since the deleted pieces in G are redundant (they map to pieces in D_t) as well as the pieces of D_t which are not in G.

Finally, let us point out that the concept type expansion and contraction operations, as defined in this chapter, do not change the BG semantics. More precisely, let G be a BG and G' be obtained from G by a type expansion or contraction. Let f be the formula translating the type definition involved in this operation. Then $\Phi(\mathcal{V}), f \models (\Phi(G) \leftrightarrow \Phi(G'))$.

Like the concept type definition, a relation type definition can be considered as equivalent to two reciprocal BG rules and similar relation type expansions and contractions can be defined.

8.6 Bibliographic Notes

In [Sow84] a maximal join operation is proposed that was implemented in the system presented in [SW86]. The definition of a maximal join between two graphs G and H in [Sow84] relies on the notion of a common generalization of G and H, say K, and "compatible" homomorphisms from K to G and H respectively, say π_1 and π_2. The join is performed according to these homomorphisms by pairwise merging each node $\pi_1(x)$ and $\pi_2(x)$ for each node x of K. However, the definition of compatibility (Definition 3.5.6 in [Sow84]) and the associated result (Theorem 3.5.7 in [Sow84]) work only if the homomorphisms are *injective* on the concept node set of K. Then, there is a bijection between the concept nodes of the joined subgraphs of G and H. Maximal joins have been used especially in natural language processing (e.g., cf. [FLDC86]).

Compatible partitions were introduced in [CM92]. Canonical conceptual graphs were introduced in [Sow84], and the characterization Theorem 8.3 as well as the recognition algorithm of canonical graphs were given in [MC93].

Type definitions as well as type expansion and contraction operations were introduced by Sowa [Sow84]. Two kinds of type expansion are defined: the minimal type expansion, which is the same as ours, and the maximal type expansion, which essentially extends the minimal type expansion with a maximal join. This second operation aims at avoiding the creation of redundant parts; but contrary to our extended type expansion it does not preserve the semantics and might strictly specialize the graph. We kept the overall idea of the type contraction operation proposed by Sowa and precisely define it with graph operations.

Although type definitions were introduced as early as 1984, their processing received little attention. Let us mention the work of Leclère (cf. [Lec95], [Lec97] and [Lec98]) extending the framework of BGs to concept and relation type definitions. In this framework, recursive type definitions (direct or indirect) are forbidden and a primitive type cannot be specialized by a defined type. The specialization relation between BGs is extended by the type expansion and contraction operations, and it is shown that soundness and completeness properties are kept, i.e., given two BGs G and H defined on a vocabulary $\mathcal{V}$, $G \succeq norm(H)$ if and only if $\Phi(\mathcal{V}), \Phi(H) \models \Phi(G)$, where $\Phi(\mathcal{V})$ is the set of formulas associated with the enriched vocabulary, which now includes formulas translating type definitions. A unique expanded form obtained by performing type expansion until stability is associated with each BG. The expanded form is a BG where all types are primitive (i.e., not defined). Thus two expanded BGs can be compared by homomorphism. Let us add that to the best of our knowledge type definitions have never been studied in conceptual graphs beyond the BG model.

Part III
Extensions

Chapter 9
Nested Conceptual Graphs

Overview

The nested conceptual graph model presented in this chapter is a direct extension of basic or simple conceptual graphs able to represent notions such as internal and external information, zooming, partial description of an entity, or specific contexts. This model also allows reasoning while taking a tree hierarchical structuring of knowledge into account. Nestings are represented by boxes. A box is an SG and, more generally, a box is a typed SG. In full conceptual graphs, a box represents the negation of the graph inside the box. Thus, for differentiating these negation boxes from the boxes used in this chapter, these boxes are usually called "positive" boxes. Nevertheless, since the only kind of boxes considered hereafter are positive boxes, we omit the term "positive."

In Sect. 9.1 different notions representable by nested conceptual graphs are presented. In Sect. 9.2, we introduce *Nested Basic Conceptual Graphs (NBGs)*, whose boxes consist of BGs. *Nested Conceptual Graphs (NGs)*, which extend NBGs with coreference links, are presented in Sect. 9.3. Coreference links can relate concept nodes of the same box (thus boxes become SGs) but also of different boxes. Coreference links in nested graphs are more difficult to manage than in simple graphs. Indeed, since boxes can represent contexts, it is generally irrelevant to merge all nodes of a coreference class into a single node. In Sect. 9.4, we define *graph types*, *typed SGs*, which are SGs with a graph type, and *Nested Typed Conceptual Graphs (NTGs)*, which generalize NBGs by typing the boxes. All of these nested graphs classes are provided with homomorphism. The FOL semantics Φ introduced for SGs is generalized to NTGs in Sect. 9.5 and a homomorphism soundness and completeness theorem is stated. As this semantics is a formula of the positive, conjunctive and existential fragment of FOL, nested and non-nested CGs are somewhat equivalent. Finally, we build a mapping *ng2bg* from nested to non-nested CGs which preserves homomorphisms. This mapping *ng2bg* shows, in another way than through logical semantics, that nested and non-nested CGs have the same descriptive power. It is easy to implement *ng2bg* and this avoids the construction of specific nested graph homomorphism algorithms.

Nevertheless, from a user viewpoint, NTGs are interesting whenever knowledge is intrinsically hierarchical, and when reasonings must follow the hierarchical structure, because in an NTG the hierarchy is explicitly and graphically represented. Nested graphs can also be interesting whenever large graphs have to be manually constructed, as the separation of levels of reasoning increases efficiency and clarity when extracting information.

9.1 Introduction

Let us give a flavor of different knowledge representation situations which are relevant to the conceptual graph model presented in this chapter. Consider, for instance, representing information about a cottage. It is possible to distinguish *internal* from *external* pieces of information about this cottage. The owner's name can be considered as an external piece of information concerning the cottage, whereas the distribution of rooms, the plan of the cottage, is internal information. In information retrieval, the ISBN number of a book can be considered as an external piece of information about that book, whereas the subject of the book is an internal piece of information. In these examples, the internal information can also be called description (of the cottage or book), and more precisely *partial description* of the given entity.

Zooming is a related notion. Let us again consider the cottage example. Having the land registry position of the cottage, one may want to zoom into the cottage, e.g., to determine the cottage plan, which can be considered as internal information about the cottage. Or, at a deeper level, having the cottage plan, one may want to know the furniture distribution in a specific room. In the book example, zooming can consist of obtaining the content of a chapter from a reference to this chapter (e.g., from the number of this chapter).

The knowledge model presented in this chapter can also be related to the *context* notion. An *informational context* can be defined as the (cultural, historical, social, geographical, etc.) surroundings or circumstances of a piece of information that are important to understand the meaning of this piece of information. For instance, in the previous cottage example, the context of the furniture distribution (in a room) is precisely the room having this furniture distribution, the context of the content of a book chapter is precisely this chapter, and so on.

The notion of context is close to the two previous notions of external versus internal pieces of information, and to zooming. Indeed, zooming takes its full meaning when one knows the origin of the zooming, and an internal piece of information takes its full meaning when this origin, which can also be an internal piece of information, is known. In the forthcoming model, only very simple contexts can be represented. Indeed, a context will be represented by a path of (nested) concept nodes. Thus, considering partial description or zooming as an intuitive meaning of what we aim to represent, seems as relevant as the context.

A partial description can be included in another partial description, thus a recursive model is proposed, and the entities are represented by a hierarchical structure. Similarly to the classical notion of "boxes within boxes" used in document processing, we define a model that consists of "graphs within graphs." Briefly said, our model consists of rooted trees of (typed) SGs, and reasoning is based on SG homomorphisms respecting the tree structure.

9.2 Nested Basic Graphs (NBGs)

Let us consider the graph G in Fig. 9.1, expressing that "a drawing has been made by the boy Paul." Suppose one wants to add two pieces of information to G. First,

Fig. 9.1 A basic conceptual graph

"the drawing is on a table," and secondly "the drawing represents a green-coloured train." These pieces of information can be considered to be of different sorts. The first one can be considered as *external* information about the drawing, i.e., it is a fact concerning the drawing taken as a whole, i.e., a black box, with what is drawn being irrelevant. On the other hand, the fact that a green train is represented on this drawing can be considered as *internal* information, or a *(partial) description*, of the drawing itself. If one wants to build a representation of these facts while keeping this difference of status, then one can add "the drawing is on a table" at the same level as G, and the fact that the drawing represents a green train can be put inside the node representing the drawing. This latter piece of information can be obtained by zooming on the drawing node, which is then considered as a glass box. After adding the fact that the drawing in Fig. 9.1 is on a table, and by zooming on the node "Drawing," the nested graph NG represented Fig. 9.2 is obtained. It can also be said that the piece of information nested in a node is relevant within the context represented by this node. Thus, the context is represented by a concept node, or more precisely by a path of concept nodes.

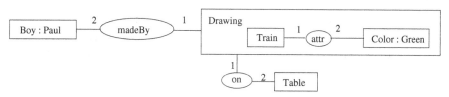

Fig. 9.2 A nested conceptual graph

Drawings of nested graphs can be quite difficult to read. A dynamic device, e.g., a graphical screen, allows nice vizualization of nested graphs by travelling level by level within the graph, i.e., by zooming in and out. For instance, for representing the nested graph in Fig. 9.2, a first image can show the graph without the description of the drawing, a graphical mark indicating that the node corresponding to the drawing has a non-empty description (as in Fig. 9.3). By zooming on a node with a non-

Fig. 9.3 A nested conceptual graph before zooming

empty description, the description graph appears, and the rest of the graph is shaded off (as in Fig. 9.4).

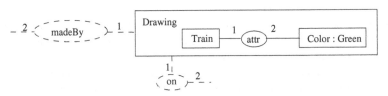

Fig. 9.4 A nested conceptual graph after zooming

We hereafter propose two equivalent definitions of nested graphs corresponding to different viewpoints of the same model. The first definition is a recursive definition obtained by adding a third field, called a *description*, to the concept node labels of a basic conceptual graph (BG), and in the second definition the recursive facet is explicitly represented by a tree. The first definition is well fitted for a graphical user interface since it corresponds to the zoom viewpoint. The second definition facilitates the drawing of nested graphs on a sheet of paper because a nested graph is basically a tree of SGs.

Definition 9.1 (Nested Basic Graph NBG).

- An *elementary* NBG G is obtained from a normal BG H by adding a third field to the label of each concept node c equal to $**$. The set of boxes of G is $boxes(G) = \{H\}$ and the complete concept node set of G is $X_G = C_H$. A trivial bijection exists between elementary NBGs and normal BGs (when no ambiguity occurs we do not distinguish between them).
- Let H be an NBG and D an elementary NBG. The graph G obtained from H by substituting D for the third field $**$ of a concept node c in X_H is an NBG. $boxes(G) = boxes(H) \cup boxes(D)$, and $X_G = X_H \cup X_D$.

- The third field of any concept node $c \in X_G$ is called the *description* of c and is denoted $Descr(c)$.

A BG appearing in the construction of an NBG G, i.e., an element of $boxes(G)$, is called a *box* of G.

An NBG is denoted $G = (C_G, R_G, E_G, l_G)$, where C_G, R_G, E_G are respectively the concept, relation and edge sets of the first elementary NBG, denoted $root(G)$, used in the construction of G. l_G is the labeling function obtained from the labeling function of $root(G)$ by adding a third field with value $Descr(c)$ to the labeling of each concept node $\in C_G$.

We distinguish the set C_G of concept nodes of an NBG G—i.e., the set of nodes of $root(G)$—from the set X_G of all concept nodes appearing in G, i.e., the set of concept nodes of all boxes of G.

We explain in Sect. 9.3 why we consider normal BGs in the previous definition. Recall that we consider normal BGs and normal SGs as identical objects (cf. Sect. 3.5).

It is important to note (for the forthcoming definitions of $Tree(G)$ and X_G) that if a BG or an NBG K is used several times in the construction of a NBG G, we consider that several copies of K (and not several times the graph K itself) are used in the construction of G. Note also that a description is not a type definition (cf. Chap. 8). First, a description applies to a specific concept node, whereas a graph defining a type applies to all concept nodes of this type. Secondly, a type definition provides a characterization of a type, i.e., it describes necessary and sufficient conditions for any object to belong to the type, whereas a description is only a partial information about an object.

Any NBG G has an associated tree $Tree(G)$ whose nodes are in bijection with the boxes of G. Before defining $Tree(G)$, let us introduce the notion of a tree of BGs.

Definition 9.2 (Tree of BGs). A *tree of BGs* is a labeled rooted tree $T = (V_T, U_T, l_T)$, such that:

- V_T, the node set of T, is in bijection with a set of pairwise disjoint normal BGs $\{G_1, \ldots, G_k\}$. For any $i = 1, \ldots, k$, the node associated with G_i is labeled G_i (the set $\{G_1, \ldots, G_k\}$ can be identified with V_T).
- For any arc (G_i, G_j) in U_T, $l_T(G_i, G_j)$ is a concept node c in G_i, and such a labeled arc is also denoted (G_i, c, G_j).
- All labels are distinct, i.e. a concept node c appears at most once as an arc label.

The mapping $Tree$ that assigns a tree of BGs to any NBG is defined as follows.

Definition 9.3 (Tree(G)). Let G be an NBG, the mapping $Tree$ is defined as follows. If G is an elementary NBG, then $Tree(G)$ is restricted to a single node labeled G. If G is the NBG obtained from a NBG H by substituting D to $**$, which is the $Descr$ of the concept node c in H, then $Tree(G)$ is built from $Tree(H)$ by adding a node labeled D successor of the node labeled K, and containing c, in $Tree(H)$. The label of (K, D) is c.

The root of $Tree(G)$ is $root(G)$. Note that for any NBG G, (J,K) is an arc in $Tree(G)$ labeled c if and only if J and K are nodes in $Tree(G)$ (i.e. are boxes of G), c is a concept node in J, and K is the label of the root of the tree associated with $Descr(c)$, i.e., the description graph of c. Otherwise said, the existence of an arc (J,c,K) in $Tree(G)$ expresses the fact that: "J and K are two boxes of G, c is a concept node in J, and K is the label of the root of $Tree(Descr(c))$."

Example. The boxes and tree of the NBG in Fig. 9.2 are represented in Fig. 9.5.

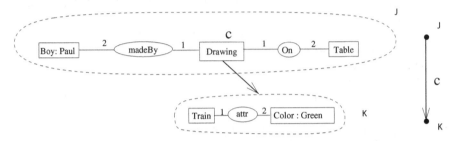

Fig. 9.5 The "tree of boxes" view of the graph in Fig. 9.2

Figure 9.6 presents a more complex example of an NBG. Its boxes are represented in Fig. 9.7, where G_1 is the root, $G_2 = Descr(d)$, $G_3 = Descr(c)$, $G_4 = Descr(e)$, $G_5 = Descr(f)$, $G_6 = Descr(i)$, $G_7 = Descr(h)$, $G_8 = Descr(l)$, $G_9 = Descr(n)$, and its tree is represented in Fig. 9.8.

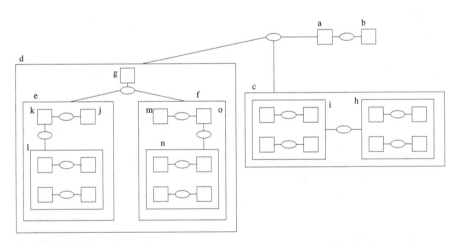

Fig. 9.6 An NBG

The following property is immediate:

Property 9.1. The mapping *Tree* is a bijection from NBGs over a vocabulary $\mathcal{V}$ to the trees of BGs over the same vocabulary $\mathcal{V}$.

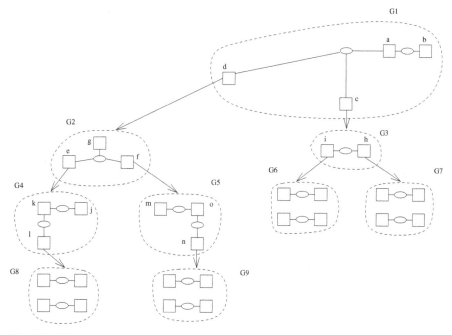

Fig. 9.7 The boxes of the graph in Fig. 9.6

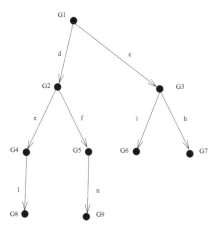

Fig. 9.8 The tree of the graph in Fig. 9.6

Thus, Definition 9.1 and Definition 9.2 can be indiscriminately used.

Definition 9.4 (Depth).

- The *depth* of an NBG G is the depth of $Tree(G)$, i.e., the maximum number of arcs of a path beginning at the root.
- The *depth* of a node x in a NBG G is the depth of the vertex of $Tree(G)$ to which it belongs, i.e., if x is in $root(G)$ then $depth(x) = 0$, and if x is in K and (J, c, K) is an arc of $Tree(G)$, then $depth(x) = depth(J) + 1$.

For instance, in Fig. 9.6 the depth of the nodes m and k is 2, whereas the depth of f and h is 1, and the depth of d and a is 0 (see also Fig. 9.7, where the depth of a node in a box is the number of arcs of a path from the root to the box containing that node).

A *complex concept node* is a node c in X_G with a non-empty $Descr(c)$, i.e., different from $**$. Such a node c is also called a context, and more precisely it is called *the context of $Descr(c)$*.concept!complex

A homomorphism from an NBG G to an NBG H is defined using the tree definition of NBGs. It is composed of an ordinary tree homomorphism π_0 from the rooted tree $Tree(G)$ to the rooted tree $Tree(H)$, and by a set of (BG) homomorphisms π_i from any box G_i of G to the box $\pi_0(G_i)$ of H.

Definition 9.5 (NBG homomorphism). Let G and H be NBGs and $Tree(G) = (V_G, U_G, l_G)$ and $Tree(H) = (V_H, U_H, l_H)$ be their associated rooted trees. An *NBG homomorphism* from G to H is a pair $\pi = (\pi_0, (\pi_{G_1}, \dots, \pi_{G_i}))$, where $V_G = \{G_1, \dots, G_l\}$, which satisfies:

- π_0 is a (tree) homomorphism from $Tree(G)$ to $Tree(H)$, i.e., a mapping from V_G to V_H, such that, if (J, K) is in U_G, then $(\pi_0(J), \pi_0(K))$ is in U_H, and $\pi_0(root(G)) = root(H)$,
- $\forall K \in V_G, \pi_K$ is a (BG) homomorphism from (the BG) K to (the BG) $\pi_0(K)$,
- if $l_G(J, K) = c$, then $l_H(\pi_0(J), \pi_0(K)) = \pi_J(c)$.

An example is given Fig. 9.9, where each arrow represents a homomorphism from the BG origin of the arrow to the BG extremity of the arrow.

A homomorphism π from an NBG G to an NBG H naturally induces a mapping (also noted π for simplicity) from the set of all nodes appearing in G to the set of all nodes appearing in H. Let x be a node in a box K of G, $\pi(x) = \pi_K(x)$.

One can extend the homomorphism definition by removing the condition that the root of the source tree has for image the root of the target tree. This extended definition is a bit more general since the root of the first tree can be mapped to any node of the target tree. In this case, only the relative depth is respected, more precisely, if the root of $Tree(G)$ is mapped to a box of depth k in $Tree(H)$ then, for any concept c of G, $depth(\pi(c)) = depth(c) + k$. Let us give an example.

Example. In Fig. 9.10, G_1 and G_2 are two facts and Q_1, Q_2 and Q_3 are three queries. Q_1 represents the question "Is there a train in the context of a thing?" Q_2 represents the question "Is there a train?" and Q_3 represents the question "Is there a train and

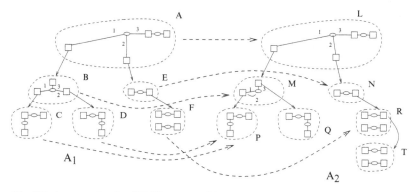

Fig. 9.9 An example of an NBG homomorphism

a boy?" With the definition 9.5, G_1 answers NO to Q_1, YES to Q_2 and YES to Q_3, and G_2 answers YES to Q_1, NO to Q_2 and NO to Q_3. With the extended definition, G_1 answers NO to Q_1, YES to Q_2 and YES to Q_3, and G_2 answers YES to Q_1, YES to Q_2 and NO to Q_3. Both definitions disagree on the answer given by G_2 to Q_2: with the first definition, the answer is NO because there is no train at the first level, and with the extended definition it is YES, and it could be added "in the context of a drawing."

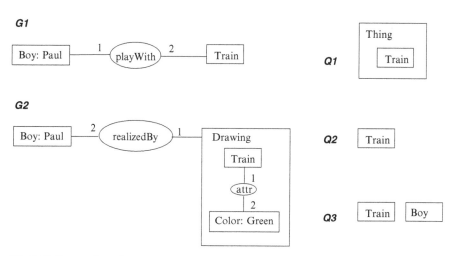

Fig. 9.10 Facts and queries

As NBG homomorphisms preserve the (relative) depth of the nodes, they allow only simple forms of reasoning. Other kinds of reasoning would be useful, especially reasonings mixing knowledge from different levels. Such reasonings can be defined by first defining transformations of graphs, then by using NG homomorphisms. The decision homomorphism problem for NBGs, NBG-HOMOMORPHISM, is NP-complete (with a BG being a particular case of NBG), and an algorithm for the NBG-HOMOMORPHISM polynomial in the complexity of an algorithm for BG-HOMOMORPHISM can be simply constructed since computing a homomorphism from a tree to a tree (or more generally to any graph) is a polynomial problem (cf. Sect. 7.2.1).

9.3 Nested Graphs (NGs)

In order to express that different nodes appearing in an NBG represent the same entity, one can add coreferences to NBGs, as we did for for BGs (cf. Definition 3.10 in Chap. 3).

The following piece of information about a scientific document that is "an article, whose subject is a wheat food product that is cooked in water, has a result, whose nutritional observation is that the vitamin content of this wheat food product decreases, whose biochemical explanation is that this wheat food product contains hydrosoluble vitamin that is dissolved, and whose nutritional evaluation is that the nutritional quality of this wheat product is deteriorated" can be represented by the graph in Fig. 9.11. In this figure, the coreferent concept nodes are represented by the named variable *x.

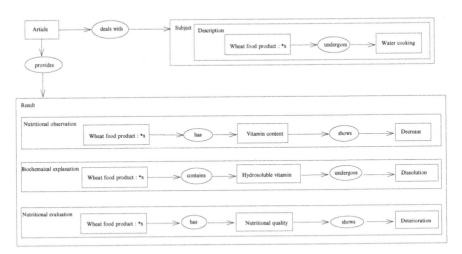

Fig. 9.11 A nested graph with coreferences represented by variables

An abstract example of a nested graph is represented in Fig. 9.12, where the nodes j, m, p, w are coreferent along with the nodes q, r. In this example, the coreference relation is represented by coreference links.

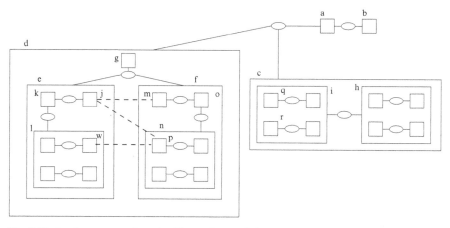

Fig. 9.12 An abstract nested graph with coreference links

Two coreferent concept nodes represent the same entity in nested graphs as in SGs. In SGs we have seen how it is possible to shrink each coreference class into a single node, thus obtaining a graph in normal form. The situation is more complicated for nested graphs. Knowledge is contextualized (by a hierarchical structure) in a nested graph, and merging two coreferent concept nodes can be done only if the contextualization is preserved. Two cases can be considered.

First, let us consider two coreferent concept nodes, say a and b, which are in the same box. These nodes may have descriptions. But, as a and b both represent the same entity (they are coreferent) in the same context (they are in the same box), one can merge a and b into a single node without altering the meaning of the graph. The description of the new node is the disjoint sum of the description of a and of the description of b.

Secondly, let us consider two coreferent concept nodes, say c and d, which are *not* in the same box, thus they are a priori *not* in the same context. The information about an entity in a context may be irrelevant in another context and merging two coreferent concept nodes which are in distinct contexts could entail inconsistencies. For instance, in Fig. 9.13, the concept nodes c and d are coreferent. They represent the same boy Paul but in two different contexts. Merging the two descriptions would state that Paul is dressed up as Zorro while this is stated only on the drawing. Nevertheless, if two coreferent nodes c and d are in the description of a and b, respectively, and a and b are coreferent nodes belonging to the same box, then after merging a and b, c and d become coreferent in the same box. In this case, they can be safely merged.

The simplest way of differentiating coreference links within a box and coreference links between two boxes is to stress that there is no coreference link within a box, i.e., each box is a normal SG. Hence the following definition:

Definition 9.6 (Nested Conceptual Graph (NG)). A *nested conceptual graph (NG) (G,coref)* is an NBG G enriched by an equivalence relation *coref* on X_G

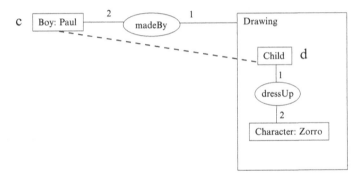

Fig. 9.13 An example of coreferent concept nodes which should not be merged

(the set of all concept nodes in all boxes of G) such that, for each box K in G, $coref$ restricted to K is the trivial equivalence, i.e., if $(c, c') \in coref$ and $c \neq c'$ then c and c' are in two different boxes.

Note that any box of an NG is a normal BG. If the box normality condition is not satisfied, i.e., if one considers a pair $(G, coref)$ where G is a NBG and $coref$ an equivalence relation which does not satisfy the condition in Definition 9.6, then G can be transformed into an intuitively semantically equivalent NG when the following normalization process is successful. One performs from the root of G a top-down (e.g., by a depth-first search or a breadth-first search) normalization of each box as follows. The normalization of a box consists of merging all coreferent nodes (in the box) into a single node, whose description is the disjoint sum of merged node descriptions. A sufficient condition for the normalization process is that each $coref$ class satisfies the condition of a coref class of an SG (cf. Definition 3.10), but this is not a necessary condition.

Definition 9.7 (NG homomorphism). A *(NG) homomorphism* from an NG G to an NG H is defined as a homomorphism of the underlying NBGs on condition that two coreferent nodes of G must have coreferent images in H.

9.4 Nested Typed Graphs

Whenever the universe of discourse can be naturally broken down into independent parts, some knowledge may possibly concern only some of these parts. In such a case, pieces of knowledge can be organized so as to respect the universe of discourse structure. In the same way, when something can be described using different viewpoints, each piece of information corresponding to a given viewpoint can be typed by this viewpoint. Let us give an example about text annotations. Concerning the content of a text, one can make a distinction between the text topic and the way this topic is presented, i.e., the text rhetoric. One can also consider the structure of the text, the word distribution, or even bibliographical data such as the author's name, the publisher or the number of pages. For each of these aspects, a graph can be constructed, with each graph being labeled by the type of annotation, e.g., topic, rhetoric, structure, bibliographic data, and so on. Thus, one introduces graph types and typed graphs, and also sets of typed graphs.

Graph types are especially useful in nested graphs when one wants to consider different sorts of nesting. A nested typed graph can be considered as an NG in which a complex concept node can have several typed descriptions. In Chap. 13 nested typed graphs are used for representing document annotations. Let us again consider the previous text annotation example. An annotation of the text identified by $T121$ and represented by a concept node c labeled (Text, T121) can be composed of an annotation concerning the text topic and another annotation concerning the text rhetoric. Then, the (global) annotation of the text is a description of c composed of a graph A of type *topic* and a graph B of type *rhetoric*, i.e., the label of c is the triple (Text, T121, { (Topic, A), (Rhetoric, B)}).

We define typed SGs (TGs) before considering nested typed graphs. A TG vocabulary $\mathcal{V}$ is an SG vocabulary supplemented by an ordered set of graph types T_G, i.e., $\mathcal{V} = (T_C, T_R, T_G, \mathcal{I})$.

Figure 9.14 presents a tree of graph types.

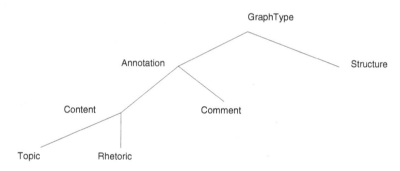

Fig. 9.14 A set of graph types

Definition 9.8 (TG and TG homomorphism). A *typed BG (resp.SG)* on a TG vocabulary $(T_C, T_R, T_G, \mathcal{I})$ is an ordered pair (g, G), where $g \in T_G$ and G is a BG (resp. SG) on $(T_C, T_R, \mathcal{I})$. A homomorphism π from G to H is a *(TG) homomorphism* from (g, G) to (h, H) if $h \leq g$.

Briefly said, a nested typed graph is a nested graph in which the boxes are no longer normal BGs but are typed normal BGs. A BG can be considered as a particular typed BG (there is only one graph type) and, in the same way, an NG can be considered as a particular nested typed graph defined as follows.

Definition 9.9 (Nested Typed Graph (NTG)).

- An *elementary* NTG G is obtained from a typed normal BG (g,H) by adding a third field to the label of each concept node c in H equal to $**$. The set of boxes of G is $boxes(G) = \{(g,H)\}$ and the complete concept node set of G is $X_G = C_H$. A trivial bijection exists between elementary NTGs and typed normal BGs (when no ambiguity occurs we do not distinguish between them).
- Let H be an NTG and $\{(g_1,H_1),\ldots,(g_k,H_k)\}$ a set of elementary NTGs, such that for any $i \neq j \in \{1,\ldots,k\}$, $g_i \neq g_j$. The graph G obtained from H by substituting $\{(g_1,H_1),\ldots,(g_k,H_k)\}$ for the third field $**$ of a concept node c in H is an NTG. $boxes(G) = boxes(H) \cup \{(g_1,H_1),\ldots,(g_k,H_k)\})$, and $X_G = X_H \cup X_{H_1} \cup \ldots \cup X_{H_k}$.
- The third field of any concept node $c \in X_G$ is called the *description* of c and is denoted $Descr(c)$.
- The set X_G of all concept nodes appearing in G is provided with an equivalence relation $coref$, such that for any box K the restriction of $coref$ to K is the trivial equivalence (i.e., if $(c,c') \in coref$ and $c \neq c'$ then c and c' are in two different boxes).

The notions defined for nested graphs (NGs) can be extended to nested typed graphs (NTGs) as follows. First, one can define a tree of typed SGs by substituting TGs for BGs in the definition of a BG tree (cf. Definition 9.2). Secondly, the transformation *Tree* for nested graphs (cf. Definition 9.3) can be extended to nested typed graphs. Thirdly, the homomorphism definition between NGs can be extended to NTGs.

Definition 9.10 (Tree of typed BGs). A *tree of TGs* is a labeled rooted tree $T = (V_T, U_T, l_T)$, such that:

- V_T, the node set of T, is in bijection with a set of pairwise disjoint typed normal BGs $\{G_1,\ldots,G_k\}$. For any $i = 1,\ldots,k$, the node associated with G_i is labeled by G_i (the set $\{G_1,\ldots,G_k\}$ can be identified with V_T).
- For any arc (G_i,G_j) in U_T, $l_T(G_i,G_j)$ is a concept node in G_i; such a labeled arc is also denoted (G_i,c,G_j).
- For all pairs of arcs (G_i,G_j) and (G_i,G_k) with same label c, the graphs G_j and G_k have a different type.

It is sometimes convenient to label an arc (G_i,c,G_j) not only by c but also by a pair $(c,type(G_j))$. In this case, the condition 3 of the previous definition becomes: All labels are distinct, i.e., if (c,g) and (d,h) are the labels of two distinct arcs, then $c \neq d$ or $g \neq h$ or both.

There are two differences between NTGs and NGs. In NTGs, the boxes are typed, and a concept node has a description composed of a set of graphs, instead of a single graph in NGs.

It is now possible to define a tree of typed BGs associated with an NTG.

Definition 9.11 (Tree(G)). The mapping *Tree* assigns to any NTG G a tree of typed SGs, denoted $Tree(G)$, which is defined as follows.

If G is an elementary NTG, then $Tree(G)$ is restricted to a single node labeled G.

If G is the NTG obtained from an NTG H by substituting $\{G_1,\dots,G_k\}$ for $**$, which is the *Descr* of the concept node c in H, then $Tree(G)$ is built from $Tree(H)$ by adding k nodes labeled G_i, $i = 1,\dots,k$ as successors of the node labeled K in $Tree(H)$ containing c. Each arc (K,G_i), for $i = 1,\dots,k$, is labeled by c.

It can immediately be checked that, as in NBGs, the mapping *Tree* is a bijection from the NTGs on a vocabulary $\mathcal{V}$ to the trees of typed BGs on $\mathcal{V}$. NTG homomorphisms are defined using the tree viewpoint.

Definition 9.12 (NTG Homomorphism). Let G and H be two NTGs with $Tree(G)$ $= (V_G, U_G, l_G)$ and $Tree(H) = (V_H, U_H, l_H)$. A *(NTG) homomorphism* from G to H is a pair $\pi = (\pi_0, \{\pi_{G_1},\dots,\pi_{G_k}\})$, where $V_G = \{G_1,\dots,G_k\}$, which satisfies:

- π_0 is an ordinary tree homomorphism from (V_G, U_G) to (V_H, U_H) which maps the root of $Tree(G)$ to the root of $Tree(H)$,
- $\forall K \in V_G$, π_K is a TG homomorphism from K to $\pi_0(K)$,
- $\forall (J, K) \in U_G$, if $l_G(J, K) = c$ then $l_H(\pi_0(J), \pi_0(K)) = \pi_J(c))$,
- two coreferent nodes of G must have coreferent images in H.

Example. The NTG in Fig. 9.15 is obtained from the NG in Fig. 9.11 by replacing the concept types *Description*, *Nutritional observation*, *Nutritional evaluation* and *Biochemical explanation* by graph types. Moreover the root is typed (here by *Annotation*).

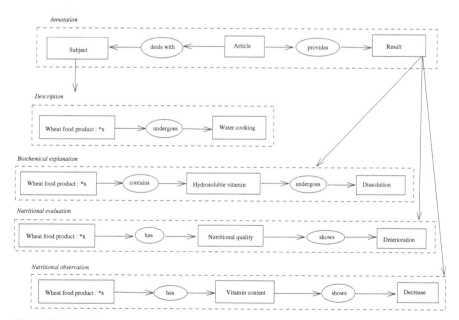

Fig. 9.15 An NTG

Example. In Fig. 9.16, it is assumed that: G is of type g, H of type h, $g \geq h$, and the graph types g_1 and g_2 are greater than or equal to g_3. The mapping π such that:

$\pi(c_1) = d_1,\ \pi(c_2) = d_2,\ \pi(c_3) = \pi(c_4) = d_3,\ \pi(r_1) = s_1,\ \pi(r_2) = s_2,\ \pi(r_3) = s_3$, is a homomorphism from (g, G) to (h, H).

(g,G)

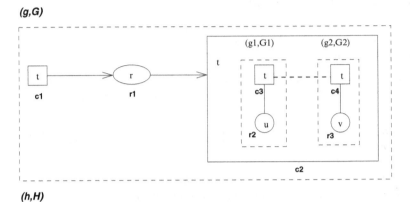

(h,H)

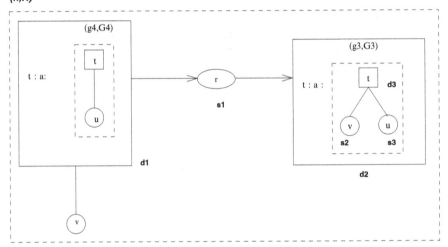

Fig. 9.16 An NTG homomorphism

9.5 The Semantics Φ_N

The FOL semantics Φ_N hereafter defined for NGs and NTGs is an extension of the FOL semantics Φ for SGs (cf. Sect. 4.2.1). The definition of Φ_N for NGs, i.e., untyped nested graphs, is based on two ideas. First, $\Phi_N(G)$ is the conjunction of a formula defining the tree structure of G with formulas associated with all boxes of G. Secondly, the formula associated with a box K is obtained from $\Phi(K)$ by adding an argument which is a variable representing K. The FOL semantics Φ_N for NTGs is built from the FOL semantics for NGs by adding for each box K of type g an atom $g(y)$ (y being the variable representing K). Thus, we consider only NTGs in this section. If graph types are useless, i.e., if NGs are considered instead of NTGs, the soundness and completeness theorem still holds with Φ_N for NGs.

After defining Φ_N, we prove that the NTG homomorphism notion is sound and complete with respect to Φ_N.

9.5.1 Definition of Φ_N

Let $\mathcal{L}$ be the FOL language associated with the vocabulary $(T_C, T_R, \mathcal{I})$ (cf. Sect. 4.2.1). The FOL language $\mathcal{L}_N$ associated with the vocabulary $\mathcal{V} = (T_C, T_R, T_G, \mathcal{I})$ on which NTGs are built is obtained from $\mathcal{L}$ by the following transformations:

- a new constant a_0 is added to the constants assigned to the individual markers,
- each predicate p of arity $n \geq 1$ in $\mathcal{L}$ is transformed into a predicate, still denoted p, of arity $n + 1$,
- a new ternary predicate $descr$ is added to $\mathcal{L}$,
- for any $g \in T_G$ a new unary predicate, also denoted g, is added to $\mathcal{L}$ (useless for NGs).

Thus, a binary predicate is associated with each concept type in T_C, and an $(n + 1)$-ary predicate is associated with each n-ary relation type. Note that the predicate $descr$, which is used for representing the tree structure of any NTG, belongs to any FOL language $\mathcal{L}_N$ associated with a vocabulary.

The following set of formulas, denoted $\Phi_N(\mathcal{V})$, is associated with $\mathcal{V} = (T_C, T_R, T_G, \mathcal{I})$:

$\forall z \forall x_1 \ldots x_n (t_1(x_1, \ldots, x_n, z) \rightarrow t_2(x_1, \ldots, x_n, z))$, for any t_1 and t_2 concept type in T_C (in this case $n = 1$) or relation type of arity $n \geq 1$ in T_R such that $t_1 \leq t_2$,

$\forall x(g_1(x) \rightarrow g_2(x))$, for all g_1, g_2 in T_G such that $g_1 \leq g_2$ (useless for NGs).

Let G be an NTG or an NG, and let $Tree(G)$ be its associated tree. A set of constants $a_1, \ldots, a_m$ is assigned to the coreference classes containing an individual concept node (the same letter is used to designate an individual marker and its associated constant), and a_0 is assigned to the root box of G. Two disjoint sets of variables are considered. First, a set of variables $x_1, \ldots, x_n$ are assigned to the generic coreference classes, and a set $y_1, \ldots, y_k$ are assigned to the k boxes of G distinct from the root box. Hereafter, we often refer to a box by its associated term, and to a coreference class by its associated term. In an NTG, a concept node c is identified by a pair (u, y), where u is the term (either a variable x_i or a constant a_i, $i \geq 1$) associated with the coreference class of c, and y is the term (either a variable y_i or a_0) assigned to the box containing c.

Example. For instance, in Fig. 9.17, the NG in Fig. 9.5 is represented with its associated variables. x_1 (resp. x_2, x_3) is the variable assigned to the generic concept node of type *Drawing* (resp. *Table*, *Train*), a_0 is the constant assigned to the root box and

y_1 is the variable assigned to the box description of the drawing. The fact that Paul is an individual concept node of type Boy in the root box a_0, is represented by the atom $Boy(Paul, a_0)$. The fact that the concept node x_3 has a *Green* attribute in the box y_1 is represented by the atom $attr(x_3, Green, y_1)$. The fact that in a_0 the concept node x_1 has y_1 for description is represented by the atom $descr(a_0, x_1, y_1)$. Finally, the fact that the type of J (resp. K) is *Proposition* (resp. *Subject*) is represented by $Proposition(a_0)$ (resp. $Subject(y_1)$).

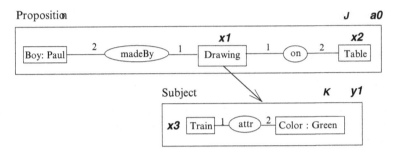

Fig. 9.17 Variables associated with concept nodes and boxes in an NTG

We require the following notation before giving the definition of Φ_N. Let us recall that, given an SG G, $\phi(G)$ is the conjunction of all atoms assigned to nodes of G without quantification, while $\Phi(G)$ is the existential closure of $\phi(G)$ (cf. Definition 4.7).

Definition 9.13 ($\phi(G, u)$). Let G be an SG and let u be a term. $\phi(G, u)$ is the formula obtained from $\phi(G)$ by adding the argument u to each atom in $\phi(G)$.

This transformation can be extended to any FOL formula. For instance, $\Phi(G, u)$ denotes the formula obtained from $\Phi(G)$ by adding the argument u to each atom in $\Phi(G)$.

Definition 9.14 ($\Phi_N(G)$). Let G be an NTG with typed boxes $(g_0, G_0), \ldots, (g_k, G_k)$, (g_0, G_0) being the root box. Let y_i, $i = 1, \ldots, k$ be the variables assigned to the G_i-s and x_i, $i = 1, \ldots, n$ be the variables assigned to the generic coreference classes. $\Phi_N(G) =$
$\exists y_1 \ldots y_k x_1 \ldots x_n (\phi(Tree(G)) \wedge g_0(a_0) \wedge \phi(G_0, a_0) \wedge \ldots \wedge g_k(y_k) \wedge \phi(G_k, y_k))$, where: $\phi(Tree(G))$ is the conjunction, for any (J, c, K) arc in $Tree(G)$, of the atoms $descr(v_j, u_i, y_k)$, where v_j is the term assigned to the box J, y_k is the variable assigned to the box K and u_i is the term assigned to the coreference class of c.

Example. The formula associated with the NTG G in Fig. 9.17 is:
$\Phi_N(G) = \exists y_1 x_1 x_2 x_3 (descr(a_0, x_1, y_1) \wedge proposition(a_0) \wedge subject(y_1) \wedge$
$Boy(Paul, a_0) \wedge Drawing(x_1, a_0) \wedge Table(x_2, a_0) \wedge madeBy(x_1, Paul, a_0) \wedge$
$on(x_1, x_2, a_0) \wedge Train(x_3, y_1) \wedge Color(Green, y_1) \wedge attr(x_3, Green, y_1))$
The formula associated with the NTG H in Fig. 9.18 is:
$\Phi_N(H) = \exists y_1 x_1 x_2 (descr(a_0, x_1, y_1) \wedge proposition(a_0) \wedge subject(y_1) \wedge Boy(x_2, a_0) \wedge$

$Drawing(x_1, a_0) \wedge madeBy(x_1, x_2, a_0) \wedge Child(x_2, y_1) \wedge Character(Zorro, y_1) \wedge$
$dressUp(x_2, Zorro, y_1))$

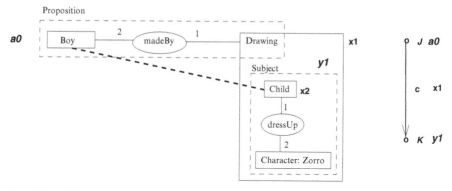

Fig. 9.18 A NTG with coreferent nodes

9.5.2 Soundness and Completeness

Let us show that the homomorphism notion between NGs is sound and complete with respect to the FOL semantics Φ_N. More precisely,

Theorem 9.1. *Let G and H be two N(T)Gs. There is a homomorphism from G to H if and only if* $\Phi_N(\mathcal{V}), \Phi_N(H) \models \Phi_N(G)$.

Note that there is no normality condition for completeness (contrary to the theorem for SGs) since we only consider normal boxes in the definition of nested graphs.

Proof. The proof is based on relationships between NTG homomorphism and $\mathcal{L}$-substitution and on the $\mathcal{L}$-substitution lemma. Ignoring the graph types, one obtains a proof for NGs.

Soundness

Let G and H be two NTGs, $Tree(G) = (V_G, U_G, l_G)$, $V_G = \{G_1, \ldots, G_k\}$, and $Tree(H) = (V_H, U_H, l_H)$, $V_H = \{H_1, \ldots, H_l\}$. Let $\pi = (\pi_0, (\pi_{G_1}, \ldots, \pi_{G_k}))$ be a homomorphism from G to H. Then (cf. Definition 9.12),

1. π_0 is a homomorphism from $Tree(G)$ to $Tree(H)$ which maps the root of $Tree(G)$ to the root of $Tree(H)$,
2. $\forall K \in V_G, \pi_K$ is a homomorphism from K to $\pi_0(K)$,
3. $\forall (J, K) \in U_G$ with $l_G(J, K) = (c, g)$, one has $l_H(\pi_0(J), \pi_0(K)) = (\pi_K(c), h)$, where h is the type of $\pi_0(K)$,

4. if c and c' are two coreferent concept nodes in G, then $\pi(c)$ and $\pi(c')$ are coreferent in H.

Let us prove that there is a $\mathcal{L}_N$-substitution from $\Phi_N(G)$ to $\Phi_N(H)$.
$\Phi_N(G) = \exists y_1 \ldots y_k x_1 \ldots x_m (\phi(Tree(G)) \wedge g_0(a_0) \wedge \phi(G_0, a_0) \wedge \ldots \wedge g_k(y_k) \wedge \phi(G_k, y_k))$, where:
$\phi(Tree(G)) = \bigwedge \{descr(y'_j, u_i, y_p) \mid (y'_j, u_i, y_p) \in Tree(G)\}$, where y'_j is either a_0 or a variable y_j.
$\Phi_N(H) = \exists z_1 \ldots z_l w_1 \ldots w_n (\phi_N(Tree(H)) \wedge h_0(a_0) \wedge \phi(H_0, a_0) \wedge \ldots \wedge h_l(z_l) \wedge \phi(H_l, z_l))$, where:
$\phi(Tree(H)) = \bigwedge \{descr(z'_j, v_i, z_p) \mid (z'_j, v_i, z_p) \in Tree(H)\}$, where z'_j is either a_0 or a variable z_j.
Let σ be the substitution defined as follows:
for any y_i, $i = 1, \ldots, k$, $\sigma(y_i) = z_{f(i)}$, where $z_{f(i)}$ is the variable assigned to the box $\pi_0(G_i)$ of H;
for any x_i, $i = 1, \ldots, m$, $\sigma(x_i) = w_{g(i)}$, where $w_{g(i)}$ is the term assigned to the coreference class of $\pi(c)$, where c is any concept node in the coreference class associated with x_i (the images of coreferent nodes are coreferent nodes, thus $w_{g(i)}$ is well defined).
Let us check that σ is a $\mathcal{L}_N$-substitution from $\Phi_N(G)$ to $\Phi_N(H)$.

- Let $descr(y'_j, u_i, y_p)$ be an atom of $\Phi_N(G)$, then $(J, K) \in U_G$, K (associated with y_p) is the description of a concept node in J (associated with y'_j), which is in the coreference class associated with u_i. Since π is a homomorphism from G to H, $(\pi_0(J), \pi_J(c), \pi_0(K))$ is in $Tree(H)$, and thus $descr(\sigma(y'_j), \sigma(u_i), \sigma(y_p))$ is an atom of $\Phi_N(H)$.
- The atom $h_0(a_0)$ of $\Phi_N(H)$ corresponds to the atom $g_0(a_0)$ of $\Phi_N(G)$.
- Let $g_i(y_i)$ be an atom of $\Phi_N(G)$, g_i the type of G_i and $\pi_0(G_i) = H_{\pi_0(i)}$. Let h be the type of $H_{\pi_0(i)}$ then $h \leq g_i$ and $h(z_{f(i)}) = h(\sigma(y_i))$ is an atom in $\Phi_N(H)$.
- Let $t(u, a_0)$ be an atom of $\Phi_N(G_0)$. It corresponds to a concept node c in G_0 which is in the coreference class associated with u. The atom $t'(u', a_0)$ is assigned to $\pi(c)$, where u' is equal to the term associated with the image by π of the coreference class u. Thus, $t'(u', a_0) = t'(\sigma(u), a_0)$, and $t' \leq t$.
- Let $t(u, y_i)$ be an atom of $\Phi_N(G_i)$. It corresponds to a concept node c in G_i, which is in the coreference class associated with u. The atom $t'(u', z_{f(i)})$ is assigned to $\pi(c)$, where u' is equal to the term associated with the image by π of the coreference class u. Thus, $t'(u', z_{f(i)}) = t'(\sigma(u), \sigma(y_i))$, and $t' \leq t$.
- Let $p(\vec{e}, y_i)$ be an atom of $\Phi_N(G_i)$ corresponding to a relation node r of G_i. The atom $p'(\vec{e}', z_{f(i)})$ is assigned to $\pi(r)$, where $p' \leq p$ and $\vec{e}'$ is the term vector associated with the nodes which are the images by π of the nodes whose term vector is $\vec{e}$, i.e. $\vec{e}' = \sigma(\vec{e})$. Thus, $p'(\pi(\vec{e}), z_{f(i)}) = p'(\sigma(\vec{e}), \sigma(y_i))$ is in $\Phi_N(H)$ and $p' \leq p$. A similar proof can be done for an atom $p(\vec{e}, a_0)$ of $\Phi_N(G_0)$ corresponding to a relation node in G_0. One concludes with the $\mathcal{L}$-substitution lemma.

$\square$

Completeness

Let G and H be two NTGs. $Tree(G) = (V_G, U_G, l_G)$, $V_G = \{G_0, G_1, \ldots, G_k\}$, and $Tree(H) = (V_H, U_H, l_H)$, $V_H = \{H_0, H_1, \ldots, H_l\}$. Let us assume that $\Phi_N(V), \Phi_N(H) \models \Phi_N(G)$, and, using the $\mathcal{L}$-substitution lemma, let σ be a $\mathcal{L}_N$-substitution from $\Phi_N(G)$ to $\Phi_N(H)$. From σ, we first build a tree homomorphism π_0 from $Tree(G)$ to $Tree(H)$ that maps the root of $Tree(G)$ to the root of $Tree(H)$; then we build a homomorphism π_i from G_i to $\pi_0(G_i)$ for any $i = 0, \ldots, k$.

One takes $\pi_0(G_0) = H_0$. For any $i = 1, \ldots, k$, $\pi_0(y_i)$ is defined by $\sigma(y_i)$. More precisely if $\sigma(y_i) = z_{f(i)}$ then $\pi_0(G_i) = H_{f(i)}$. The substitution σ associates to $g_i(y_i)$ an atom $h(\sigma(y_i))$ with $h \leq g_i$. Thus the type of $\pi_0(G_i)$ is $\leq$ type of G_i.

Let us now check that π_0 is a tree homomorphism from $Tree(G)$ to $Tree(H)$. If $k = 0$, then $Tree(G)$ is restricted to a single node and π_0 is a tree homomorphism. Let us assume that $k \geq 1$. Let us consider an arc (J, K) in $Tree(G)$ with $l_G(J, K) = c$. An atom $descr(y_j', u_i, y_p)$ in $\Phi_N(G)$ corresponds to this arc and $descr(\sigma(y_j'), \sigma(u_i), \sigma(y_p))$ is an atom of $\Phi_N(Tree(H))$. Thus, $(\pi_0(J), \pi_0(K))$ is an arc of $Tree(H)$ and $l_H(\pi_0(J), \pi_0(K)) = (d, type(\pi_0(K))$ where d is the (single) concept node in $\pi_0(J)$ associated with the term $\sigma(u_i)$ (all boxes are normal). This proves that π_0 is a homomorphism from $Tree(G)$ to $Tree(H)$ (and it maps the root of $Tree(G)$ to the root of $Tree(H)$).

Let us show that σ is an $\mathcal{L}$-substitution from $\Phi(G_i)$ to $\Phi(\pi_0(G_i))$, for $i = 0, \ldots, k$. Let $p(\vec{e})$ be an atom of $\Phi(G_i)$; then there is an atom $q(\sigma(\vec{e}))$ in $\Phi(\pi_0(G_i))$ such that $q \leq p$. Indeed, for any $i = 0, \ldots, k$ the atoms of $\Phi_N(G_i)$ are mapped by σ to atoms of $\Phi_N(\pi_0(G_i))$ with a decrease of the predicate. Thus, the result holds for $i = 0$. For any $i \in \{1, \ldots, k\}$, $p(\vec{e}, y_i)$ is an atom of $\Phi(G_i, y_i)$; therefore it is an atom of $\Phi_N(G)$ and there is an atom $q(\sigma(\vec{e}), \sigma(y_i))$ of $\Phi_N(H)$ with $q \leq p$. $q(\sigma(\vec{e}), \sigma(y_i))$ is an atom of $\Phi(\pi_0(G_i), \sigma(y_i))$. Thus $q(\sigma(\vec{e}))$ is an atom of $\Phi(\pi_0(G_i))$ with $q \leq p$. As σ is an $\mathcal{L}$-substitution from $\Phi(G_i)$ to $\Phi(\pi_0(G_i))$, for $i = 0, \ldots, k$, then with Property 4.7 there is a homomorphism π_i from G_i to $\pi_0(G_i)$ such that for any concept c in G_i associated with the term u in $\Phi(G_i)$, the concept $\pi_i(c)$ is associated with $\sigma(u)$ in $\Phi(\pi_0(G_i))$. The last condition of an NTG homomorphism (cf. Definition 9.12) is satisfied since if $l_G(J, K) = c$ (associated with u), then $l_H(\pi_0(J), \pi_0(K)) = d$ where d is the single concept in $\pi_0(J)$ associated with $\sigma(u)$ and $d = \pi_J(c)$. $\square$

9.6 Representation of Nested Typed Graphs by BGs

In this section, we define an injective mapping $ng2bg$ which assigns a BG to a nested graph. We will see that this mapping preserves the homomorphisms.

A BG (or SG) vocabulary $\mathcal{V}' = (T_C, T_R \cup T_G \cup \{cont, descr\}, \mathcal{I} \cup \{a_0\})$ is assigned to a typed graph vocabulary $\mathcal{V} = (T_C, T_R, T_G, \mathcal{I})$, where the elements in T_G are now considered as unary relation types, and $cont$, $descr$ and $coref$ are three new binary relations and a_0 is a new individual marker.

Let $(G, coref_G)$ be an NTG on $\mathcal{V}$, with $Tree(G) = (V_G, U_G, l_G)$ and $V_G = \{G_1, \ldots, G_k\}$, G_1 being the root. First, for $i = 1, \ldots, k$, each G_i (on the vocabulary $\mathcal{V}$) is transformed into an BG G'_i (on the vocabulary $\mathcal{V}'$) as follows. A new concept node c_i is added to G_i and, if $i = 1$, i.e., G_i is the root of G, then its label is $(\top, a_0)$. Otherwise its label is $(\top, *)$; c_i is linked to a new unary relation node labeled by g_i (the type of G_i). For each concept node c in G_i, a new relation node labeled $cont$ is created with c_i as its first neighbor and c as its second neighbor.

Secondly, the tree structure has to be coded by the relation $descr$ used as follows. Whenever G_j is the description of c in G_i, i.e., (G_i, c, G_j) is an arc of $Tree(G)$, then a new relation node labeled $descr$ is added, which links c in G'_i to c_j in G'_j.

Thirdly, $coref_G$ is translated by relation nodes of type $coref$. For each pair $(c_i, c_j) \in coref_G$, we add $coref(c_i, c_j)$ (as $coref_G$ is an equivalence relation, we also have $coref(c_j, c_i)$, $coref(c_i, c_i)$ and $coref(c_j, c_j)$).

Example. For instance, let G be the NTG in Fig. 9.18, then $ng2bg(G)$ is represented in Fig. 9.19.

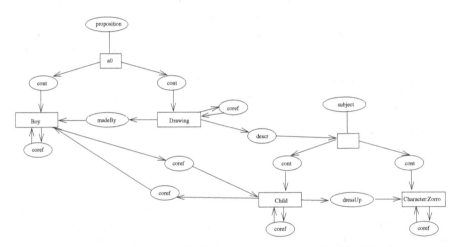

Fig. 9.19 The SG $ng2bg(G)$ associated with the nested typed graph G in Fig. 9.18

Theorem 9.2. ng2bg *is an injective mapping from the set of NTGs on* $\mathcal{V}$ *to the set of BGs on* $\mathcal{V}'$. *Furthermore, there is a bijection between the set of (NTG) homomorphisms from G to H and the set of (BG) homomorphisms from ng2bg(G) to ng2bg(H).*

Proof. Let G be an NTG on $\mathcal{V}$, with $Tree(G) = (V_G, U_G, l_G)$ and $V_G = \{G_1, \ldots, G_k\}$, G_1 being the root and $G' = ng2bg(G)$. For $i = 1, \ldots, k$, c_i is the node in G' associated with G_i and G'_i is the subBG of G' corresponding to G_i. Let H be an NTG on $\mathcal{V}$, with $Tree(H) = (V_H, U_H, l_H)$ and $V_H = \{H_1, \ldots, H_l\}$, with H_1 being the root and $H' = ng2bg(H)$. For $i = 1, \ldots, l$, d_i is the node in H' associated with H_i and H'_i is the subBG of H' corresponding to H_i. Let $\pi = (\pi_0, \{\pi_1, \ldots, \pi_k\})$ be a homomorphism from G to H where π_i is a homomorphism from G_i to $H_{f(i)}$. A homomorphism π' from G' to H' is built as follows (cf. Fig. 9.20).

- For any $i = 1,\dots,k$ and any node x in G_i', $\pi'(x) = \pi_i(x)$.
- For any c_i, $i = 1,\dots,k$, $\pi'(c_i) = d_{f(i)}$.
- For any unary relation node r (of type $l_G(G_i)$) incident to c_i, $\pi'(r)$ is the unary relation node (of type $l_H(H_{f(i)}) \le l_G(G_i)$) incident to $\pi'(c_i) = d_{f(i)}$.
- Let r be a binary relation node of type $cont$ in G' and (c_i, c) its neighbor list. $\pi'(r)$ is the relation node of type $cont$ in $H_{f(i)}$ having for neighbor list $(\pi'(c_i), \pi'(c))$.
- Let r be a binary relation node of type $descr$ in G' and (c, c_j) its neighbor list. Let us assume that c is in G_i'. $\pi'(r)$ is the relation node of type $descr$ in H' having for neighbor list $(\pi'(c), d_{f(j)})$.
- Let r be a binary relation node of type $coref_G$ in G' and (u_1, u_2) its neighbor list. Let us assume that u_1 and u_2 are in G_i'. $\pi'(u_1)$ and $\pi'(u_2)$ are in $H_{f(i)}'$, and there is a binary relation node r' of type $coref_H$ in $H_{f(i)}'$. One takes $\pi'(r) = r'$.

One can check that π' is a homomorphism from G' to H', and this mapping from the (NTG) homomorphisms from G to H to the (BG) homomorphisms from G' to H' is injective and surjective. $\square$

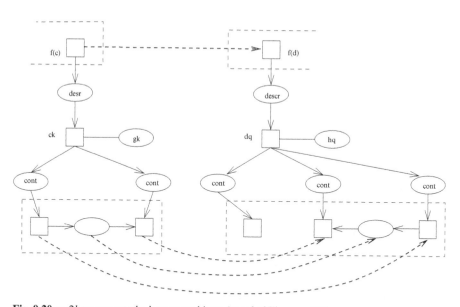

Fig. 9.20 $ng2bg$ preserves the homomorphisms (proof of Theorem 9.2)

The previous correspondence allows us to transport notions and properties of SGs to NTGs and, in a way, the nested typed graphs (NTGs) are not more expressive than simple graphs (SGs). However, even if the transformation $ng2bg$ is simple, the SG obtained seems more difficult to understand for a human being than the initial NTG.

9.7 Bibliographic Notes

It is generally agreed that knowledge has a contextual component, and different formalizations have been proposed to deal with knowledge contexts (e.g., cf. [BP83], [Guh91] and [McC93]).

History of nested conceptual graphs. Nested conceptual graphs were introduced in [Sow84]. Concept nodes representing a proposition can have a referent that is a graph; the intuitive semantics is that this graph is asserted by, or describes, the surrounding proposition. After that, several variants of nested conceptual graphs have been proposed to represent contexts, knowledge description by increasing level of detail, objects in object oriented programming, or related notions. Natural language processing is a main application domain (e.g., [Sow92] [Naz93] [Dic94]).

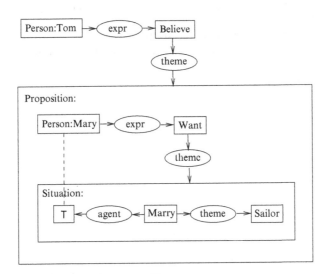

Fig. 9.21 Nested Graph representing a natural language sentence

For instance, the Fig. 9.21 (due to Sowa, e.g., [Sow99])) represents the sentence "Tom believes that Mary wants to marry a sailor." Tom is the experiencer (expr) of the concept [Believe], which is linked by the relation (theme) to a proposition that Tom believes. This proposition is described by another graph, which states that Mary is the experiencer of [Want], whose theme is a situation that Mary hopes will occur. That situation is described by another graph, which says that Mary (represented by the concept [⊤]) marries a sailor. The coreference link shows that the concept [⊤] in the situation concept refers to the same individual as the concept [Person: Mary] in the proposition concept.

We introduced a combinatorial structure in [CM97], provided with a homomorphism called "Graph of graphs," and showed that the different kinds of nested conceptual graphs found in the CG literature are instantiations of this generic notion, and more precisely are trees of specific graphs. Relationships with category theory are also pointed out in this paper. Indeed, a kind of graphs provided with a morphism is a concrete category and transformations like *ng2bg* are functors between two categories. Note however that this framework did not take coreference links into account. It was only concerned with the recursive structure of nested graphs.

The model of nested graphs (without coreference) presented in this chapter was introduced in [MC96] and developed in [CM97]. The latter paper also introduces typed nestings, to specify relationships between the surrounding node and one of its descriptions. Note that, in the model presented in this chapter, we shifted nesting types to graph types. This allows us to type graphs at the first level, and not only as descriptions of a concept node. This model was motivated by two applications in knowledge representation, one concerning simulation (cf. [BBV97]) and the other concerning document indexing (cf. [Gen00]) Since then it has been used in other works (e.g., [GC05], [MLCG07] and [TML06] from which the example of Fig. 9.11 and Fig. 9.15 are extracted). In [CMS98] and [PMC98]) nested graphs with coreference links are processed and two sound and complete logical semantics are defined (see below).

Different logical semantics. Sowa extends the mapping Φ (as defined on SGs) by introducing a special binary predicate *descr* whose first argument is a term and second argument is a formula. For each complex concept node c with referent a non-empty graph D, $\Phi(G)$ contains the atom $descr(e, \Phi(D))$, where e is the term assigned to c. This logical semantics is similar to the semantics of contexts in [Guh91] [McC93], with *descr* playing the same role as the *ist* predicate. Note that the predicate *descr* takes a formula as argument, thus this logical semantics goes beyond FOL.

A non-classical logical semantics, with a calculus based on Gentzen-like sequents, equivalent to NG homomorphism, was presented in [PMC98]. In [Sim98] [CMS98], two alternative semantics in classical FOL are given to NGs. The first one extends Φ (as defined on SGs). For homomorphism completeness, the target NG has to be in so-called *k-normal* form, where k is the depth of the source NG. Every graph can be put in *k-normal* form. The drawback is that putting a graph into *k-normal* form may involve increasing the depth of its tree structure. The second semantics extends Ψ (as defined on SGs, cf. Chap. 4). As for SGs, there is no normality condition on the target graph. None of these two semantics are entirely satisfying because they do not translate the notion of context. With Φ, a concept node is simply represented by the term associated with its coreference class without a term representing its context. Thus properties attached to a concept node can be equivalently attached to any coreferent concept node, even if this node is in another context. With Ψ, the opposite approach is taken: Two coreferent nodes are translated by distinct terms (plus a term translating the fact that they represent the same entity); consequently, properties attached to one node cannot be equivalently attached to the second node, even if these nodes are in the same context. The logical semantics Φ_N defined in this chapter distinguishes between two kinds of coreference, depending on whether it is intra-context or inter-contexts, and thus can be seen as a combination of previous semantics Φ and Ψ. The logical semantics for typed nested graphs is new.

Other works on nested graphs. In [Bag99], a generalization of nested graphs (boxed graphs) is defined in which each box contains a BG, and relation nodes can link concept nodes belonging to different boxes. This extension was mainly done to reify coreference links. The idea is to replace the coreference relation (or to supplement it) by different relation types translating different coreference relations. Each type is provided with rules which describe the properties of this coreference relation. More generally, a boxed graph can be seen as a nested graph with several roots, added with a set of relation nodes linking concept nodes of different boxes (these relation nodes are said to be out of context). The boxed graph homomorphism does not necessarily map roots to roots, and a relation node out of context can be mapped to a node in or out of context, provided that edges are preserved. A transformation from boxed graphs to simple graphs is given, which, as a side effect, gives another proof of equivalence between simple and nested graphs (see also [Bag01] (in French)).

In [Ker01], descriptions are named. However, this framework does not provide operations on nested graphs. A formal declarative semantics is given by an embedding into a nested structure, with an associated notion of truth, from which a notion of deduction on nested graphs is defined. A transformation from these nested graphs to simple graphs is exhibited, which preserves deduction.

In [Pre98a] and [Pre00], the NGs studied are without generic nodes. A semantics (called contextual semantics) in connection with situation theory [BP83] and formal concept analysis [GW99] is proposed. Triadic power context families, i.e., a kind of formal context in the sense of formal concept analysis, are defined. The notion of a standard model built on a triadic power context family and the related notion of semantic entailment are introduced. An NG G_2 is semantically entailed by an NG G_1 if and only if G_2 is valid in the standard model of G_1. Finally, the framework is provided with eight sound and complete elementary rules which are sound and complete with respect to semantic entailment.

Chapter 10
Rules

Overview

In this chapter, we present a strict generalization of basic conceptual graphs (and nested conceptual graphs). A rule expresses knowledge such as: "If H is present then C can be added," where H and C are two graphs with a correspondence between some of their concept nodes. H is called the hypothesis of the rule, and C its conclusion. A rule frequently represents implicit or general knowledge, which can be applied to particular entities, thus making it explicit on these entities. The following is a typical use. Let F be a basic graph representing facts and let R be a rule representing general knowledge. F can be enriched by applying R: Each time the hypothesis of R can be mapped to F (F contains a specialization of the rule hypothesis), then its conclusion can be added to F according to the mapping.

In the first section, we give definitions and logical semantics of the kind of rules that are studied in this chapter. The second section is devoted to rule application in forward chaining. After defining a derivation mechanism, it is proven that this rule derivation mechanism is sound and complete with respect to the logical semantics. The third section is devoted to rule application in backward chaining. Backward mechanism is similar to Prolog resolution. But, since a graph rule is more general than a definite clause, the situation is more complex than with Prolog. The graph structure is used in the backward mechanism proposed. The fourth section deals with computational complexity results. It is stated that the deduction problem is non-decidable (i.e., it is only semi-decidable) and decidable cases are exhibited.

10.1 Definition and Logical Semantics of a Rule

Several definitions of a rule can be found in the conceptual graph literature. Below we introduce the classical definition of a rule as an ordered pair of lambda-BGs. We shall also give the definition of a rule as a bicolored graph, which is in some sense less expressive, but is interesting for visualization purposes. Let us first define λ-BGs.

Definition 10.1 (λ-BG).

- A λ-BG, $L = (c_1 \ldots c_n) \, G$, $n \geq 0$, consists of a BG G and n distinguished generic concept nodes of G, $c_1 \ldots c_n$. A BG can be considered as a lambda-BG with $n = 0$. L is said to be normal if G is normal.
- The logical semantics of a λ-BG $(c_1 \ldots c_n) \, G$ is obtained from the logical semantics of G by removing the existential quantification of all variables associated with $c_1 \ldots c_n$ (which are thus free variables), i.e., $\Phi((c_1 \ldots c_n) \, G) = \exists y_1 \ldots \exists y_k \phi(G)$, where the variables y_i-s are associated with the generic concept nodes distinct from the c_i-s and $\phi(G)$ is the conjunction of atoms associated with all nodes in G (cf. definition 4.7).

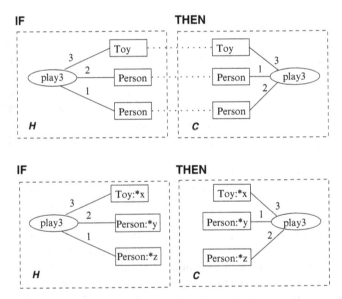

Fig. 10.1 Two drawings of the same rule

Definition 10.2 (λ-rule).

- A λ-*rule* R is an ordered pair of λ-BGs $(H = (c_{1_1} \ldots c_{1_n}) \, H', C = (c_{2_1} \ldots c_{2_n}) \, C')$. There is a bijection from the distinguished nodes of H to the distinguished nodes of C which associates c_{1_i} with c_{2_i} for all $i = 1, \ldots, n$.
- H is called the *hypothesis* of R, and C is its *conclusion*.
- The nodes c_{1_i} and c_{2_i} are called the *connection* nodes of R. A λ-rule is also denoted $R = (c_1 \ldots c_n) H \Rightarrow C$ or simply $R = H \Rightarrow C$.

When there is no ambiguity, we simply note c_i for c_{1_i} and c_{2_i}. In drawings, distinguished concept nodes are often labeled by named generic markers of form $*x_1 \ldots *x_n$. The bijection between them may also be represented by links connecting each c_{1_i} with c_{2_i}. Figure 10.1 presents two drawings of the same rule which states that the relation $play_3$ is symmetrical on its two first arguments of type Person: "If a person z and a person y play together with a toy x, then y and z also play together with x." This rule is very simple, since its conclusion is very simple: All concept nodes are connection nodes and corresponding concept nodes in the hypothesis and the conclusion have the same type. Its application to a fact only adds a new relation node between already existing concept nodes. Other kinds of rules are presented in Fig. 10.2. Rule R_1 states that "If a Person x has for uncle a Person y, then there is a Person who is a parent of x and a brother of y." Its application adds not only relation nodes but also a new generic node. Rule R_3 states that "If a Person has for uncle the Person x then x is a Man." Its application possibly restricts the type of an existing concept node.

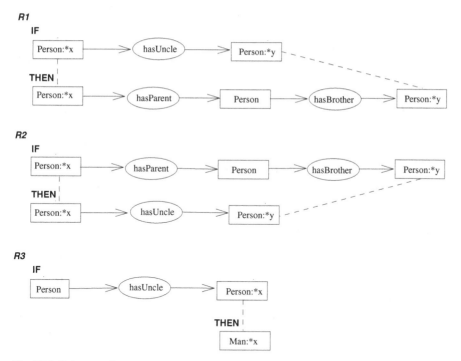

Fig. 10.2 Rule examples

10.1.1 Logical Semantics of a Rule

Briefly, the FOL semantics of a rule states that, for all variables associated with the connection nodes, if the hypothesis holds true, then the conclusion also holds true.

Definition 10.3 (FOL semantics of a λ-rule). Let $R = H \Rightarrow C$ be a λ-rule with $H = (c_{1_1} \ldots c_{1_q}) H'$ and $C = (c_{2_1} \ldots c_{2_q}) C'$. $\Phi(R) = \forall x_1 \ldots \forall x_q (\Phi(H) \rightarrow \Phi(C))$ where, for $i = 1, \ldots, q$, x_i is the variable assigned to the connection nodes c_{1_i} and c_{2_i}, and $\Phi(H)$ and $\Phi(C)$ are FOL semantics of the λ-BGs H and C (by definition, $x_1 \ldots x_q$ are free variables in $\Phi(H)$ and $\Phi(C)$).

Note that, if for some i, c_{1_i} and c_{2_i} have the same type, the atom $type(x_i)$ associated with c_{2_i} in $\Phi(C)$ can be suppressed.

Let $x_{q+1}, \ldots, x_{l_1}$ denote the variables assigned to the generic nodes in H that are not connection nodes, and $y_1, \ldots, y_{l_2}$ denote the variables assigned to generic nodes in C that are not connection nodes. Then, $\Phi(R)$ can be written as follows:
$\Phi(R) = \forall x_1 \ldots \forall x_q (\exists x_{q+1} \ldots \exists x_{l_1} \phi(H) \rightarrow \exists y_1 \ldots \exists y_{l_2} \phi(C))$, where $\phi(H)$ and $\phi(C)$ are the conjunction of atoms associated with all nodes in H and C, respectively. This formula can equivalently be rewritten as:
$\Phi(R) = \forall x_1 \ldots \forall x_{l_1} (\phi(H) \rightarrow \exists y_1 \ldots \exists y_{l_2} \phi(C))$, where the only existentially quantified variables are the variables associated with generic nodes of the conclusion that are not connection nodes.

Example. The formula assigned to the rule in Fig. 10.1 is
$\forall x \forall y \forall z (Toy(x) \wedge Person(y) \wedge Person(z) \wedge play_3(z, y, x) \rightarrow play_3(y, z, x))$.
The formulas assigned to the rules R_1 and R_2 in Fig. 10.2 are:
$\Phi(R_1) = \forall x \forall y (Person(x) \wedge Person(y) \wedge hasUncle(x, y) \rightarrow \exists z (Person(z) \wedge hasParent(x, z) \wedge hasBrother(z, y)))$.

For R_2, which is the reciprocal rule of R_1, one has:
$\Phi(R_2) = \forall x \forall y (\exists z (Person(x) \wedge Person(y) \wedge Person(z) \wedge hasParent(x, z) \wedge hasBrother(z, y)) \rightarrow hasUncle(x, y))$, or equivalently:
$\Phi(R_2) = \forall x \forall y \forall z (Person(x) \wedge Person(y) \wedge Person(z) \wedge hasParent(x, z) \wedge hasBrother(z, y) \rightarrow hasUncle(x, y))$.

For R_3 one has: $\Phi(R_3) = \forall x (\exists y (Person(x) \wedge Person(y) \wedge hasUncle(y, x)) \rightarrow Man(x))$,
or equivalently $\Phi(R_3) = \forall x \forall y (Person(x) \wedge Person(y) \wedge hasUncle(y, x) \rightarrow Man(x))$.

The logical formula assigned to a rule can generally not be represented by an equivalent clause (i.e., a universally quantified disjunction of positive or negative literals) because of the existentially quantified variables in the conclusion (e.g. $\Phi(R_1)$). In the other direction, a clause can be represented by a rule only if it is a definite clause (i.e., it possesses exactly one positive literal) with no function. A definite clause D is translated into a rule whose logical formula is of form $\forall x_1 \ldots \forall x_q (\top(x_1) \wedge \ldots \wedge \top(x_q) \wedge conj_N \rightarrow conc)$, where: $x_1 \ldots x_q$ are the variables of the clause, $conj_N$ is the conjunction of atoms that appear negatively in D and $conc$ is the positive atom in D. The atoms $\top(x_i)$ are used to type each variable. For instance, a clause $p(x, y, z) \vee \neg r(x, y) \vee \neg r(x, z)$ is translated into a rule logically interpreted by $\forall x \forall y \forall z (\top(x) \wedge \top(y) \wedge \top(z) \wedge r(x, y) \wedge r(x, z) \rightarrow p(x, y, z))$.

Before entering the processing of rules, let us define another representation of rules.

10.1.2 Rule as a Bicolored Graph

Bicolored basic graphs defined hereafter can be used to define rules and provide nice visualizations of them. However, they slightly restrict the expressiveness of rules (in a sense that we will specify).

Definition 10.4 (Bicolored BG). A bicolored basic graph (G,l) is a BG G provided with a coloring l of its nodes with two colors, say 0 and 1.

Definition 10.5 (Rule as a bicolored BG).

- A bicolored rule $R = (G,l)$ is a bicolored basic graph, which satisfies: The subgraph H induced by the color 0 nodes is a subBG of G (i.e., if a relation node is colored by 0 then all its neighbors must also have color 0).
- H is called the *hypothesis* of the rule.
- Concept nodes of color 0 with at least one neighbor outside the hypothesis part are the *frontier* nodes.
- Nodes with color 1 together with the frontier nodes induce a subgraph C called the *conclusion* of the rule.

The condition on the hypothesis of the rule prevents the rule application mechanism (as it is defined in definition 10.6) for producing BGs that would be syntactically incorrect. The frontier nodes are shared by the rule hypothesis and conclusion. They correspond to the connection nodes of λ-rules, which are generic nodes, and to shared individual nodes.

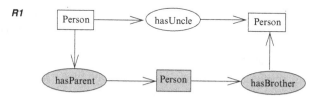

Fig. 10.3 Colored graph representation of the rule R_1 in Fig. 10.2

In drawings, the hypothesis is shown in white, and the conclusion, except frontier nodes, in gray. Figure 10.3 shows a bicolored representation of the rule R_1 represented as a pair of λ-graphs in Fig. 10.2.

The logical semantics of these rules is as follows: Let $\phi(H)$ and $\phi(C)$ be the conjunctions of atoms associated with the hypothesis H and the conclusion C. Let $x_1 \ldots x_n$ be the variables assigned to generic frontier nodes; let $y_1 \ldots y_p$ (resp. $z_1 \ldots z_q$)

be the variables assigned to the other generic concept nodes of H (resp. C). Then,
$\Phi(R) = \forall x_1 \ldots \forall x_n (\exists y_1 \ldots \exists y_p \, \phi(H) \rightarrow \exists z_1 \ldots \exists z_q \, \phi(C))$.

A bicolored rule with hypothesis H and conclusion C can be seen as a λ-rule $((c_1...c_n)H, (c_1...c_n)C)$, where $(c_1 \ldots c_n)$ is any total ordering of the generic frontier nodes. In the other direction, the λ-rules corresponding to bicolored rules have the specific property that all c_{i_1} and c_{i_2} have the same type. Intuitively, a bicolored rule might add new relation and concept nodes, but it cannot specialize the type of an existing node. This is not a real loss of expressivity from a logical viewpoint since concept types can be seen as unary relations. In other words, a knowledge base consisting of facts and λ-rules can be translated into a logically equivalent knowledge base consisting of facts and bicolored rules with a change of vocabulary. Let us call t this transformation. Let $\mathcal{V}$ be the vocabulary; $t(\mathcal{V})$ is the vocabulary built by translating all concept types, except the universal type $\top$, into unary relation types, while keeping the same partial order. The concept type set of $t(\mathcal{V})$ is composed of the single type $\top$. Any BG on $\mathcal{V}$ is transformed into a BG on $t(\mathcal{V})$ as follows: Each concept node with label $l = (t_1...t_k, m)$ is translated into a concept node with label $(\top, m)$ and, if the type in l is not equal to $\top$, for each t_i in l, one incident relation node with label (t_i). The following property is immediate:

Property 10.1. Let G and H be two BGs. There is a homomorphism π_1 from G to H if and only if there is a homomorphism π_2 from $t(G)$ to $t(H)$, such that π_1 and π_2 coincide on concept nodes, i.e., for all $c \in C_G, t(\pi_1(c)) = \pi_2(t(c))$.

Facts and λ-rules can thus be translated into facts and bicolored rules, while preserving not only the logical semantics of the KB, but also homomorphisms between objects of the KB.

10.2 Forward Chaining

Generally speaking, there are two classical ways of using rules: forward and backward chaining. *Forward chaining* starts from the facts and applies rules to the facts in order to produce new facts; this cycle is repeated until either no rule can be applied (all factual knowledge deducible from the base has been made explicit) or a specified state is reached (for instance a specialization of a goal, or an answer to a question has been found). *Backward chaining* starts from a question, classically called a goal, and tries to build a derivation leading to an answer to this goal in a backward manner. The idea is that, if a rule conclusion "matches" the goal, i.e., an application of this rule could possibly answer part of the goal, then the corresponding rule application can be used in a derivation answering the goal, provided that a new goal, composed of the remaining part of the original goal and the hypothesis of the rule, is answered; the process stops when the new goal is empty (in this case, the last rule considered has an empty hypothesis—it is a fact—and the goal is completely mapped to this fact) or all possibilities have been explored.

10.2.1 Rule Application

Definition 10.6 (Rule application). A λ-rule $R = H \Rightarrow C$ is applicable to a BG G if there is a homomorphism π from the hypothesis H of R to G. In this case, the result of the *application of R to G according to π* is the BG $G' = (R, \pi)G$ obtained from G and the conclusion C of R by merging each c_{2_i} of C with $\pi(c_{1_i})$. If normalization is required, the possible new individual nodes added by C are merged with the possible individual nodes with the same marker in G.

The node obtained by merging c_{2_i} and $\pi(c_{1_i})$ has the same marker as $\pi(c_{1_i})$. If c_{1_i} and c_{2_i} have the same type (i.e., R could be considered as a bicolored rule), then its type is exactly the type of $\pi(c_{1_i})$.

Otherwise, its type is the conjunction of both types. This conjunction can be a banned type: In this case, despite the fact that the rule is applicable to G according to π, the result is a BG inconsistent with respect to the vocabulary. The knowledge base is thus inconsistent and has to be repaired (otherwise, from a logical viewpoint, any BG could be deduced from it, which is not desirable from a knowledge representation viewpoint).

Notes on Particular Cases

Basic vocabularies can also be considered instead of conjunctive vocabularies. In this case, the type of the obtained node is a maximal common subtype of the type of c_{2_i} and the type of $\pi(c_{1_i})$. Thus, if the concept type hierarchy is not an inf-semi-lattice, then there might be several possible new labels, or none if the types have no common subtype. Let us recall that the set of conjunctive types is a lattice (however, some of the types might be banned). If bicolored rules are considered instead of λ-rules, the structure of the concept type set does not matter. Indeed, the rule application process simply adds the conclusion according to the images of the frontier nodes and cannot specialize the label of existing nodes. In terms of elementary specialization operations, it makes the disjoint sum of G and the conclusion C of R, restricts the label of each frontier node c in C to the label of its image $\pi(c)$, then joins each c and $\pi(c)$.

Examples. See rule R in Fig. 10.1 and graphs G and G' in Fig. 10.4 for the simplest example: There is only one way of applying R to G, and G' is the resulting graph. Let us now consider the abstract example of Fig. 10.5. There are three concept types t_1, t_2 and t_3, where t_1 is greater than t_2 and t_3, which are both incomparable. Rule R states that "for all entity x, if x is of type t_1 and $r(x)$ holds true, then x is of type t_3." R is applicable to G, with the connection node of type t_1 being mapped to the node of type t_2, which leads to a node of type $\{t_2, t_3\}$. G' is obtained. Note that $\{t_2, t_3\}$ might be banned. Figure 10.6 shows a more complex rule application: Rule R_4 (which is the conjunction of rules R_1 and R_3 in Fig. 10.2) is applied to the fact F yielding F'. It is assumed that *Painter* $\leq$ *Person* and *Man* $\leq$ *Person*.

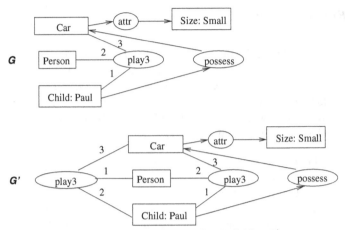

The rule in Fig. 10.1 is applied to G yielding G'

Fig. 10.4 A simple rule application

Fig. 10.5 Type restriction by rule application

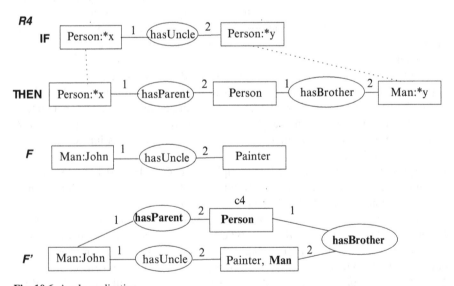

Fig. 10.6 A rule application

In Fig. 10.7, the (bicolored) rule R represents the following statement: "If a person x is an employee of a company z, which is managed by a person y, then x gets a salary s and y decides on s." Fact F states that B manages company C. It is assumed that *manager* is a specialization of *employee*. There is a homomorphism π from the hypothesis of R to F, which maps both connection nodes x and y to the same node B, and z to C. The application of R to F according to π produces the fact F', which states that "B, manager of company C, gets a salary and decides on it."

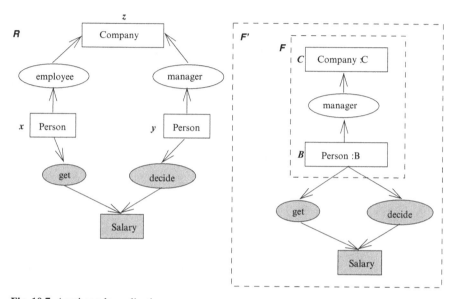

Fig. 10.7 Another rule application

10.2.2 Derivation and Deduction

Let $\mathcal{R}$ be a set of rules. If G' is obtained from a BG G by application of a rule R in $\mathcal{R}$, we say that G' is *immediately derived* from (G, R) or, from $(G, \mathcal{R})$. G' is said to be *derived* from $(G, \mathcal{R})$ if there is a sequence of BGs $G_0(=G), G_1, \ldots, G_k(=G')$ $(k \geq 0)$ such that each G_{i+1} $(0 \leq i < k)$ is immediately derived from $(G_i, \mathcal{R})$. Such a sequence is called an $\mathcal{R}$-*derivation* (or simply a derivation) from G. When the KB is composed of facts and rules, the deduction problem, denoted $\mathcal{FR}$-DEDUCTION (cf. Chap. 11), asks if there is a sequence of rule applications that enrich the facts such that the goal can be reached, i.e., a derivation from the facts leading to a graph to which the goal can be mapped. In what follows, a set of facts $\mathcal{F}$ is identified with a BG F.

Definition 10.7 (Deduction with facts and rules). $\mathcal{FR}$-DEDUCTION takes a KB $\mathcal{K} = (F, \mathcal{R})$ as input, where F is a set of facts and $\mathcal{R}$ is a set of rules, and a goal Q, and asks if Q is deducible from $\mathcal{K}$, i.e., if there is an $\mathcal{R}$-derivation from F to a BG F' such that $Q \succeq F'$.

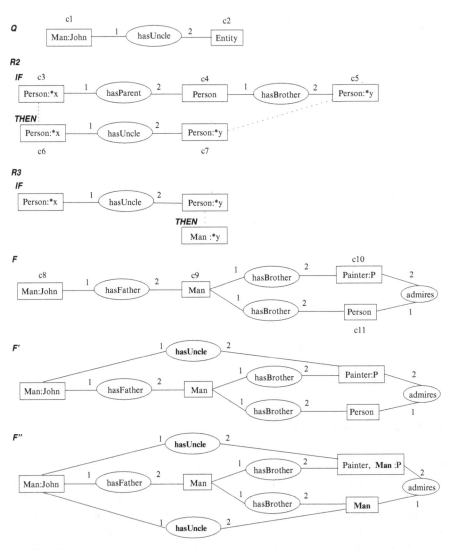

Fig. 10.8 Graph derivation and deduction

Example. In Fig. 10.8, $\mathcal{R} = \{R_2, R_3\}$. F' is immediately derived from $(F, \mathcal{R})$ by an application of R_2 which maps the sequence of nodes (c_3, c_4, c_5) to (c_8, c_9, c_{10}). F'' is derived from $(F, \mathcal{R})$ by a derivation of length 4 (R_2 and R_3 are each used twice).

Q cannot be deduced from F, but it can be deduced from $(F, \mathcal{R})$: Indeed, there is a homomorphism from Q to F' (and *a fortiori* to F'').

Note on the expressivity of bicolored rules with respect to λ-rules. Every instance of $\mathcal{FR}$-DEDUCTION can be transformed into an instance of $\mathcal{FR}$-DEDUCTION solely using bicolored rules by way of transformation t detailed in Sect. 10.1.2. A rule R is applicable to a fact F if and only if $t(R)$ is applicable to $t(F)$. Moreover, it can be immediately checked that the following three graphs are logically equivalent: F' immediately derived from (F, R), $t(F')$ and graph F'' immediately derived from $(t(F), t(R))$. They have identical sets of concept nodes. F'' may contain redundant unary relation nodes, which are easily suppressed, in the same way as a conjunction of concept types is minimized. The only true difference between F' and $t(F')$ is that F' can be inconsistent due to a possible banned concept type occurring in a label. Thus, if the knowledge base is consistent, i.e., no inconsistent graph can be derived from it, then λ-rules and bicolored rules can be said to have the same expressivity.

The forward chaining scheme is presented in Table 25. The algorithm explores the tree of all derivations from fact F in a breadth-first manner. One step (i.e., an iteration of the "while true" loop) consists of finding all *new* homomorphisms from rule hypotheses to the current F, and performing all rule applications corresponding to these homomorphisms. The order in which these rule applications are considered does not matter: The same graph (up to isomorphism) is obtained at the end of the step for all possible orders. At the beginning of each step, it is checked whether Q can be mapped to F. If yes, Q is deducible from the knowledge base and the algorithm returns true. Otherwise the set of all *new* homomorphisms is computed. Given a rule $R = H \Rightarrow C$, a homomorphism from H to F is new if has not been already computed at a previous step. A simple method for recognizing new homomorphisms at step i consists of memorizing the set of nodes N_{i-1} added to F, or modified (by a type restriction) at step $i - 1$; then a homomorphism from H to F is new if it maps at least one node of H to a node of N_{i-1}. If the set of new homomorphisms is empty, Q is not deducible from the knowledge base and the algorithm returns false. Otherwise, F is enriched by all rule applications according to the new homomorphisms.

If the algorithm returns true, Q is indeed deducible from the KB since a derivation has been built, which ends with a graph to which Q can be mapped. As the algorithm explores the space of all possible rule applications in a breadth-first manner, and at a given step, the number of (new) rule applications to F is finite, it is ensured that whenever there are some derivations allowing us to deduce Q the algorithm will build one of these derivations. The algorithm thus returns true if and only if Q is deducible from the base. If the algorithm returns false, Q is indeed not deducible from the base: New rule applications can no longer be done; thus the current F represents all information deducible from the KB (see the notion of a closed graph in Sect. 10.4.2), and Q cannot be mapped to it. However, the tree of all derivations may have infinite paths: Thus, when Q is not deducible from the base, the algorithm might not end. See Sect. 10.4.2 for further details.

Algorithm 25: ForwardChaining

Input: a knowledge base $\mathcal{K} = (F, \mathcal{R})$ in normal form, and a BG Q
Output: true if and only if Q can be deduced from $\mathcal{K}$ otherwise false or infinite calculus
begin
    while *true* **do**
        if *there is a homomorphism from Q to F* **then**
            └ **return** *true*
        ToApply $\leftarrow \emptyset$
        forall $R: H \Rightarrow C \in \mathcal{R}$ **do**
            forall new *homomorphism* $\pi : H \to F$ **do**
                └ ToApply $\leftarrow$ ToApply $\cup \{(R, \pi)\}$
        if *ToApply* $= \emptyset$ **then**
            └ **return** *false*
        forall $(R, \pi) \in$ *ToApply* **do**
            $F \leftarrow$ apply R to F according to π
            └ Normalize F if needed
end

10.2.3 Non-Redundant Rule Application

Obviously, it is irrelevant to apply a rule to a graph if this rule does not bring new information. A trivial case is that if a rule can be applied once, then it can be applied indefinitely in the same way to the obtained graph. These further applications do not bring anything new. They are said to be *trivially useless*. The notion of *new* homomorphism in the forward chaining algorithm (Table 25) prevents trivially useless rule applications. Another case of trivial redundancy is that of twin relation nodes, i.e., relations with the same label and exactly the same arguments in the same order. Consider, for instance, the rule "if $r(x,y)$ then $r(x,y)$." This rule can be applied indefinitely according to a new homomorphism each time, but all applications produce twin relation nodes. More generally, the application of a rule R brings new information to a graph G if G and $(R, \pi)G$ are not equivalent. Determining whether two graphs are equivalent is an NP-complete problem (see Chap. 2). The next property allows us to test whether a rule application would add new information with a local check. Instead of checking whether $(R, \pi)G$ can be mapped to G, it is indeed sufficient to check whether the part that would be added to G could be "folded" onto G.

Property 10.2 (Redundant application). Let G be an irredundant BG, $R = H \Rightarrow C$ be a rule, and let π be a homomorphism from H to G. Then $(R, \pi)G$ is equivalent to G if and only if there is a homomorphism π' from C to G, such that for each connection node c_i, $\pi'(c_{2_i}) = \pi(c_{1_i})$.

Proof. Assume that there is a homomorphism from $(R, \pi)G$ to G. Property 2.9 states that, if G' is a redundant graph and K is an irredundant subgraph of G' equivalent to G', then there is a homomorphism π'' from G' to K, which keeps every node in

K invariant. If we consider $G' = (R, \pi)G$ and $K = G$, we define π' as the restriction of π'' to the subgraph of G' corresponding to C; π' satisfies the condition of the property. Conversely, if there is a homomorphism π' satisfying the condition of the property, then there is a homomorphism from $(R, \pi)G$ to G built as follows: Each node coming from C is mapped to $\pi'(C)$; each other node is mapped to itself. Since there is also a homomorphism from G to $(R, \pi)G$ (the identity mapping), G and $(R, \pi)G$ are equivalent. $\square$

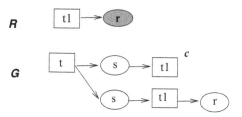

Fig. 10.9 G needs to be irredundant in Property 10.2

If G is redundant, only the $\Leftarrow$ direction of Property 10.2 is true. In this case, we only have a sufficient condition for a rule application to be useless. See Fig. 10.9, where R is pictured as a bicolored rule ("for all x, if $t_1(x)$ then $r(x)$") and G is redundant; the application of R to G at node c would yield a graph equivalent to G; however, there is no homomorphism, from the conclusion of R to G that maps the concept node of R to c.

Property 10.2 prevents uninteresting rule applications. However, it would be more interesting to accurately characterize the part of the rule conclusion that adds new information in order to keep an irredundant graph. Indeed, even if G is irredundant, the previous property does not guarantee that $G' = (R, \pi)G$ will be irredundant. G' will not be equivalent to G but may contain redundant parts. See for instance Fig. 10.10, in which $t_1 \geq t_2$: G is not redundant; R can be applied at b and at c; the application at b is avoided because it is a redundant application; the application at c is not redundant but the obtained graph G' is redundant (there is a homomorphism from G' to one of its strict subgraphs, that maps b to c).

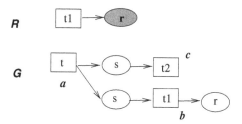

Fig. 10.10 Property 10.2 does not prevent redundancy ($t_1 \geq t_2$)

10.2.4 Soundness and Completeness of Forward Chaining

In this section, we prove the logical soundness and completeness of the forward mechanism based on rule application, i.e., given a KB $\mathcal{K} = (F, \mathcal{R})$, any BG Q is deducible from $\mathcal{K}$ by a rule derivation if and only if its logical translation is deducible from the logical translation of $\mathcal{K}$. As the involved operations rely on homomorphism, completeness is submitted to a normality condition. A KB is said to be normal if the set of facts considered as a single graph (F) is normal.

Theorem 10.1 (Forward chaining soundness and completeness). *Let $\mathcal{K} = (F, \mathcal{R})$ be a KB, where F is a set of facts and $\mathcal{R}$ a set of rules, and let Q be a BG, with all being relative to a vocabulary $\mathcal{V}$.*

- *(soundness) if Q is deducible from $\mathcal{K}$ (i.e. there is a BG F' that can be derived from $\mathcal{K}$ such that Q can be mapped to F') then $\Phi(\mathcal{V})$, $\Phi(F)$, $\Phi(\mathcal{R}) \models \Phi(Q)$*
- *(completeness) provided that F is normal and the graph obtained at each derivation step is put into normal form: if $\Phi(\mathcal{V})$, $\Phi(F)$, $\Phi(\mathcal{R}) \models \Phi(Q)$ then Q is deducible from $\mathcal{K}$.*

As a knowledge base consisting of facts and λ-rules can be translated into a logically equivalent knowledge base consisting of facts and bicolored rules (cf. the last remark in Sect. 10.1.2), only bicolored rules are considered in the proof.

Proof of soundness

Lemma 10.1. *Let F and Q be BGs and let R be a rule over $\mathcal{V}$, such that Q is immediately derived from F by an application of R, i.e., $Q = (R, \pi)F$, then: $\Phi(\mathcal{V}), \Phi(F), \Phi(R) \models \Phi(Q)$.*

Notations
   $\Phi(F) = \exists x_1 \ldots x_{l(F)}(P_1 \wedge \ldots P_{n(F)})$.
The connection nodes of R are $\{c_{11}, \ldots, c_{1q}\}$ and $\{c_{21}, \ldots, c_{2q}\}$, x_i is the term associated with c_{1i} and c_{2i}. H is the hypothesis of R and C its conclusion.
   $\Phi(R) = \forall x_1 \ldots \forall x_q (\forall x_{q+1} \ldots \forall x_{l_1} Q_{11} \wedge \ldots \wedge Q_{1m_1} \rightarrow (\exists y_1 \ldots y_{l_2} Q_{21} \wedge \ldots \wedge Q_{2m_2}))$.
   The clausal forms of $\Phi(F)$ and $\Phi(R)$ are:
   $C(\Phi(F)) = \{\rho(P_i) \mid i = 1, \ldots, n(F)\}$ where ρ is the substitution $\rho = \{(x_1, a_1), \ldots, (x_{l(F)}, a_{l(F)})\}$ with a_i-s being skolem constants;
   $C(\Phi(R)) = \{\neg Q_{11}(x_1, \ldots, x_{l_1}) \vee \ldots \vee \neg Q_{1m_1}(x_1, \ldots, x_{l_1})$
   $\vee Q_{2i}(x_1, \ldots, x_{l_1}, f_1(x_1, \ldots, x_{l_1}), \ldots, f_{l_2}(x_1, \ldots, x_{l_1})) \mid i = 1, \ldots, m_2'\}$
where the f_i-s are skolem functions. If the atoms Q_{2i}, for $i = m_2' + 1, \ldots, m_2$, are atoms associated with connection nodes of the conclusion of R then the clause of $\Phi(R)$ corresponding to Q_{2i} is a tautology and it can be deleted.
   $\Phi(Q(=(R, \pi)F)) = \exists x_1 \ldots x_{l(F)}(P_1 \wedge \ldots P_{n(F)}) \wedge \exists y_1 \ldots y_{l_2}(\eta(Q_{21}(x_1, \ldots, x_{l_1}, y_1, \ldots, y_{l_2})) \wedge \ldots \wedge \eta(Q_{2m_2'}(x_1, \ldots, x_{l_1}, y_1, \ldots, y_{l_2})))$
where η is the substitution which assigns to each variable x_i associated with the connection node c_{1i} the term assigned to $\pi(c_{1i})$ by Φ, i.e.,

$\eta = \{(x_1, term(\pi(c_{11}))), \ldots, (x_q, term(\pi(c_{1q})))\}$.
cf. Fig. 10.11.

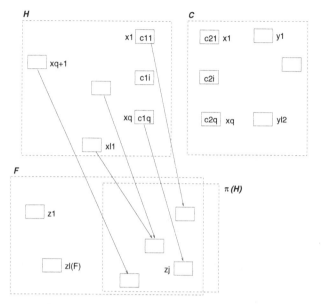

Fig. 10.11 Terms associated with nodes in forward application of a rule

Lemma Proof technique

We show, by the resolution method, that $\{\Phi(\mathcal{V}), \Phi(F), \Phi(R), \neg\Phi((R, \pi)F)\}$ is inconsistent. More precisely, we show that there is a substitution λ such that any $\lambda(A)$ where $A \in atoms((R, \pi)F)$ can be obtained by resolution from $\lambda(E)$, where $E = \{\Phi(\mathcal{V}), \Phi(F), \Phi(R)\}$.

Proof. Since π is a homomorphism from H to F, there is a $\mathcal{L}$-substitution σ from $\Phi(H)$ to $\Phi(F)$, which satisfies (cf. Property 4.6):
For each concept node c in H, $\sigma(term(c)) = term(\pi(c))$,
for all $i = 1, \ldots, m_1$ $\sigma(Q_{1i}) \in Aug(atoms(\Phi(F)))$.
Let μ be the composition of ρ and σ, then μ satisfies:

- For all concept nodes c in H, $\mu(term(c)) = \rho(\sigma(term(c))) = \rho(term(\pi(c)))$,
- for all $i = 1, \ldots, m_1$, since $\sigma(Q_{1i}) \in Aug(atoms(\Phi(F)))$ one has $\mu(Q_{1i}) \in Aug(C(\Phi(F)))$.

Let λ be the substitution $\lambda = \rho \cup \mu'$ where $\mu' = \{(y_1, f_1(\mu(x_1), \ldots, \mu(x_{l1}))), \ldots, (y_{l2}, f_{l2}(\mu(x_1), \ldots, \mu(x_{l1})))\}$. cf. Fig. 10.12

Let $A \in atoms(\Phi(Q(=(R, \pi)F)))$. Let us check that any $\lambda(A)$ can be obtained by resolution from $\lambda(E)$.

If A is associated with a node in F then $\lambda(A)$ is in $C(\Phi(F))$ thus it is in $\lambda(E)$. Otherwise, A is associated with a node, not equal to a connection node, in C and

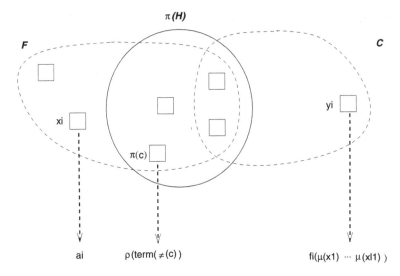

Fig. 10.12 The substitution λ

$\lambda(A) = \mu'(Q_{2m})$. We know that for all $i = 1,\dots,m_1$ $\mu(Q_{1i}) \in Aug(C(\Phi(F)))$. Therefore, using the clause in $\Phi(R)$ containing Q_{2m} all the negative atoms can be eliminated, and one obtains $\mu'(Q_{2m})$. $\quad\square$

The soundness is proven by recurrence on the number of rule applications using Lemma 10.1.

Proof of completeness
Notations

$\Phi(F) = \exists x_1 \dots x_{l(F)}(P_1 \wedge \dots P_{n(F)})$,
$C(\Phi(F)) = \{\rho(P_i) \mid i = 1,\dots,n(F)\}$ where ρ is the substitution $\rho = \{(x_1,a_1),\dots,(x_{l(F)},a_{l(F)})\}$ with a_i-s being skolem constants;
For any $R \in \mathcal{R}$:

$$\Phi(R) = \forall x_1^R \dots \forall x_{q(R)}^R (\forall x_{q(R)+1}^R \dots \forall x_{l_1(R)}^R Q_{11}^R \wedge \dots \wedge Q_{1m_1(R)}^R \rightarrow$$
$$(\exists y_1^R \dots y_{l_2(R)}^R Q_{21}^R \wedge \dots \wedge Q_{2m_2(R)}^R)),$$

$$C(\Phi(R)) = \bigwedge_{i=1}^{m_2'(R)}(\neg Q_{11}^R(x_1^R,\dots,x_{l_1(R)}^R) \vee \dots \vee \neg Q_{1m_1(R)}^R(x_1^R,\dots,x_{l_1(R)}^R)$$
$$\vee Q_{2i}^R(x_1^R,\dots,x_{l_1(R)}^R,f_1(x_1^R,\dots,x_{l_1(R)}^R),\dots,f_{l_2(R)}(x_1^R,\dots,x_{l_1(R)}^R))))$$

where the f_i^R-s are skolem functions.
Let $\mathcal{E}$ denote the set of ground (or Herbrand) clauses associated with the set of clauses $\{C(\Phi(\mathcal{V})),C(\Phi(H)),C(\Phi(\mathcal{R})),C(\neg\Phi(Q))\}$ and built on $I \cup \{a_i \mid i = 1,\dots,l(F)\} \cup \{f_i^R \mid i = 1,\dots,l_2(R), R \in \mathcal{R}\}$.
$Sub(U)$ is the set of ground substitutions $\{(x_1,t_1),\dots,(x_p,t_p)\}$ such that the t_i-s are pairwise distinct and do not belong to I. For such a substitution, if the x_i-s are associated with generic nodes in a graph, then there is a bijection between these nodes and the t_i-s without mix-up with the individual nodes.

Proof.

Proof steps

The main steps of the proof are as follows. Since $\Phi(\mathcal{V})$, $\Phi(F)$, $\Phi(\mathcal{R}) \models \Phi(Q)$, $\mathcal{E}$ is unsatisfiable and thus there is $\mathcal{E}' \subseteq \mathcal{E}$ such that $\mathcal{E}'$ is finite and unsatisfiable.

Starting from $A_0 = Aug(C(\Phi(F)))$ a sequence $A_0 \subseteq A_1 \subseteq \ldots \subseteq A_s = A_{s+1}$ of clause sets is built. A sequence $(H_n, \rho_n, R_n, \pi_n)$ is associated with sequence A_n as follows.

$H_0 = F$, $\rho_0 = \rho$ and R_n is a rule, π_n is a homomorphism from the hypothesis of R_n to H_n, H_{n+1} is the normal form of $(R_n, \pi_n)H_n$ and ρ_{n+1} is a substitution in $Sub(U)$ obtained from ρ_n and π_n.

Let H_s be the last graph obtained by this construction. One proves that:

1. There is a clause B in $C(\neg\Phi(Q))$ with all its atoms in A_s,
2. let $C'(\Phi(H_s)) = \{\rho_s(A) \mid A \in atoms(\Phi(H_s))\}$, then $A_s = Aug(C'(\Phi(H_s)))$.

Therefore, there is a homomorphism from Q to H_s (and this proves the completeness). Indeed, let $p(u_1, \ldots, u_k)$ be any atom of $\Phi(Q)$, and let σ be the substitution defined by the clause B, i.e., $p(\sigma(u_1), \ldots, \sigma(u_k))$ is an atom of B. There is an atom $q(v_1, \ldots, v_k)$ in $\Phi(H_s)$ such that $q \leq p$ and $\rho_s(v_i) = \sigma(u_i)$ for all $i = 1, \ldots, k$. Therefore, there is a $\mathcal{V}$-substitution of $\Phi(Q)$ to $\Phi(H_s)$, and by using Property 4.6 Q maps to H_s.

Let us now build the sequences A_n and $(H_n, \rho_n, R_n, \pi_n)$, for $n = 0, \ldots, s$ and let us prove their properties.

Construction of the sequence A_n
$$A_0 = Aug(C(\Phi(F)))$$
for all $n \geq 0$, if:

- There is $R \in \mathcal{R}$,
- there is a ground substitution $\mu =$
$$\{(x_1^R, t_1), \ldots, (x_{l_{1(R)}}^R, t_{l_{1(R)}}), (y_1^R, f_1^R(t_1, \ldots, t_{l_{1(R)}})), \ldots, (y_{l_{2(R)}}^R, f_{l_{2(R)}}^R(t_1, \ldots, t_{l_{1(R)}}))\},$$
- there is $m \in \{1, \ldots, m_2'(R)\}$ such that $\mu(\neg Q_{11}^R \vee \ldots \vee \neg Q_{1m_{1(R)}}^R \vee Q_{2m}^R) \in \mathcal{E}'$
and, for all $i = 1, \ldots, m_1^R$, $\mu(Q_{1i}^R) \in A_n$ and $\mu(Q_{2m}^R) \notin A_n$

then $A_{n+1} = A_n \cup Aug(\{\mu(Q_{2i}^R) \mid i = 1, \ldots, m_2(R)\}$, otherwise $A_{n+1} = A_n$.

At each step a new clause in $\mathcal{E}'$ is chosen and $\mathcal{E}'$ is finite. Thus the construction stops at a rank s.

Proof of the first property

Let us prove the first property, i.e., there is a clause B in $C(\neg\Phi(Q))$ with all its atoms in A_s. A_s is a set of ground atoms, therefore we can define an interpretation v as follows: $v(A) = true$ if and only if $A \in A_s$. We show by reduction to the absurd that if the property is not satisfied, i.e., if any clause in $C(\neg\Phi(Q))$ has an atom that is not in A_s, then $v(\mathcal{E}') = true$ and this is absurd since $\mathcal{E}'$ is unsatisfiable.

Let D be a clause in $\mathcal{E}'$.

- Let $D \in C(\neg \Phi(Q))$, then D has at least one atom that is not in A_s, thus $v(D) = true$.
- Let $D \in C(\Phi(F))$, then D is in A_0; thus it is also in A_s and $v(D) = true$.
- Let $D \in C(\mathcal{V})$. $D = \neg A \vee A'$ and $A' \in Aug(A)$. If $A \in A_s$ then $A' \in A_s$ because $Aug(A_s) = A_s$. Therefore, $v(A') = true$ and $v(D) = true$. Otherwise $v(A) = false$ and $v(D) = true$.
- Let $D = \mu(\neg Q^R_{11} \vee \ldots \vee \neg Q^R_{1m_1(R)} \vee Q^R_{2m})$. If for all $i = 1, \ldots, m^R_1$ $\mu(Q^R_{1i}) \in A_s$ then $\mu(Q^R_{2m})$ is also in A_s, therefore $v(\mu(Q^R_{2m})) = true$ and $v(D) = true$, otherwise there is i, $1 \le i \le m_1(R)$ with $\mu(Q^R_{1i}) \notin A_s$ and $v(D)$ is still equal to $true$.

Construction of the sequence $(H_n, \rho_n, R_n, \pi_n)$

$H_0 = F$ and $\rho_0 = \rho$.

For all $n = 0, \ldots, s - 1$ let us consider the clause $\mu(\neg Q^R_{11} \vee \ldots \vee \neg Q^R_{1m_1(R)} \vee Q^R_{2m})$ used for building A_{n+1} from A_n, then $R_n = R$ and $\mu_n = \mu$.

There is a homomorphism π from hypothesis H of R to H_n such that $\mu(term(c)) = \rho_n(term(\pi(c)))$ for any c concept node in H. Let $\pi_n = \pi$.

$H_{n+1} = (R_n, \pi_n)H_n$, and any variable $y^{R_n}_i$ associated with the generic concept nodes in the conclusion of R_n is renamed y^n_i in order to use different variables for different nodes (a rule can be used several times).

$\rho_{n+1} = \rho_n \cup \{(y^n_1, f^R_1(\mu_n(x^R_1), \ldots, \mu_n(x^R_{l_1(R)}))), \ldots, (y^n_{l_2(R)}, f^R_{l_2(R)}(\mu_n(x^R_1), \ldots, \mu_n(x^R_{l_1(R)})))\}$.

Let $C'(\Phi(H_n)) = \{\rho_n(B) \mid B \in atoms(\Phi(H_n))\}$. Then, one shows by recurrence on n that: $A_n = Aug(C'(\Phi(H_n)))$ and $\rho_n \in Sub(U)$.

This ensures that if $n < n_0$, then π_n exists by Property 4.6, and therefore H_{n+1} can be constructed.

For $n = 0$, one has $A_0 = Aug(C(\Phi(F))) = Aug(C'(\Phi(H_0)))$ and $\rho_0 = \rho \in Sub(U)$.

Let us assume that the sequence is constructed until n and that $A_n = Aug(C'(\Phi(H_n)))$ and $\rho_n \in Sub(U)$. Let us consider $n + 1$.

$A_{n+1} = A_n \cup Aug(\{\mu_n(Q^R_{2i}) \mid i = 1, \ldots, m_2(R)\}$
$= Aug(C'(\Phi(H_n))) \cup Aug(\{\mu_n(Q^R_{2i}) \mid i = 1, \ldots, m_2(R)\}$
$= Aug(C'(\Phi(H_n)) \cup \{\mu_n(Q^R_{2i}) \mid i = 1, \ldots, m_2(R)\})$

Lemma 10.2 proved hereafter gives
$\{\rho_{n+1}(B) \mid B \in atoms(\Phi((R_n, \pi_n)H_n))\} = C'(\Phi(H_n)) \cup \{\mu_n(Q^R_{2i}) \mid i = 1, \ldots, m_2(R)\}$
then $A_{n+1} = Aug(\{\rho_{n+1}(B) \mid B \in atoms(\Phi((R_n, \pi_n)H_n))\}$
and $A_{n+1} = Aug(\{\rho_{n+1}(B) \mid B \in atoms(\Phi(H_{n+1}))\}) = Aug(C'(\Phi(H_{n+1})))$.

It remains to show that for any $n \ge 0$ one has $\rho_n \in Sub(U)$. $\rho_0 = \rho$. Thus the property holds for $n = 0$. Let us assume that $\rho_n \in Sub(U)$, and let us consider $n + 1$. The terms associated with the generic concept nodes in H_n are pairwise disjoint and are not in I since $\rho_n \in Sub(U)$. The terms associated with the generic concepts of C^R, which are not connection nodes, are $f^R_1(\mu_n(x^R_1), \ldots, \mu_n(x^R_{l_1(R)})), \ldots, f^R_{l_2(R)}(\mu_n(x^R_1), \ldots, \mu_n(x^R_{l_1(R)}))$. They are pairwise disjoint and are not in I. Let us show that if $(z, t) \in \rho_n$, then t is not equal to a $f^R_i(\mu_n(x^R_1), \ldots, \mu_n(x^R_{l_1(R)}))$. Indeed, otherwise a clause

$\mu_n(\neg Q_{11}^R \vee \ldots \vee \neg Q_{1m_1(R)}^R \vee Q_{2m}^R)$ would have been chosen before in the construction of the sequence A_n, and this is impossible. $\square$

Lemma 10.2. *Let:*

- $\Phi(F) = \exists z_1 \ldots z_{l(F)}(P_1 \wedge \ldots P_{n(F)})$,
- $\rho = \{(z_1,t_1),\ldots,(z_{l(F)},t_{l(F)})\}$ *where all t_i-s are in $Sub(U)$ (the t_i-s are thus in bijection with the generic concept nodes in F)*,
- $C'(\Phi(F)) = \{\rho(P_i) \mid i = 1,\ldots,n(F)\}$,
- *R be a rule (H,C) with $\Phi(R) = \forall x_1 \ldots \forall x_q^R(\forall x_{q+1} \ldots \forall x_{l_1} Q_{11} \wedge \ldots \wedge Q_{1m_1}^R \rightarrow \exists y_1 \ldots y_{l_2}^R Q_{21} \wedge \ldots \wedge Q_{2m_2})$*

  $C(\Phi(R)) = \bigwedge_{i=1}^{m_2'}(\neg Q_{11}(x_1,\ldots,x_{l_1}) \vee \ldots \vee \neg Q_{1m_1}(x_1,\ldots,x_{l_1})$
  $\vee Q_{2i}(x_1,\ldots,x_{l_1},f_1(x_1,\ldots,x_{l_1}),\ldots,f_{l_2}(x_1,\ldots,x_{l_1})))$,
- *π be a homomorphism from the hypothesis of R to F and μ, such that: For any concept node c in H, $\mu(term(c)) = \rho(term(\pi(c)))$ and $\forall m = 1,\ldots,m_1$ one has $\mu(Q_{1m}) \in Aug(C'(\Phi(F)))$,*
- *λ be the substitution $\lambda = \rho \cup \mu'$ where $\mu' = \{(y_1, f_1(\mu(x_1),\ldots,\mu(x_{l1}))),\ldots, (y_{l2}, f_{l2}(\mu(x_1),\ldots,\mu(x_{l1})))\}$.*

Then, $\{\lambda(A) \mid A \in atoms(\Phi((R,\pi)F))\} = C'(\Phi(F)) \cup \{\mu(Q_{2i}) \mid i = 1,\ldots,m_2'\}$.

Proof. Let A be an atom of $\Phi((R,\pi)F)$. If A is associated with a node of F then $\lambda(A) = \rho(A) \in C'(\Phi(F))$. Conversely, any atom in $C'(\Phi(F))$ is equal to a $\lambda(A)$. Let A be associated with a concept node c in C which is not a connection node. If c is an individual node with marker i and type t, then $\lambda(A) = t(i)$, and there is $j \in \{1,\ldots,m_2'\}$ with $Q_{2j} = t(i) = \mu(Q_{2j})$. If c is a generic concept node having type t and with y_j as its variable, then $\lambda(A) = t(f_j(\mu(x_1),\ldots,\mu(x_{l1})))$ which is equal to $\mu(Q')$, where Q' is the atom Q_{2j} associated with c.

Let A be associated with a relation node r in C. Let us consider an argument u of A. If it corresponds to a concept node which is not a connection node then $\lambda(u) = \mu(u)$. Otherwise it is equal to a connection node c_{2j}. $\lambda(u) = \rho(term(\pi(c_j)))$ and $\mu(x_j)$ is precisely equal to $\rho(term(\pi(c_j)))$. Finally, one can check that any $\mu(Q_{2i})$ for $i = 1,\ldots,m_2'$ is equal to a $\lambda(A)$ and thus: $\{\lambda(A) \mid A \in atoms(\Phi((R,\pi)F))\} = C'(\Phi(F)) \cup \{\mu(Q_{2i}) \mid i = 1,\ldots,m_2'\}$. $\square$

10.3 Backward Chaining

We describe a backward chaining mechanism for bicolored rules in this section. Moreover, it is assumed that every individual marker always occurs with the same concept type in concept nodes. As a consequence of these restrictions, a rule application to a fact F never restricts labels of existing concept nodes in F (either directly, or indirectly in the normalization step). Its sole effect is to add new nodes. The backward mechanism detailed in this section is based on a graph notion, that of a *piece*, which is strongly related to this property. To take general λ-rules into account, the piece notion as well as the *unification* notion would have to be extended, which has not yet been done. The above restrictions do not lead to a loss of expressivity (see the note in Sect. 10.2.2).

10.3.1 Outline of the Backward Chaining Mechanism

The global backward mechanism relies on the same ideas as the classical goal resolution, as performed by Prolog for instance. It starts from the initial goal Q. An elementary step consists of first determining a part of the current goal Q that could possibly be obtained by a rule R: If there is such a part, say q, a "unification" of q and the conclusion of R is computed. Secondly, a new goal is built from Q, say Q', by deleting q and adding the hypothesis of R. Q is said to be "erased" by Q'. Q' is such that, if one applies R to Q', a specialization of Q is obtained. In other words, if Q' is proven to be deducible from the KB, so is Q. A fact is considered as a rule with an empty hypothesis. It allows us to delete part of the current goal without adding a new part to prove. The process ends successfully if the last produced goal is empty (and, in this case, the last operation is a unification of the entire goal with a fact).

Let us illustrate the global process of backward chaining on a very simple example pictured in Fig. 10.13. The knowledge base is composed of a fact F and a rule $R = H \Rightarrow C$ (pictured as a λ-rule). Q is the goal. Q is a consequence of the knowledge base since a specialization of Q can be obtained by one application of R to F (see previous Fig. 10.8: an application of $R = R_2$ to F yields F' which answers Q). To proceed backward, a naive idea could be to check if there is a homomorphism from the goal, or part of the goal, to the conclusion of the rule; if yes, this part of the goal could be replaced by the hypothesis of the rule (according to the idea that, if the hypothesis is proven, then a specialization of Q is obtained by applying the rule). However, homomorphism is not sufficient (indeed, check that there is no homomorphism from Q to C). We have to look for a *unifier* of Q and C, which unifies part of Q and part of C; more specifically, this unifier defines specializations of Q and C with respect to a compatible pair of C-partitions (cf. Definition 8.12) on subsets of the concept node sets in Q and C. Here we consider $(\{c_1\}, \{c_2\})$ for Q and $(\{c_6\}, \{c_7\})$ for C; c_1 is thus associated with c_6 and c_2 with c_7. The greatest lower bound of the labels of c_1 and c_6 is $(Man, John)$; for c_2 and c_7, it is $(Person, *)$; this compatible pair of C-partitions leads to specializations of C and of Q, which in this example are the same: See $u(C) = u(Q)$ in Fig. 10.13 (the name u is used for "unifier"). It remains to prove the hypothesis H of R, and more precisely to prove a specialization H' of H according to the unifier: See H' in Fig. 10.13, which is obtained from H by specializing the label of c_3 into $(Man, John)$. It is the new goal Q'; there are two unifiers of Q' and F. The first one associates c_5 in Q' with c_{10} in F: Their labels are restricted to $(Painter : P)$; the second one associates c_5 in Q' with c_{11} in F and no label restriction has to be done. In both cases, Q' is completely proven by F: Since F is a rule without hypothesis, the new goal is replaced by the empty graph in both cases, which yields two proofs that Q can be deduced from the knowledge base. In general, backward chaining involves more complex operations. Namely, unification does not consider a whole goal but a part of it, and the kind of specialization performed may be based on a more complex compatible pair of C-partitions, where classes are not restricted to just a single node, and thus may lead to the merging of concept nodes of the same graph.

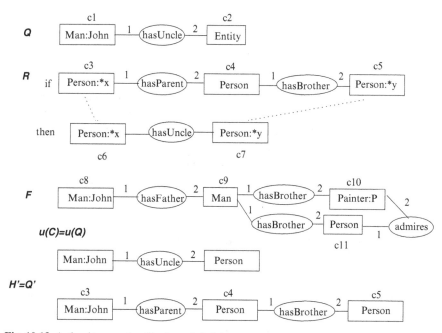

Fig. 10.13 A simple example of backward chaining

The differences between the graph backward mechanism and classical goal resolution come from the structure of graph rules, which is more complex than the structure of definite clauses, since the conclusion can be any BG and may contain new generic nodes. The mechanism exploits the graph structure of rules and goals, and tries to process subgraphs as large as possible without decomposing them into simpler structures. It relies on the notion of a *piece*, which is a subgraph of a rule conclusion (cf. Definition 8.22 for a general definition of a piece). A *cut point* of a rule conclusion C is either a frontier node or an individual node (that is not shared with the hypothesis). The pieces of C are defined by its cut points: Two nodes in C belong to the same piece if there is a path between them that does not go through a cut point.

The idea behind the piece notion is that it represents a "unit of knowledge" brought by a rule application. More precisely, each rule R with k pieces in its conclusion can be decomposed into an equivalent set of k rules, each with the same hypothesis as R, and a conclusion composed of a single piece. The conclusions of these rules cannot be further decomposed while keeping the semantics of R. A rule application to a fact necessarily adds at least one piece to this fact. In backward chaining, the unification between a goal and a rule is thus guided by the pieces of the rule conclusion, and a unification is eligible only if part of the goal corresponding to a piece unifies with an entire piece of the rule conclusion (otherwise this unification would be useless). Thus, the goal is erased by subgraphs corresponding to pieces.

10.3.2 Piece Resolution

Following our main thread (i.e., the maximal use of graph theoretical notions), instead of splitting the goal into trivial subgraphs (stars) composed of one relation and its arguments and erasing each of them, subgraphs as large as possible which can be processed as a whole are determined. Such subgraphs are pieces (cf. Definition 8.22), whose definition is reviewed hereafter.

Definition 10.8 (Rule cut points and conclusion pieces). Let $R = H \Rightarrow C$ be a rule.

1. A *cut point* in H is a frontier node.
2. A *cut point* in C is either a frontier node or an individual concept node in C that is not shared with H.
3. The *pieces* of C are defined with respect to the cut points in C.

Two nodes of C belong to the same piece if and only if there is a path between them that does not go through any cut point (however, a cut point can be an extremity of such a path). In particular, the arguments of a relation node belong to the same piece. If C has no cut point, it is itself a piece. Since a fact is a rule without a frontier node, its pieces are determined by its individual concept nodes.

In the rule R in Fig. 10.13 (pictured as a λ-rule), the cut points are the connection nodes, and the hypothesis as well as the conclusion have only one piece. Figure 10.14 presents a more complex rule (pictured as a bicolored rule). The cut points of the hypothesis H are the frontier nodes a and b, the cut points of the conclusion C are a, b and the individual node e; H has one piece, and C has four pieces denoted P_1, P_2, P_3 and P_4.

The following notion of a unifier determines part of the current goal Q that could be answered by an application of a rule R (i.e., a part that could be mapped to a fact obtained by an application of R). As a rule application may involve merging several nodes of C and specializing their labels, one looks for a homomorphism from a subgraph of Q to a *specialization* of C, but not to C itself, with this specialization concerning the cut points of C. This homomorphism determines "pieces" in Q: The cut points of Q are the nodes mapped to cut points of C. Moreover, this homomorphism has to map at least one piece of Q. Otherwise the unification would be useless for the backward proof, as previously explained.

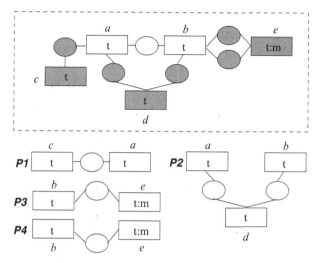

Fig. 10.14 Cut points and pieces of a rule

The following definitions use several notions introduced in Chap. 8, i.e., specialization of a graph with respect to a compatible pair of C-partitions (cf. Definition 8.13) and join of two graphs with respect to a compatible pair of C-partitions (cf. Definition 8.14).

Definition 10.9 (Unifier). Let Q be a goal and $R = H \Rightarrow C$ be a rule. A *unifier* $u(Q,R)$ of Q and R is a triple (P_C, P_Q, π_u) such that:

- $P_C = (p_1^1 \dots p_k^1)$ and $P_Q = (p_1^2 \dots p_k^2)$ is a (possibly empty) compatible pair of C-partitions of a subset C' of cut points of C and a subset C'' of concept nodes of Q respectively; let $u(C) = s(C, P_Q, P_C)$ denote the specialization of C with respect to these partitions (if P_C and P_Q are empty, $u(C) = C$);
- let us call cut points of Q the concept nodes of C'',
- π_u is a homomorphism from a subgraph of Q to $u(C)$, which satisfies the following conditions:

  1. it maps at least one piece of Q to $u(C)$ (only entirely mapped pieces of Q will be of interest in the backward chaining),
  2. it maps all concept nodes appearing in P_Q to their corresponding node in $u(C)$: each concept node in a class p_i^1 of P_Q is mapped to the concept node of $u(C)$ built from p_i^2 in P_C.

As pointed out in the above definition, the partitions P_C and P_Q may be empty. Assuming that Q is connected, this is the case when π_u is a homomorphism from the whole Q to the conclusion C, such that no concept node of Q is mapped to a cut point of C. This implies in particular that Q has no individual concept node (otherwise this node would be mapped to an individual concept node in C, which would be a cut point).

If R is a fact, its cut points are its individual concept nodes, and the label of these cut points cannot be restricted in $u(C)$ (because of the unique type assumption on individual markers). In the previous definition, $u(C)$ is thus obtained from C by merging individual concept nodes with the same label, i.e., $u(C)$ is a generalization (or is equal to) the normal form of C. Hence, the following property holds true:

Property 10.3. Let $R = \varepsilon \Rightarrow C$ be a rule with an empty hypothesis (i.e., a fact), Q a goal and $u(Q,R) = (P_C, P_Q, \pi_u)$ a unifier of Q and R. Then, π_u is a coref-homomorphism from a subBG of Q to C; thus, it yields a homomorphism from a subBG of Q to the normal form of C.

Let Q be a goal and R be a rule with conclusion C. A backward application of R to Q can be done if there is a unifier $u(Q,R)$ of Q and R. Roughly said, a new goal Q' is built by removing the part of Q that has been unified with the conclusion of R and adding the hypothesis of R, which has been specialized according to the unifier.

Definition 10.10 (Backward chaining elementary step). Given a unifier $u(Q,R = H \Rightarrow C) = (P_C, P_Q, \pi_u)$, Q is replaced by Q' which is built as follows:

1. Build $u(Q) = s(Q,P_Q,P_C)$ the specialization of Q with respect to P_Q and P_C; the cut points of $u(Q)$ are nodes coming from the cut points of Q, and the pieces of $u(Q)$ are defined by its cut points (thus, they correspond to the pieces of Q specialized on their cut points);
2. build $u(Q)'$ by removing from $u(Q)$ the pieces mapped by π_u to $u(C)$, except for the concept nodes whose removal would lead to a subgraph that is not a correct BG: For all piece p of Q mapped by π_u, first remove from $u(Q)$ all the relation nodes of p and all concept nodes of p that are not cut points in $u(Q)$; then remove a cut point of p only if all pieces it belongs to are removed;
3. let P'_C be obtained from P_C by removing the classes containing an individual node not shared with H (this node does not belong to the frontier of the rule); let $P'_Q = (p_1^{2'} \ldots p_l^{2'})$ be obtained from P_Q by removing the classes corresponding to suppressed classes of P_C (P'_Q and P'_C thus constitute a compatible pair of C-partitions); let $P'_H = (p_1^3 \ldots p_l^3)$ be the partition on a subset of the cut points of H built in a similar way: Let C'_H be the subset of cut points in H (i.e., frontier nodes) that appear in P'_C; P'_H is the partition of C'_H induced by P'_C: Two nodes of C'_H are in the same class if and only if their corresponding nodes in C are in the same class of P'_C (P'_H and P'_Q constitute a compatible pair of C-partitions);
4. build $u(H) = s(H,P'_Q,P'_H)$ the specialization of H with respect to P'_Q and P'_H;
5. join $u(Q)'$ and $u(H)$ with respect to P'_H and P'_Q. We obtain Q'.

The possible introduction of new individuals in the conclusion of the rule implies special steps. The removal of parts of Q that unify with C is performed according to the partition P_C on some cut points in C. If these cut points contain individuals not shared with H, the corresponding classes in the partitions P_C and P_Q are suppressed, yielding P'_C and P'_Q. The partition P'_H is computed with respect to P'_C (or equivalently to P'_Q) and not to P_C (or P_Q). The new goal is obtained by joining the remaining part of Q and hypothesis H according to P'_H and P'_Q. Step 3 is thus needed only if C introduces new individuals. Otherwise, in step 4, $P'_Q = P_Q$ and $P'_H = P_H$.

If R is a fact, steps 3 and 4 are not needed. Indeed, P_C is composed of individual concept nodes; step 3 computes empty P'_C and P'_H. As H is empty, so is $u(H)$. Finally $Q' = u(Q)'$, which has been computed at step 2. The following property is immediate:

Property 10.4. The graph Q', produced by a backward chaining elementary step with a unifier $u(Q, R = H \Rightarrow C)$, is empty if and only if R is a fact (i.e., H is empty) and there is a homomorphism from Q to $u(C)$ (thus to the normal form of C).

Definition 10.11 (Piece resolution).
Let $\mathcal{K} = (F, \mathcal{R})$ be a KB, where F is a set of facts and $\mathcal{R}$ a set of rules, and let Q be a BG. A *piece resolution* of Q in $\mathcal{K}$ is a sequence $Q = Q_0, Q_1 \ldots Q_p, p \geq 0$ such that for all $1 \leq i \leq p$, Q_i is obtained by a backward chaining elementary step with a unifier $u(Q_{i-1}, R)$, where $R \in \mathcal{R} \cup \{F\}$. The piece resolution ends successfully if $Q_p = \emptyset$ (and in this case $u(Q_{p-1}, R)$ is such that $R = F$).

Note that the knowledge base does not need not to be in normal form. However, if it is in normal form, the unification with facts can be performed more efficiently because it is based on a homomorphism instead of a coref-homomorphism (cf. Proposition 10.3).

Example. Figure 10.15 shows a rule R (that was already used in the rule application example of Fig. 10.7), part of a fact F, and a goal Q (which can be read as: "Does B possess a vehicle, does she/he get a salary and who decides it?") Let Q be the initial goal. Q is unified with R by (P_C, P_Q, π_u), where $P_C = (\{c_1\}, \{c_2\})$, $P_Q = (\{c_4\}, \{c_6\})$; $u(C)$ is obtained from C by restricting the label of c_4 to $(Person : B)$; let us keep the same names for the nodes in $u(C)$ as in C; $\pi_u = \{(c_4, c_1), (c_5, c_2), (c_6, c_3), (r_3, r_1), (r_4, r_2)\}$. The new goal is Q'. Q' is unified with F by $(P_C, P_{Q'}, \pi_u)$, where $P_C = (\{c, d\})$, $P_{Q'} = (\{c_7, c_8\})$; let e be the node of $u(C)$ obtained by merging c and d. $\pi_u = \{(c_6, a), (c_7, e), (c_8, e), (c_9, b), (r_5, r_8), (r_6, r_8), (r_7, r_9)\}$. See that π_u is a homomorphism from Q' to $norm(F)$. The empty goal is obtained.

A backward chaining scheme is presented in Table 26. It maintains a set, called *Unifiers*, of all not yet applied unifiers . At each step (i.e., an iteration of the "while true" loop), it computes unifications for the current goal and adds them to Unifiers. If Unifiers is empty, it returns false (there is no successful resolution). Otherwise, it removes an element in Unifiers and performs the corresponding elementary backward chaining step. If the obtained goal is empty, it returns true (a successful derivation has been built). Whereas the forward chaining scheme presented in Table 25 builds derivations by a breadth-first exploration of the graph space, no exploration strategy is imposed here. Backward chaining algorithms usually try to reconstruct derivations in a depth-first manner, with the objective being to decide quicker when the answer is yes. However, the price to pay is that the algorithm may enter into an infinite resolution sequence without being able to try others. The strategy is determined by the way of managing the set Unifiers. If it is managed as a stack, we have

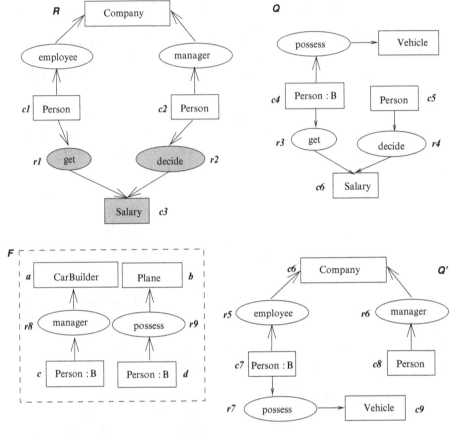

Fig. 10.15 Piece resolution

a depth-first exploration. If it is managed as a queue, we have a breadth-first exploration, which ensures that no resolution is missed (this guarantees that the algorithm
returns true if Q can be deduced), but it will likely be slower.

10.3.3 Soundness and Completeness of Backward Chaining

The following result is obtained:

Theorem 10.2 (Backward chaining soundness and completeness). *Let* $\mathcal{K}=(F,\mathcal{R})$
*be a KB, where F is a set of facts (not necessarily in normal form) and $\mathcal{R}$ a set of
rules, and let Q be a BG. There is a piece resolution of Q that ends successfully if
and only if $\Phi(\mathcal{K}) \vDash \Phi(Q)$ (where $\Phi(\mathcal{K}) = \Phi(\mathcal{V}) \cup \Phi(\mathcal{R}) \cup \Phi(F)$).*

Algorithm 26: BackwardChaining

Input: a knowledge base $\mathcal{K} = (F, \mathcal{R})$, and a BG Q
Output: true, false or infinite calculus; if true, then Q can be deduced from $\mathcal{K}$ and if false,
      then Q cannot be deduced from $\mathcal{K}$

begin
    Unifiers $\leftarrow \emptyset$
    while *true* **do**
        forall $R \in \mathcal{R} \cup \{F\}$ **do**
            forall *unifier* $u(Q,R)$ **do**
                Unifiers $\leftarrow$ Unifiers $\cup\, u(Q,R)$

        if *Unifiers = $\emptyset$* **then**
            return *false*
        $u(Q,R) \leftarrow$ remove an element from Unifiers
        $Q' \leftarrow$ elementary backward chaining step from Q according to $u(Q,R)$
        if *Q' is the empty graph* **then**
            return *true*
        $Q \leftarrow Q'$
end

The soundness and completeness are not proven directly, but rely on the soundness and completeness of the forward chaining: It is shown that Q and R unify, producing the new goal Q', if and only if there is an application of R to Q' that produces a specialization of Q. Then, if Q is not the empty graph, and considering that rules in $\mathcal{R}$ have a non-empty hypothesis, a piece resolution of Q that ends successfully necessarily uses F in the end. Let us consider a piece resolution in which F is unified with Q_{p-1} to produce the empty graph Q_p , and let $R_{i_0} \ldots R_{i_{p-2}}$ be the rules (including facts) used to produce $Q_1 \ldots Q_{p-1}$. Then there is a derivation sequence that applies the rules in reverse order, $R_{i_{p-2}} \ldots R_{i_0}$, to F and produces a graph to which Q can be mapped. Reciprocally, a subsequence can be extracted from a derivation sequence, from which a piece resolution can be built by considering the rule applications in reverse order.

Let us now give the main steps of the proof by decomposing it into several lemmas.

Lemma 10.3. *Let Q be a goal and $R = H \Rightarrow C$ be a rule (possibly with an empty hypothesis). Let Q' be obtained by a backward chaining elementary step with a unifier $u(Q,R)$. Then R can be applied to Q' to produce a specialization of Q.*

Proof. Since $H \succeq u(H)$ and $u(H) \succeq Q'$, there is a homomorphism from H to Q', say π_1. π_1 can be chosen such that, for each concept node c in H, if c is in a class P_i^3 of P_H', $\pi_1(c)$ is the node obtained from P_i^3, otherwise $\pi_1(c)$ is the same node in Q'. Let $Q_1 = (R, \pi_1)Q'$ be the graph obtained by applying R to Q'. The following mapping from the nodes in Q to the nodes in Q_1 is a homomorphism, say π, from Q to Q_1: For all nodes x, if x belongs to $u(Q)'$, then $\pi(x)$ is the corresponding node in Q_1 (either the node itself, or the node obtained from its class in P_Q if x is a cut point in P_Q that belongs to at least a non-erased piece of Q); otherwise, x is a node in Q that has been erased: Then, x belongs to an erased piece of Q, and we take $\pi(x) = \pi_u(x)$ (x is thus mapped to a node in C). □

Lemma 10.4. *Let $\mathcal{K} = (F, \mathcal{R})$ be a KB, where F is a set of facts and $\mathcal{R}$ a set of rules, and let Q be a BG. If there is a piece resolution of Q that ends successfully then there is an $\mathcal{R}$-derivation from $\text{norm}(F)$ ending with a BG G' such that G' is a specialization of Q.*

Proof. By induction on the number of unifications. If the resolution has one step, the property is true: There is a homomorphism from Q to $\text{norm}(F)$, thus a derivation $F_0 = F$. Assume this is true for any $1 \leq p < n$. Let us consider a successful piece resolution of length n. The first step unifies Q with a conclusion C producing Q'. With the derivation from Q' being of length strictly less than n, the induction hypothesis can be applied: There is an associated rule derivation leading from F to a G' to which Q' can be mapped. Let $\pi_{Q'}$ be a homomorphism from Q' to G'. By applying the rule $R = H \Rightarrow C$ of the first step to Q' as in Lemma 10.3, one would obtain a specialization of Q. Let π_H be a homomorphism from H to Q'. Let π_Q be a homomorphism from Q to $(R, \pi_H)Q'$. R can be applied to G' by the homomorphism $\pi_{Q'} \circ \pi_H$. Let G'' be the graph produced. G'' is obtained from F by an $\mathcal{R}$-derivation. Check that Q can be mapped to G'' by the following homomorphism π: For all nodes x in Q, if $\pi_Q(x)$ comes from a node y in Q' then $\pi(x) = \pi_{Q'}(y)$; otherwise, $\pi(x)$ is the node coming from C corresponding to $\pi_Q(x)$. □

Corollary 10.1. *Backward chaining is sound.*

Proof. From Lemma 10.4 and the soundness of forward chaining (Theorem 10.1).

Lemma 10.5. *Let Q be a goal, $R = H \Rightarrow C$ be a rule with a non-empty hypothesis, and let G be a BG to which R can be applied, producing G'. If there is a homomorphism from Q to G' such that at least one node in Q is mapped to a concept node of G' that does not come from G, then there is a piece unification of Q and R that produces a goal Q' with $Q' \succeq G$.*

Proof. Let π_H be the homomorphism from H to G used in the rule application. Let π be a homomorphism from Q to G' that satisfies the condition. Let π_u be obtained from π by restricting its domain to the part of Q mapped to the subgraph of G' coming from C. A unifier $u(Q, R) = (P_C, P_Q, \pi_u)$ can be built, where: P_Q is the partition of the set of nodes x which are in the domain of π_u such that $\pi_u(x)$ is a either a $\pi_H(y)$ (i.e., it corresponds to a frontier node of the rule) or an individual node coming from C; two nodes are in the same class of the partition if and only if they have the same image by π_u; P_C is the compatible partition of the subset of cut points in C corresponding to nodes in G' that are images of nodes in P_Q. There is a bijection from P_Q to P_C, which maps a class in P_Q whose nodes are mapped to a node z by π_u, to the class in P_C, whose nodes are mapped to z by the rule application. Let us now check that π_u maps pieces of Q. By definition of $u(Q, R)$ and of cut points, if, for a node x in a piece of Q, $\pi_u(x)$ is in C, then all nodes of this piece are mapped by π_u to C. By hypothesis, there is at least such a node x, thus at least a piece of Q is mapped to C by π_u. Let Q' be the graph obtained with $u(Q, R)$. There is a homomorphism π' from Q' to G built as follows. For each node x coming from $u(Q)$, $\pi'(x) = \pi(x)$. For each node y coming from $u(H)$, its image is $\pi'(x) = \pi_H(x)$. If a node comes from both $u(Q)$ and $u(H)$, we have $\pi(x) = \pi_H(x)$, hence π' is a homomorphism from Q' to G.   □

Lemma 10.6. *Let* $\mathcal{K} = (F, \mathcal{R})$ *be a KB, where F is a set of facts and $\mathcal{R}$ a set of rules, and let Q be a BG. If there is a derivation from F to F', a specialization of Q, then there is a piece resolution of Q that ends successfully.*

Proof. By induction of the length of the derivation. The property is true for a derivation of length 0: If Q can be mapped to F by a homomorphism π, then an elementary backward chaining step with a unifier built on $\pi_u = \pi$ produces the empty graph. Assume the property is true for any $1 \leq p < n$. Let us consider a derivation of length n: $F_0 = F, ..., F_n = F'$. Let π be a homomorphism from Q to F_n and let $R = H \Rightarrow C$ be the rule applied to go from F_{n-1} to F_n. If $\pi(Q)$ is entirely contained in the subgraph of F_n coming from F_{n-1}, then π yields a homomorphism from Q to F_{n-1} and, by induction hypothesis, there is a successful piece resolution of Q associated with the derivation $F_0 = F, ..., F_{n-1}$. In this case, the rule application leading from F_n to F_{n-1} is useless to deduce Q. Otherwise, by Lemma 10.5, there is a piece unification of Q and R producing $Q' \succeq F_{n-1}$. By induction hypothesis, there is a successful piece resolution of Q' associated with the derivation $F_0 = F, ..., F_{n-1}$; there is thus a successful piece resolution of Q.   □

Corollary 10.2. *Backward chaining is complete.*

Proof. From Lemma 10.6 and the completeness of forward chaining (Theorem 10.1).   □

Corollaries 10.1 and 10.2 prove the theorem.

No mechanism is better than another in the general case. Backward chaining is much more complex to implement than forward chaining. It is interesting to note that the rule base can be compiled in such way that forward chaining can be guided by the goal in the same way as backward chaining (Sect. 10.4.2.2).

10.4 Computational Complexity of $\mathcal{FR}$-DEDUCTION with Rules

10.4.1 Semi-Decidability of $\mathcal{FR}$-DEDUCTION with Rules

In the example of Fig. 10.8, all $\mathcal{R}$-derivations from F are finite. But it is not true in general. More precisely, let $\mathcal{K} = (F, \mathcal{R})$ be a KB, where F is a set of facts and $\mathcal{R}$ is a set of rules, and let Q be a goal. If the answer to the question "Is Q deducible from $\mathcal{K}$" is positive, the forward chaining algorithm (Table 25) will produce a specialization of Q after a finite number of rule applications. But if the answer is no, there is no guarantee that the algorithm will ever stop. And no algorithm can be guaranteed to stop in all cases: Indeed, $\mathcal{FR}$-DEDUCTION is undecidable. More precisely, it is semi-decidable as we will show it in this section: An answer can be computed in finite time for all positive instances of $\mathcal{FR}$-DEDUCTION but not for all negative ones. Thus, there is no chance to find an algorithm that always stops when the answer is no.

We have an algorithm that runs in finite time if Q is deducible from $\mathcal{K}$ (see Table 25). Thus, to prove the semi-decidability of $\mathcal{FR}$-DEDUCTION, we only have to exhibit a reduction from a problem known to be semi-decidable. Several such reductions have been exhibited (see bibliographical notes for details). Here we give a reduction from *word problem in a semi-thue system*. The proof is interesting in itself since it proves that, even when rules are of the form "if *path* p_1 then *path* p_2," where p_1 and p_2 share origin and extremity, the problem remains semi-decidable.

Property 10.5. $\mathcal{FR}$-DEDUCTION is semi-decidable.

Proof. When the answer to the problem is "yes," a breadth-first search of the tree of all derivations from $\mathcal{K}$ provides the answer in finite time (Table 25). Let us prove that $\mathcal{FR}$-DEDUCTION is not decidable with a reduction from the WORD PROBLEM IN A SEMI-THUE SYSTEM, which is expressed as follows: Given the two words m and m' and $\Gamma = \{g_1, \ldots, g_k\}$ a set of rules, with each rule g_i being a pair of words (α_i, β_i), the question is whether there is a derivation from m to m'. There is an immediate derivation from m to m' (we note $m \to m'$) if, for some g_j, $m = m_1 \alpha_j m_2$ and $m' = m_1 \beta_j m_2$. A *derivation* from m to m' (we note $m \rightsquigarrow m'$) is a sequence $m = m_0 \to m_1 \to \ldots \to m_p = m'$.

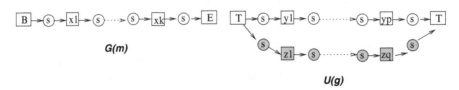

G(m)

U(g)

Fig. 10.16 Transformation from the WORD PROBLEM into $\mathcal{FR}$-DEDUCTION

Let $(m, m', \Gamma = \{g_1, \ldots, g_k\})$ be an instance of the previous word problem. A concept type x_i is assigned to each letter x_i. There are three other concept types: B (for "begin"), E (for "end") and $\top$ (for "anything"). $\top$ is the greatest concept type and all other types are pairwise incomparable. There is one relation type s (for "has successor"). BG $G(m)$ is assigned to any word $m = x_1 \ldots x_k$, and the rule $U(g)$ is assigned to any rule $g = (y_1 \ldots y_p, z_1 \ldots z_q)$, as represented in Fig. 10.16. Let $\mathcal{U}(\Gamma) = \{U(g_1), \ldots, U(g_k)\}$. By a straightforward proof (a recurrence on the smallest derivation length), we obtain that to every path from the node typed B to the node typed E ("begin" to "end") in a BG derived from $(G(m), U(\Gamma))$, corresponds a word (and not a subword) derivable from m, and reciprocally. It follows that $m \rightsquigarrow m'$ if and only if $G(m')$ is deducible from $(G(m), U(\Gamma))$. $\quad\square$

$\mathcal{FR}$-DEDUCTION is not only semi-decidable, it is a computation model, as can be proven by a reduction from the *halting problem of a Turing machine*. This reduction indeed shows that every Turing machine can be simulated by the forward chaining mechanism. The knowledge base built in the proof has only one rule, which proves that $\mathcal{FR}$-DEDUCTION remains non-decidable even with one rule (see the bibliographical notes).

10.4.2 Decidable Cases

Let us now summarize the results on decidable cases.

10.4.2.1 Finite Expansion Rule Sets

In what follows, we assume that the rule application mechanism prevents the generation of trivial redundancies: A rule is never applied twice according to the same homomorphism from its hypothesis to a fact, and the generation of twin relation nodes is avoided (cf. Sect. 10.2.3).

Let us say that an irredundant graph H is *full* with respect to a set of rules $\mathcal{R}$ if every graph that can be obtained by applying one of these rules to H is equivalent to H. In other words, no rule application can bring new information to the graph. Assuming that F is an irredundant graph and that graphs obtained by a rule application are put into irredundant form, if a full graph can be derived from F then it is unique (up to isomorphism).

A weaker stopping condition for forward chaining, with F and a set of rules $\mathcal{R}$, can be to stop when a graph is obtained to which no rule can be applied in an original way (i.e., all possible rule applications have been performed). We call it a *closed* graph. More formally, F' is closed with respect to $\mathcal{R}$ and with respect to an $\mathcal{R}$-derivation $F_0 (= F),\ldots,F_k (= F')$ where F_i $(1 \leq i \leq k)$ is the graph obtained by application of a rule of $\mathcal{R}$ to F_{i-1} according to a homomorphism π_i, if for every rule $R = H \Rightarrow C$ of $\mathcal{R}$, for every homomorphism π from H into F', there is a homomorphism π_i from H to F_{i-1} $(1 \leq i \leq k)$, such that $\pi = \pi_i$.

Given a set of rules $\mathcal{R}$ and a graph F, if a closed graph is $\mathcal{R}$-derivable from F, then it is unique (up to isomorphism). Moreover, if this graph is derivable with n rule applications, then n is the maximal length of an $\mathcal{R}$-derivation, and all derivations of length n lead to it. When it exists, we call it the *closure* of F with respect to $\mathcal{R}$, which we denote $F_{\mathcal{R}}^*$.

If the closure of F exists, its irredundant form is equivalent to the full graph of F. However, there are cases where the full graph exists but not the closure. Consider, for instance, a bicolored rule "if H then C" without frontier node (thus H and C are disconnected), and such that C is a specialization of H. Applying the rule to a fact consists in adding a new connected component C to this fact. Let F' be the graph obtained after the first rule application. As C is a specialization of H, the rule can be applied once again by mapping its hypothesis to the newly added C component, and this indefinitely. All applications after the first one produce a graph equivalent to F'. Thus, the irredundant form of F' is full.

Informally, the notion of a closed graph translates the fact that nothing can be added that has not been already added, whereas the notion of a full graph says that nothing can be added that really adds new information to the graph.

Definition 10.12 (Finite expansion set). A set of rules $\mathcal{R}$ is called a *finite expansion set* if, for every BG F, there is a (finite) $\mathcal{R}$-derivation $F \ldots F'$ such that the irredundant form of F' is full with respect to $\mathcal{R}$. This full graph is denoted $F_{\mathcal{R}}$.

If $\mathcal{R}$ is a finite expansion set, $\mathcal{FR}$-DEDUCTION becomes decidable. Indeed, in order to determine whether a BG Q is deducible from a KB $(F, \mathcal{R})$, it suffices to compute $F_{\mathcal{R}}$, then to check the existence of a homomorphism from Q to $F_{\mathcal{R}}$.

Note however that the finite expansion set property is not a *necessary* condition for decidability. See for instance Fig. 10.17: The set of rules $\{R\}$ is not a finite expansion set, because each application of R leads to a new application of R. Nevertheless, $\mathcal{FR}$-DEDUCTION is decidable for $\{R\}$: For any set of facts F, for any goal Q, the answer can be given after k steps of forward chaining, where k is the maximal number of relation nodes in a directed path of Q consisting of relations with labels less or equal to (*hasParent*).

R   

Fig. 10.17 A non-f.e.s of rules

Let us now focus on particular cases of finite expansion set A λ-rule is said to be *range restricted* if its application does not create any generic concept node. Equivalently, its conclusion is restricted to cut points (connection nodes or individual nodes) and relation nodes. We use this expression by analogy with the so-called rules in Datalog, where all variables of the head must appear in the body. Such rules are also called *safe* in the literature. Consider for instance the rules of Fig. 10.2: R_2 and R_3 are range-restricted, while R_1 is not.

A range restricted rule $R = H \Rightarrow C$ can be decomposed into an equivalent set of rules with a conclusion restricted to at most one node outside connection nodes. There is at most one rule for each node in the conclusion of R: For each connection node c_{2_i} in C such that the type of c_{2_i} differs from the type of c_{1_i}, one rule with hypothesis H and a conclusion restricted to c_{2_i}; for each individual node c in C, one rule with hypothesis H and a conclusion restricted to c; for each relation node r, one rule whose hypothesis is the disjoint union of H and all individual concept nodes of the conclusion of R, and conclusion is r, with the same neighbors as in C. The logical interpretation of such rules are (function free) range restricted Horn rules. If a BG Q is deducible from a set of range restricted rules $\mathcal{R}$, then it is deducible from the set of their decompositions, and reciprocally.

Note that, contrary to general finite expansion sets, the existence of the closure and of the full graph are equivalent notions in the case of range restricted rules.

Property 10.6. A set of range restricted rules is a finite expansion set.

Proof. Since facts are maintained in normal form, an individual marker appears at most once in them. The number of individual nodes created by the set of rules is bounded by $M = |\mathcal{R}| \times max_{R \in \mathcal{R}} |C_R|$, where $|C_R|$ is the size of the conclusion of the rule R. The number of relation nodes added to a graph F (no twin relation nodes are created) is thus bounded by $N = \sum_{n=1}^{k} P_n (|C_F| + M)^n$, where P_n is the number of relation types with a given arity n appearing in a rule conclusion, k is the greatest arity of such a relation type and C_F is the set of concept nodes in F. Hence, the closure of a graph can be obtained with a derivation of length $L \leq N + M$. We thus obtain $F_{\mathcal{R}}^*$ (or $F_{\mathcal{R}}$) in finite time. $\square$

It follows from the proof of Property 10.6 that the length of a derivation from F to $F_{\mathcal{R}}^*$ or $F_{\mathcal{R}}$ is in $O(n^{k+1})$, where n is the size of $(F, \mathcal{R})$ and k is the greatest arity of a relation type appearing in a rule conclusion. This rough upper bound could be refined but it is sufficient to obtain the following property, which will be used throughout the proofs of complexity results involving range restricted rules (cf. Property 10.8 and results in Chap. 11).

Property 10.7. Under the assumption that the maximum arity of a relation type is a constant, given a range restricted set of rules $\mathcal{R}$, the length of an $\mathcal{R}$-derivation from F is polynomially related to the size of $(F, \mathcal{R})$.

Property 10.8 (Complexity with range restricted rules). Under the assumption that the maximum arity of a relation type is a constant, when $\mathcal{R}$ is a set of range restricted rules, $\mathcal{FR}$-DEDUCTION is NP-complete.

Another example of a finite expansion set is that of disconnected rules. A λ-rule $R = (H, C)$ is *disconnected* if its set of connection nodes is empty. Equivalently, a bicolored rule is disconnected if its hypothesis and conclusion are disjoint. We obtain the same complexity property as for range restricted rules: Under the assumption that the maximum arity of a relation type is a constant, when $\mathcal{R}$ is a set of disconnected rules, $\mathcal{FR}$-DEDUCTION is NP-complete.

Finally, let us point out that the union of two finite expansion sets is not necessarily a finite expansion set. Indeed, in the transformation of Proposition 10.5, each rule can be decomposed into an equivalent set of two rules, with one of these rules being range restricted and the other one being a disconnected rule (see bibliographic notes). The next section will present conditions under which finite expansion sets can be safely used together (cf. Proposition 10.10).

10.4.2.2 Conditions on the Rule Dependency Graph

The notion of a graph representing dependencies between rules of a given set is classical in rule-based systems. The basic idea is that a rule R_2 depends on a rule

R_1 if there can be a fact, say F, such that an application of R_1 to F leads to new applications of R_2 to the resulting fact, say F'. In other words, there are more homomorphisms from the hypothesis of R_2 to F' than to F. As BG rules are more complex than classical definite clauses, the dependency notion will be more complex than in classical rule-based systems.

Definition 10.13 (Neutral rule). Let $R_1 : H_1 \to C_1$ and $R_2 : H_2 \to C_2$ be two rules. R_1 is *neutral* for R_2 if for every BG F on $\mathcal{V}$, for every homomorphism $\pi : H_1 \to F$, let $F' = (R, \pi)F$, the set of homomorphisms from H_2 to F' is the same as the set of homomorphisms from H_2 to F. Otherwise, R_2 *depends on* R_1.

The next theorem gives an effective characterization of rule dependency. The framework considered in this result is the same as in the backward chaining mechanism (see Sect. 10.3).

Theorem 10.3. $R_1 : H_1 \to C_1$ and $R_2 : H_2 \to C_2$ be two rules on the same vocabulary, with a non-empty hypothesis and conclusion. Then, R_2 depends on R_1 if and only if H_2 and R_1 are unifiable in the sense of definition 10.9 (where H_2 and R_1 respectively play the role of Q and R).

Proof: cf. the bibliographical notes.

Let $\mathcal{R}$ be a set of rules. The *graph of rule dependencies* of $\mathcal{R}$, notation $D(\mathcal{R})$, is the directed graph whose nodes are the rules of $\mathcal{R}$, and, for all pair of nodes (R_1, R_2), there is an arc (R_1, R_2) if R_2 depends on R_1. This graph may contain loops, since a rule may depend on itself. Facts can be integrated with $D(\mathcal{R})$. By definition, a fact does not depend on any rule, i.e., cannot have incoming arcs. We thus consider a fact F as a rule with an empty hypothesis, and look for the rules that depend on it. A query can be integrated as well: by definition, no rule depends on it, thus it cannot have outgoing arcs. We thus consider a query Q as a rule with an empty conclusion and look for the rules it depends on. To solve $\mathcal{FR}$-DEDUCTION for a given Q, only the rules which are on a path from F (where F stands for all facts) to Q are useful. The subgraph of $D(\mathcal{R})$ induced by these rules is called the *simplified graph of rule dependencies* and denoted by $D(\mathcal{R}, Q)$.

Deciding whether a rule depends on another is an NP-complete problem (indeed a unification is a polynomial certificate and, for particular cases of rules, a unification is a homomorphism). Building the graph of rule dependencies thus requires $|\mathcal{R}|^2$ calls to an algorithm solving an NP-complete problem. However, it has two interests: for algorithmic efficiency in practice and for decidability results.

The dependency graph is costly to compute, but it can be used as a compilation of the knowledge base: It is computed once, independently from solving $\mathcal{FR}$-DEDUCTION, thus the cost of its construction is not the point. Only Q has to be classified for each use.

A simple use of the dependency graph in the forward chaining mechanism is as follows. In the first step, all rules that are successors of the source F in the simplified dependency graph $D(\mathcal{R}, Q)$ are considered. Then, at step i, $i > 1$, the *only* rules considered are those depending on a rule that has been applied at step $i - 1$.

The dependency graph can be further used to improve forward chaining (See the bibliographic notes for further details).

When $D(\mathcal{R}, Q)$ has no circuit, the number of steps is thus bounded by $(l + 1)$ where l is the length of a longest path in $D(\mathcal{R})$. Hence the following decidability result:

Property 10.9. If the simplified dependency graph $D(\mathcal{R}, Q)$ has no circuit (including no loop), then $\mathcal{FR}$-DEDUCTION is decidable.

In this case, the algorithm can be improved by considering rules according to their *level* in $D(\mathcal{R}, Q)$: the rule level is the maximal length of a path from the source to it. As $D(\mathcal{R}, Q)$ is without a circuit, the maximal length of a path is less than the number of rules. In the first step, the rules of level 1 are considered (thus the successors of the source). More generally, the rules of level i are considered at step $i, i \geq 1$. Thus, each rule is checked only once for applicability.

Combining finite expansion sets and rule dependency leads to a stronger decidability result:

Property 10.10. If each strongly connected component of the simplified dependency graph $D(\mathcal{R}, Q)$ is a finite expansion set of rules, then $\mathcal{FR}$-DEDUCTION is decidable.

Let us recall that two vertices x and y of a directed graph are in the same strongly connected component if there is a path from x to y and a path from y to x. A strongly connected component of the dependency graph represents a maximal set of rules that mutually depend (directly or indirectly) on each other. Let $C_1...C_p$ be the strongly connected components of the simplified dependency graph. Consider the reduced graph corresponding to the components: there is one vertex c_i for each strongly connected component C_i, and there is an arc $c_i c_j$ if there is an arc from a rule in C_i to a rule in C_j (i.e., a rule of C_j depends on a rule C_i). Associate with each c_i its level (as explained just above). The rules can be processed by levels: at step i, each c_j of level i is processed; processing the set of rules represented by c_j consists of applying these rules until the closed (or the full) graph is obtained (and it exists because these rules form a finite expansion set).

10.5 Bibliographic Notes

The notion of a λ-BG is introduced in [Sow84] (under the name of λ-*abstraction*). A CG rule is traditionally seen as a couple of basic conceptual graphs, H and C, with co-reference links between some nodes of H and some nodes of C (e.g., [Sow84], [GW95], [SM96]). [BGM99b, BGM99a] defined rules and constraints as bicolored basic graphs. [SM96] (see also [Sal98]) defined backward chaining operations and proved associated soundness and completeness results. Note that, albeit rules are defined as an ordered pair of basic graphs, it is compulsory in this framework that corresponding distinguished nodes have the same type, thus these rules are equivalent to bicolored rules.

Let us emphasize the originality of piece resolution. It is a really graph-based mechanism, in the sense that it processes goals and rules by subgraphs as large as possible, i.e., the pieces. Other resolution mechanisms known for CG rules are more or less translated from logical resolution. To the best of our knowledge, only one is (and has been proven) sound and complete [GW95]. In this mechanism, the goal is split into trivial subgraphs, with each composed of one relation node and its neighbors. Then, unification is performed on each trivial subgraph. This mechanism is thus very similar to classical logical resolution mechanisms, which erase the goal one atom after another.

Note that pieces are trivial in Prolog, and in rules equivalent to definite or Horn clauses in general. A practical comparison between piece resolution and Prolog resolution was made in [CS98]. More precisely, the rule and piece notions were translated into the logical framework, and a straightforward implementation of the obtained piece resolution mechanism was compared to the SLD resolution of Prolog. Skolemization was used to translate CG rules, or, more precisely from their logical translation, into Prolog rules. The comparison showed an important gain, sometimes huge, in the number of backtracks. One reason for this gain is obviously that part of the backtracks made by Prolog was hidden in the piece unification operation. But there are other reasons. Due to the piece notion, the graph resolution mechanism can perform unifications which are guided by the structure of the goal and rules. Then substitutions that are not piece unifications are rejected, whereas Prolog uses them blindly and detects a failure. Of course, if pieces have no more than one relation node, the piece mechanism is not interesting: In this case, it involves a complex machinery for no gain at all, since unifications are the same as in Prolog.

The fact that the logical translation of a CG rule is similar to a tuple-generating dependency (TGD) in databases was first noted by Rousset (oral communication, 1997). Indeed, the only differences between a TGD t (expressed as a logical formula, cf. for instance [AHV95]) and the formula $\Phi(R)$ associated with a rule R, are on one hand that t may contain equality atoms, and on the other hand that $\Phi(R)$ goes with a set of logical formulas translating the partial orders on types in the vocabulary. But t can be (polynomially) transformed into an equivalent set of formulas without equality (cf. [CS98] based on a result of [BV84]), and we can incorporate the vocabulary in the graphs (cf. the mapping *flat* in Chap. 5). The following *implication problem for dependencies* has been shown to be semi-decidable on TGDs: given a set T of TGDs, and a TGD t, is t deducible from T? [Var84]. Coulondre and Salvat used this result to prove that $\mathcal{FR}$-DEDUCTION is semi-decidable [CS98]. Let us give here this reduction since it is very simple. Consider an instance of the implication problem for dependencies. Replace in the TGD t each universally quantified variable by a new constant not appearing in t and T. Let t' be the TGD obtained. Then t is deducible from the set T if and only if the conclusion of t' is deducible from T and the hypothesis of t'. Now, translating TGDs into rules (with the conclusion and hypothesis of t' being actually translated into BGs), one obtains an instance of $\mathcal{FR}$-DEDUCTION.

In his thesis [Bag01], Baget proved the semi-decidability of $\mathcal{FR}$-DEDUCTION with a reduction from the *halting problem of a Turing machine*, i.e., he proved that every Turing machine can be simulated in the $\mathcal{FR}$ model. This proves that deduction with rules is a computation model. The reduction given from the *word problem in a semi-thue system* in the present chapter is from [BM02]. The word problem is introduced in [Thu14] and is proven to be semi-decidable in [Pos47, reduction to his correspondence Problem].

The complexity of deduction with rules and constraints is studied in detail in [BM02]). Finite expansion sets are defined in this paper. In particular, it is proven that the union of two finite expansion sets is not necessarily a finite expansion set. Decidability results concerning the graph of rule dependencies are from [Bag04]. The graph of rule dependencies is first introduced in this paper, and studied in more depth in [BS06]. In particular, the equivalence between the notion of rule dependency and a unifier (as defined in the backward chaining mechanism) is proven in this latter paper. Uses of the graph of rule dependencies to improve the efficiency of forward chaining as well as backward chaining are also studied in the same paper.

Chapter 11
The BG Family: Facts, Rules and Constraints

11.1 Overview of the BG Family

In this chapter, we present a family of reasoning models which generalizes the models seen so far. The DEDUCTION problem, expressed in a generic way, takes a knowledge base $\mathcal{K}$ and a goal Q as input, and asks whether Q can be deduced from $\mathcal{K}$. Depending on the knowledge base composition, different models are obtained. If $\mathcal{K}$ is composed of facts only, we obtain the BG (or SG) model studied in the first chapters. When rules are added, we obtain the model studied in the preceding chapter. Now we will also consider graph *positive and negative constraints*, which introduce another problem, i.e., CONSISTENCY ("Is the knowledge base consistent with respect to the constraints?") Deduction from inconsistent bases is forbidden. Depending on how rules and constraints are combined, rules have different semantics and become *inference rules* or *evolution rules*. Inference rules represent implicit knowledge and evolution rules represent possible actions transforming factual knowledge (which is possibly enriched by applications of inference rules). Finally, the general form of a KB is $(\mathcal{F}, \mathcal{R}, \mathcal{E}, \mathcal{C})$, where $\mathcal{F}$ is a set of facts, $\mathcal{R}$ and $\mathcal{E}$ are respectively a set of *inference rules* and a set of *evolution rules*, and $\mathcal{C}$ is a set of *constraints*. When $\mathcal{C}$ is empty, $\mathcal{R}$ and $\mathcal{E}$ can be confused. The $\mathcal{BG}$ family is thus composed of the following six models.

- The $\mathcal{F}$ model for $\mathcal{K} = (\mathcal{F}, \emptyset, \emptyset, \emptyset)$, in which DEDUCTION asks if Q can be deduced from $\mathcal{F}$, i.e., if there is a homomorphism from Q to $\mathcal{F}$;
- the $\mathcal{FR}$ model for $\mathcal{K} = (\mathcal{F}, \mathcal{R}, \mathcal{E}, \emptyset)$, in which DEDUCTION asks if Q can be deduced from $(\mathcal{F}, \mathcal{R} \cup \mathcal{E})$, i.e., if there is an $\mathcal{R} \cup \mathcal{E}$-derivation from $\mathcal{F}$ to a BG F_k such that Q can be deduced from F_k; as the set of constraints is empty, $\mathcal{R}$ and $\mathcal{E}$ can be confused, hence the name of the model;
- the $\mathcal{FC}$ model for $\mathcal{K} = (\mathcal{F}, \emptyset, \emptyset, \mathcal{C})$, in which DEDUCTION asks if $\mathcal{F}$ is consistent with respect to $\mathcal{C}$ and if Q can be deduced from $\mathcal{F}$;
- the $\mathcal{FRC}$ model for $\mathcal{K} = (\mathcal{F}, \mathcal{R}, \emptyset, \mathcal{C})$, in which DEDUCTION asks if $(\mathcal{F}, \mathcal{R})$ is consistent with respect to $\mathcal{C}$ and if Q can be deduced from $(\mathcal{F}, \mathcal{R})$;

- the $\mathcal{FEC}$ model for $\mathcal{K} = (\mathcal{F}, \emptyset, \mathcal{E}, \mathcal{C})$, in which DEDUCTION asks if there is an $\mathcal{E}$-derivation from $\mathcal{F}$ to a BG F_k, involving only consistent BGs with respect to $\mathcal{C}$, and such that Q can be deduced from F_k;
- the $\mathcal{FREC}$ model for $\mathcal{K} = (\mathcal{F}, \mathcal{R}, \mathcal{E}, \mathcal{C})$, in which DEDUCTION asks if there is an $\mathcal{RE}$-evolution (this notion will be defined later, let us just now say that it is a particular derivation involving $\mathcal{R}$ and $\mathcal{E}$) from $\mathcal{F}$ to a BG F_k, such that (1) for each F_i in this derivation, $(F_i, \mathcal{R})$ is consistent with respect to $\mathcal{C}$, and (2) Q can be deduced from $(F_k, \mathcal{R})$.

The hierarchy of these models is represented in Fig. 11.1. The complexity of the DEDUCTION problem is mentioned when it is decidable. If rules are involved, DEDUCTION is no longer decidable, as seen in the preceding chapter. We will study how particular cases of rules and constraints affect the decidability property and classify the complexity of DEDUCTION when appropriate. Table 11.1 at the end of the chapter summarizes all complexity results.

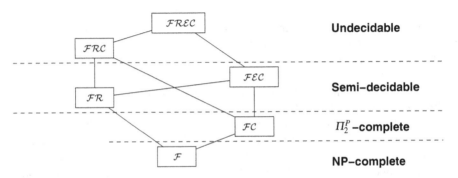

Fig. 11.1 The BG family

The examples in this chapter are inspired from a resource allocation problem, where the global aim is to assign offices to persons of a research group while fulfilling some constraints. $\mathcal{F}$ describes information about office locations, persons and the group organization. A set $\mathcal{R}$ of inference rules describing implicit knowledge about the domain is added. $\mathcal{C}$ represents allocation constraints. $\mathcal{F}$ and $\mathcal{R}$ can be seen as describing a world, which is consistent if it fulfills the constraints in $\mathcal{C}$. $\mathcal{E}$ is the set of evolution rules, allowing to go from one world to another. In this particular example, evolution rules are used to assign an office to a person. Finally, given such a KB, the allocation problem can be modeled as a deduction problem. The goal Q represents a situation where each person of the group has an office. It can be deduced from the base if there is a world, obtainable from the initial world by the evolution rules while satisfying the constraints, such that Q can be mapped to this world.

11.2 $\mathcal{FC}$: **Facts and Constraints**

Let us first introduce *constraints*, which are used to validate knowledge, i.e., to check its consistency. In the simpler model, the checked knowledge is a set of facts. In further sections, we will see more complex uses of constraints when the knowledge base includes rules.

11.2.1 Positive and Negative Constraints

A *constraint* can be positive or negative, expressing a knowledge of form "if *A* holds, so must *B*", or "if *A* holds, *B* must not." It has the same syntactical form as a rule. Intuitively, if the constraint is positive, it is satisfied by a graph *G* if each time "A holds" in *G* then "*B* also holds" in *G*. As usual, "A holds" in *G* means that there is a homomorphism from *A* to *G*. Thus *G* satisfies the constraint if each homomorphism from *A* to *G* can be extended to a homomorphism from *B* to *G*. Symmetrically, *G* satisfies a negative constraint if no homomorphism from *A* to *G* can be extended to a homomorphism from *B* to *G*.

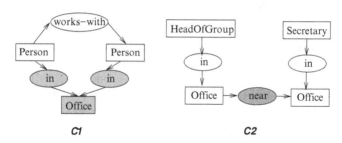

Fig. 11.2 Constraints

Figure 11.2 represents two constraints (seen as bicolored graphs). The negative constraint C_1 expresses that "two persons working together should not share an office." The positive constraint C_2 expresses that "the office of a head of a group must be near the offices of all secretaries." The BG *G* in Fig. 11.3 does not satisfy C_1 because "there is a researcher who works with researcher K" (homomorphism from the condition of *C*1) "and they share office #124" (extension of this homomorphism to a homomorphism from the whole *C*1). *G* satisfies C_2 because there is no homomorphism from the condition of C_2 to *G* (with the natural assumption that *Researcher* is not a subtype of *HeadOfGroup*).

When the KB is composed of a set of facts and a set of constraints, the role of constraints is to validate the fact base. The KB is said to be consistent if all constraints are satisfied. Deduction is defined only on a consistent base. Provided that the base is consistent, deduction is performed by a simple homomorphism check

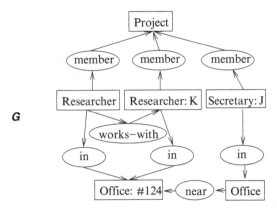

Fig. 11.3 *G* does not satisfy constraint C_1 in Fig. 11.2

as in the BG model. Note that even if they have the same form, constraints and rules

Fig. 11.4 Rule versus Constraint

play a different role. Consider, for instance, the bicolored graph *R* in Fig. 11.4: As
a rule, it says that every researcher is a member of a project. Take the fact $\mathcal{F} =$
[Researcher:K] and the query *Q*= [Researcher:K] → (member) → [Project] ("is K
member of a project?"). If *Q* is asked on $\mathcal{K} = (\mathcal{F}, \mathcal{R} = \{R\})$, the answer is "yes."
Now see *R* as a positive constraint *C*. It says that every researcher *must* be a member
of a project. $\mathcal{K} = (\mathcal{F}, \mathcal{C} = \{C\})$ is inconsistent, thus nothing can be deduced from it,
including *Q*. The KB has to be repaired first.

Definition 11.1 (Constraints). A *positive* (resp. *negative*) *constraint C* is a bicolored BG. $C_{(0)}$, the subgraph induced by the nodes of color 0, is called the *condition* of constraint, $C_{(1)}$, the subgraph induced by the nodes of color 1 and the frontier nodes, is called its *obligation* (resp. *interdiction*). A BG *G satisfies* a positive
(resp. *negative*) constraint *C* if every (resp. no) homomorphism from the condition
of *C* to the irredundant form of *G* (resp. to *G*) can be extended to a homomorphism
from *C* as a whole. A BG *G* π-*violates* a positive (resp. negative) constraint *C* if π is
a homomorphism from the condition of *C* to the irredundant form of *G* (resp. to *G*)
that cannot be extended (resp. that can be extended) to a homomorphism from *C* as
a whole. *G violates C* if it π-violates *C* for some homomorphism π, in other words
G does not satisfy *C*.

Nothing prevents the condition or the obligation/interdiction of a constraint to
be empty. If the obligation/interdiction is empty, the constraint plays no role. If the
condition is empty, satisfaction can be expressed in a simpler way since the empty

graph can be mapped to any graph. A BG G satisfies a positive constraint C^+ with an empty condition if and only if there is a homomorphism from $C_{(1)}^+$ to the normal form of G. Similarly, G satisfies a negative constraint C^- with an empty condition if and only if there is no homomorphism from $C_{(1)}^-$ to G.

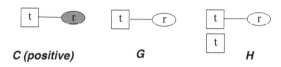

C (positive)   **G**   **H**

Fig. 11.5 Redundancy and constraint violation

Let us point out the importance of the irredundancy condition on the graph to be validated by positive constraints: Should we omit this condition, there may be two equivalent BGs, such that one satisfies a positive constraint and the other does not. Figure 11.5 shows an example of such graphs. G satisfies C, but the equivalent (redundant) graph H, obtained by making the disjoint union of G and the condition of C, does not. A way of avoiding different consistency values for equivalent graphs is to define positive constraint satisfaction with respect to the irredundant form of a BG. This problem does not occur with negative constraints. Indeed, let G_1 and G_2 be two equivalent graphs and assume that G_1 π-violates a negative constraint C; since there is a homomorphism from G_1 to G_2, say π_1, $\pi_1 \circ \pi$ is a homomorphism from C to G_2, thus G_2 also violates C.

In the previous definition, a constraint is defined as a bicolored graph. It can also be defined in the more general form of a couple of λ-BGs with very slight changes: The condition is the first λ-graph and the obligation/interdiction is the second one. A BG G satisfies a positive constraint $C : \lambda (c_1 \ldots c_n) C_c \rightarrow O_c$ if, for every homomorphism π_1 from C_c to the irredundant form of G, there is a homomorphism π_2 from O_c to the irredundant form of G, such that for all $i = 1 \ldots n$, $\pi_1(c_{1_i}) = \pi_2(c_{2_i})$. G π_1-violates C if there is π_1 from C_c to the irredundant form of G with no π_2 satisfying the previous condition.

All results presented here for bicolored constraints hold true for λ-contraints.

Let us give a very simple example to show what λ-constraints may add. Suppose one wants to define the signature of relations by constraints (i.e., for each relation symbol, define the maximal type of each argument of a relation node typed by this symbol). The constraint to express is of form "if there is a relation node r of type t_r, then for all i ranging from 1 to the arity p of t_r, the i-th neighbor of r has to be of type less or equal to a certain type t_i." The condition part is made of one relation node $r = (t_r)$ with p neighbors $[\top : *]$, with each concept node being a distinguished node; the obligation is restricted to the distinguished nodes $[t_1 : *] \ldots [t_p : *]$, where the ith node corresponds to the ith neighbor of r (see Fig. 11.6). Bicolored graphs cannot express this constraint (for this particular case, where all distinguished nodes in the condition are of form $[\top : *]$, they could express it if we removed the requirement that the condition part of a constraint has to be a well-formed BG; such an extension could be safely done for constraints, whereas it could not for rules since the application of a rule could then lead to a graph that is not well-formed).

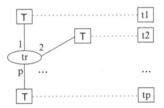

Fig. 11.6 A λ-constraint representing a relation signature

Two constraints C_1 and C_2 are said to be *equivalent* if any graph that violates C_1 also violates C_2, and conversely. The following property on negative constraints is immediate.

Property 11.1. Any negative constraint is equivalent to the negative constraint obtained by coloring all its nodes by 1.

Furthermore, negative constraints can be considered as a particular case of positive ones: Consider the positive constraint C' obtained from a negative constraint C by coloring all nodes of C by 0, then adding a concept node colored by 1, with type *NotThere*, where *NotThere* is incomparable with all other types and does not appear in any graph of the KB, except in constraints. Then a graph G violates the negative constraint C if and only if it violates the positive constraint C'. Positive constraints strictly include negative constraints, in the sense that the associated consistency problems are not in the same complexity class, as will be shown in Sect. 11.2.3 (cf. Theorems 11.3 and 11.4).

Property 11.2. Unless $\Pi_2^P = $ co-NP, positive constraints are a strict generalization of negative constraints.

Since negative constraints can be seen as a particular case of positive ones, we will now, unless indicated otherwise, let "a set of constraints" denote a set of positive constraints: Some of them can be equivalent to negative ones.

Definition 11.2 (Consistency/Deduction in $\mathcal{FC}$). A KB $\mathcal{K} = (\mathcal{F}, \mathcal{C})$ is *consistent* if $\mathcal{F}$ satisfies all constraints of $\mathcal{C}$. A BG Q can be *deduced* from $\mathcal{K}$ if $\mathcal{K}$ is consistent and Q can be deduced from $\mathcal{F}$. We call $\mathcal{FC}$-CONSISTENCY and $\mathcal{FC}$-DEDUCTION the associated decision problems.

11.2.2 Translation to FOL

Note first that constraints lead to non-monotonic deduction. Indeed, consider a knowledge base with a consistent set of facts $\mathcal{F}$ and a BG Q deducible from $\mathcal{F}$.

If one adds information to $\mathcal{F}$, one may create a constraint violation; since nothing can be deduced from an inconsistent knowledge base, previous deductions no longer hold. In particular Q cannot be deduced anymore. This is why as soon as constraints are involved it is impossible to obtain results of form "Q can be deduced from the KB $\mathcal{K}$ if and only if $\Phi(\mathcal{K}) \models \Phi(Q)$." However, if deduction cannot be translated to FOL, consistency (or inconsistency) can, and this is the point studied in this section.

For negative constraints, the correspondence between consistency and logical deduction is immediate, and relies on homomorphism soundness and completeness with respect to the semantics Φ (theorem 4.3). Intuitively, a BG G violates a negative constraint C^- if and only if the knowledge represented by C^- is deducible from the knowledge represented by G.

Theorem 11.1. *A BG G violates a negative constraint C iff $\Phi(\mathcal{V}), \Phi(G) \models \Phi(C)$, where $\Phi(C)$ is the logical formula assigned to the BG underlying C.*

A first way of translating consistency with respect to to positive constraints into logics consists of translating the "homomorphism" notion into the notion of $\mathcal{L}_\mathcal{V}$-substitution on formulas associated with graphs, where $\mathcal{L}_\mathcal{V}$ is the ordered logical language assigned to $\mathcal{V}$ (cf. Sect. 4.2.1).

As a corollary of Theorem 4.4 establishing the equivalence between homomorphism and $\mathcal{L}_\mathcal{V}$-substitution, we obtain:

Property 11.3. A graph G π-violates a positive constraint C if and only if the $\mathcal{L}_\mathcal{V}$-substitution σ from $\Phi(C_{(0)})$ to $\Phi(G)$ associated with π cannot be extended to a $\mathcal{L}_\mathcal{V}$-substitution from $\Phi(C)$ to $\Phi(G)$, where $\Phi(C)$ is the logical formula assigned to the BG underlying C.

Another bridge can be built using rules. Indeed, a graph G satisfies a positive constraint C if and only if, considering C as a rule, all applications of C to G produce a graph equivalent to G. Or, more specifically:

Property 11.4. A BG G π-violates a positive constraint C iff, considering C as a rule, the application of C to G according to π produces a graph that is not equivalent to G.

Proof. Let C and G such that G satisfies C. If π_0 is a homomorphism from $C_{(0)}$ to G, let us consider the graph G' obtained by the application of C (considered now as a rule) to G according to π_0. Let us now build the following homomorphism π' from G' to G: for each node v, $\pi'(v) = v$ if v belongs to G; otherwise, v is a copy of a node w of $C_{(1)}$, and let π be one of the homomorphisms from C to G that extends π_0, we have $\pi'(v) = \pi(w)$. Then π' is a homomorphism from G' to G, and since G trivially maps to G', they are thus equivalent. This proves the $\Leftarrow$ part of Property 11.4.

For the $\Rightarrow$ part, we use Property 2.9, which states that, for any BG and any of its irredundant subgraphs equivalent to it, there is a folding from the BG to this subgraph. Suppose now that G π-violates C. Since constraint violation is defined with respect to the irredundant form of a graph, we can consider, without loss of generality, that G is irredundant. We denote by G' the graph obtained by the application of C (again, considered now as a rule) to G according to π. We prove that "G' equivalent to G" leads to a contradiction. If G' is equivalent to G, then there is a homomorphism from G' to G. And since G is an irredundant subgraph of G', there is a folding f from G' to G. Consider now π' the homomorphism from C to G defined as follows: for any node x in C, if x is in $C_{(0)}$ then $\pi'(x) = \pi(x)$, otherwise let x' be the copy of x in G', and let $\pi'(x) = f(x')$. Since for all x in $C_{(0)}$, and in particular for the frontier nodes, $f(\pi(x)) = \pi(x)$, π' is a homomorphism and it extends π. This contradicts the "G π-violates C" hypothesis. Thus G' is not equivalent to G. $\square$

Property 11.5. If a BG G satisfies a positive constraint C, then any graph in a $\{C\}$-derivation of G is equivalent to G.

Proof. Let $G_0(= G)$,..., G_k be a $\{C\}$-derivation of G. From Property 11.4, each G_i, $1 \leq i \leq k$, is equivalent to G_{i-1}, thus by transitivity, is equivalent to G. $\square$

Using soundness and completeness of the deduction in $\mathcal{FR}$, and Properties 11.4 and 11.5, one obtains the following relation with FOL deduction.

Theorem 11.2. *A BG G violates a positive constraint C if and only if there is a BG G' such that $\Phi(\mathcal{V}), \Phi(G), \Phi(C) \vDash \Phi(G')$ and not $\Phi(\mathcal{V}), \Phi(G) \vDash \Phi(G')$, where $\Phi(C)$ is the translation of C considered as a rule.*

11.2.3 Complexity of Consistency and Deduction

This section is devoted to the complexity of $\mathcal{FC}$-CONSISTENCY and $\mathcal{FC}$-DEDUC-TION. The complexity is first established in the general case, then for particular constraints, namely negative and disconnected constraints.

Theorem 11.3 (Complexity of consistency in $\mathcal{FC}$). *Consistency in $\mathcal{FC}$ is Π_2^P-complete.*

Proof. Without change of complexity, one can consider that $\mathcal{C}$ is composed of only one positive constraint, say C. First recall that deciding whether a BG F satisfies C is done on the irredundant form of F. We may consider two ways of integrating the complexity of making a graph irredundant in the complexity of $\mathcal{FC}$-CONSISTENCY. One way of doing this is to assume that the irredundant form of F is computed before the consistency check. This can be achieved with a number of calls to a homomorphism oracle linear in the size of F (cf. Sect. 2.6). However, since we then have to solve a function problem (compute the irredundant form of F) instead of a decision problem (is F irredundant?), we prefer to integrate irredundancy into the consistency check: Then, for a homomorphism π_0 from the condition of C to F, the homomorphism from C to F we look for does not necessarily extend π_0, but it extends the composition of a homomorphism from F to one of its subgraphs (possibily equal to F itself) and π_0.

$\mathcal{FC}$-CONSISTENCY belongs to Π_2^P since it corresponds to the language $L = \{x \mid \forall y_1 \, \exists y_2 \, R(x, y_1, y_2)\}$, where x encodes an instance $(F, \{C\})$ of the problem and $(x, y_1, y_2) \in R$ if and only if y_1 encodes a homomorphism π_0 from $C_{(0)}$ to F and y_2 encodes a homomorphism π_F from F to one of its subgraphs and a homomorphism π from C to F such that $\pi[C_{(0)}] = \pi_F \circ \pi_0$. Note that if F is in irredundant form, then π_F is an automorphism.

Now let us consider the B_2^c problem: Given a boolean formula E, and a partition $\{X_1, X_2\}$ of its variables, is it true that for any truth assignment for the variables in X_1 there is a truth assignment for the variables in X_2 such that is E true? This problem is Π_2^P-complete, since its complementary B_2 is shown to be Σ_2^P-complete in [Sto77]. In order to build a polynomial reduction to $\mathcal{FC}$-CONSISTENCY, we use a restriction of this problem to k-CNF, i.e., conjunctions of disjunctions with at most k literals per clause. Let us call 3-SAT_2^c the special case where E is a 3-CNF, in other words an instance of 3-SAT. Then 3-SAT_2^c is also Π_2^P-complete. Indeed, in the same paper (Theorem 4.1) it is shown that B_2 with E restricted to a 3-disjunctive normal form (3-DNF) remains Σ_2^P-complete. Since the negation of a 3-DNF is a 3-CNF, it follows that the complementary problem B_2^c with E restricted to a 3-CNF is Π_2^P-complete.

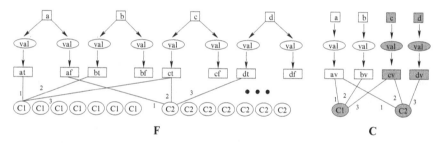

Fig. 11.7 Example of transformation from 3-SAT_2^c to $\mathcal{FC}$-CONSISTENCY

Let us now reduce 3-SAT_2^c to $\mathcal{FC}$-CONSISTENCY. The transformation used is very similar to that from 3-SAT to $\mathcal{BG}$-DEDUCTION (proof of Theorem 2.5 illustrated in Fig. 2.25). Let E be an instance of 3-SAT. Let $H(E)$ and $G(E)$ be the BGs obtained by the transformation described in the proof of Theorem 2.5. Let $F = H(E)$. The constraint C is obtained by adding a coloration to $G(E)$: All relation nodes obtained from a clause (nodes typed C_i) and all nodes obtained from variables in X_2 (concept nodes typed x or xv and relation nodes typed val) are colored by 1 (i.e., belong to the obligation of C). Once again, having clauses of bounded size leads to a polynomial transformation. The graph F and the positive constraint C presented in Fig. 11.7 are obtained from the 3-SAT formula $(a \vee b \vee \neg c) \wedge (\neg a \vee c \vee \neg d)$ and the partition $\{X_1 = \{a, b\}, X_2 = \{c, d\}\}$.

Each truth assignment of the variables in E such that E is true naturally yields a homomorphism from C to F, and reciprocally (as indicated in the proof of Theorem 2.5). Furthermore, any truth assignment for the variables in X_1 naturally yields a homomorphism from $C_{(0)}$ to F, and reciprocally. Thus, the question "is it true that for any truth assignment for the variables in X_1 there is a truth assignment for the variables in X_2 such that is E true?" is equivalent to the question "is it true that for any homomorphism π_0 from $C_{(0)}$ to F there is a homomorphism from C to F extending π_0?" □

Corollary 11.1. *Deduction in* $\mathcal{FC}$ *is* Π_2^P-*complete.*

The next theorem shows a complexity decrease when only negative constraints are considered.

Theorem 11.4 (Complexity with negative constraints). *If all constraints are negative,* $\mathcal{FC}$-CONSISTENCY *is co-NP-complete and* $\mathcal{FC}$-DEDUCTION *is DP-complete.*

Proof. Co-NP-completeness of $\mathcal{FC}$-CONSISTENCY: from NP-completeness of homomorphism checking (theorem 2.5).

DP-completeness of $\mathcal{FC}$-DEDUCTION: this problem can be expressed as "Is it true that Q maps to F and that no constraint of C maps to F?" Thus it belongs to DP. Now let us consider that C contains only one constraint. A reduction from 3-SAT to BG-HOMOMORPHISM (cf. for instance the proof of theorem 2.5) provides a straightforward reduction from SAT/UNSAT (cf. for instance [Pap94]) to $\mathcal{FC}$-DEDUCTION, thus the DP-completeness. □

It would be interesting to exhibit particular cases of constraints, more general than negative ones, that make this complexity fall into intermediary classes (e.g., DP and Δ_2^P for $\mathcal{FC}$-CONSISTENCY). Some syntactic restrictions we defined for rules are good candidates: Though a finite expansion set of constraints makes no sense, let us consider *range restricted constraints*, i.e., (positive) constraints whose obligation does not contain any generic concept node. Let us also define *disconnected constraints* as (positive) constraints where the condition and obligation are disjoint. The following property highlights the relationships of these particular cases with negative constraints:

Property 11.6. Negative constraints are a particular case of both range-restricted constraints and disconnected constraints.

Proof. As noted previously, a negative constraint is equivalent to a positive constraint whose obligation is composed of one concept node of type *NotThere*, where *NotThere* is incomparable with all other types and does not appear in the knowledge base except in C (it is thus a disconnected constraint). Without loss of generality this node can be labeled by an individual marker (which, as *NotThere*, appears only in C), thus leading to a constraint which is both disconnected and range-restricted. □

Theorem 11.5 (Complexity with disconnected constraints). *When* C *contains only disconnected constraints,* $\mathcal{FC}$-CONSISTENCY *and* $\mathcal{FC}$-DEDUCTION *become DP-complete.*

Proof. $\mathcal{FC}$-INCONSISTENCY belongs to DP, since we must prove that for one constraint there is a homomorphism from its condition and no homomorphism from its obligation. Since DP = co-DP, $\mathcal{FC}$-CONSISTENCY belongs to DP, and DP-completeness is proven with a reduction from SAT/UNSAT (as in proof of theorem 11.10). For $\mathcal{FC}$-DEDUCTION, we have to *independently* solve the $\mathcal{FC}$-CONSISTENCY (DP-complete) and the $\mathcal{F}$-DEDUCTION (NP-complete) problems, hence the result. $\square$

To the best of our knowledge, there is currently no result about the complexity of $\mathcal{FC}$-CONSISTENCY with range-restricted constraints. The difficulty in classifying this problem relies on the integration of irredundancy.

11.3 Combining Rules and Constraints

Two different kinds of rule and constraint combinations are considered in this section. With the first one, yielding the model $\mathcal{FRC}$, rules are considered as *inference rules* representing implicit knowledge. Facts and inference rules can be seen as describing a "world," and a rule application modifies the explicit description of the world (i.e., the facts). Constraints define the consistency of this world. In the second one, yielding the model $\mathcal{FEC}$, another semantics is given to rules, which are now called *evolution rules*: Rules represent possible transitions from facts to other facts, and all facts have to be consistent. Finally, the model $\mathcal{FREC}$ integrates both kinds of rules. In short, facts and inference rules define a world that has to be consistent, and evolution rules define possible transitions from one world to other worlds, that have to be consistent.

11.3.1 $\mathcal{FRC}$: Constraints and Inference Rules

In the $\mathcal{FRC}$ model, a KB is composed of a set of facts $\mathcal{F}$, a set of inference rules $\mathcal{R}$, and a set of constraints $\mathcal{C}$. The notion of consistency now has to take rules into account.

Consider, for instance, the KB $\mathcal{K} = (\mathcal{F} = \{G\}, \mathcal{R} = \{R_1, R_2\}, \mathcal{C} = \{C_2\})$ where G is the upper graph in Fig. 11.8, R_1 and R_2 are the rules in Fig. 11.9, and C_2 is the positive constraint in Fig. 11.2. The fact G alone does not satisfy constraint C_2 (there are two violations of C_2, because none of the offices #2 and #3 are known to be near office #1). But it does after a certain number of rule applications. Thus $\mathcal{K}$ is said to be consistent. In this case, it is easy to define and check consistency because the world description can be completely explicited by a finite BG, the full graph of G with respect to $\mathcal{R}$, represented as H in Fig. 11.8, we just have to check whether this graph is consistent. In the general case, consistency relies on whether each "constraint violation" can be repaired by rule applications. Moreover, applying a rule to $\mathcal{F}$ can create inconsistency, but a further application of a rule may restore

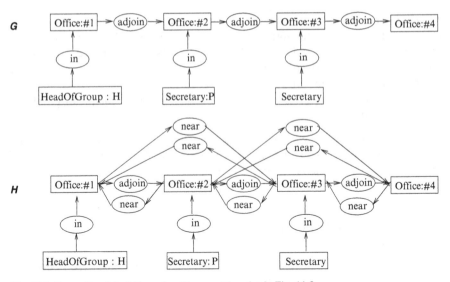

Fig. 11.8 Facts: *G* and its full graph with respect to rules in Fig. 11.9

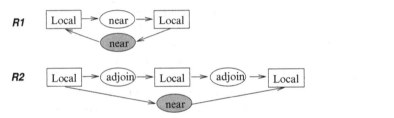

Fig. 11.9 Rules

consistency. Let us formalize this notion of *consistency restoration*. Suppose there is a π-violation of a positive constraint C in $\mathcal{F}$; this violation (C, π) is said to be $\mathcal{R}$-*restorable* if there is an $\mathcal{R}$-derivation from $\mathcal{F}$ to a BG F' such that F' does not π-violates C (in other words, there is a homomorphism π' from F' to $irr(F')$, the irredundant form of F', such that the homomorphism $\pi' \circ \pi$ from the condition of C to $irr(F')$ can be extended to a homomorphism from C as a whole). Note that the $\mathcal{R}$-restoration can create new violations that must themselves be proven $\mathcal{R}$-restorable.

Definition 11.3 (Consistency/Deduction in $\mathcal{FRC}$). A KB $\mathcal{K} = (\mathcal{F}, \mathcal{R}, \mathcal{C})$ is *consistent* if, for any BG F' that can be $\mathcal{R}$-derived from $\mathcal{F}$, for every constraint $C \in \mathcal{C}$, for every π-violation of C in F', (C, π) is $\mathcal{R}$-restorable. A BG Q can be deduced from $\mathcal{K}$ if $\mathcal{K}$ is consistent and Q can be deduced from $(\mathcal{F}, \mathcal{R})$. We call $\mathcal{FRC}$-CONSISTENCY and $\mathcal{FRC}$-DEDUCTION the associated decision problems.

Let us give another example. Consider a KB composed of one fact $F=$ [Person:Mary], one rule and one constraint, both represented by the same bicolored graph shown in Fig. 11.10. The rule says that "every person has a father and a

Fig. 11.10 A rule/constraint

mother," and the constraint says that this must be the case. F violates the constraint but a rule application can repair this violation, while creating another violation. Finally, every constraint violation can be repaired by a rule application, thus the KB is consistent.

Note that the violation of a negative constraint can never be restored. The consistency of a KB with negative constraints only can thus be defined in a simpler way.

Property 11.7 (Consistency/Deduction in $\mathcal{FRC}$ with negative constraints). A KB $\mathcal{K} = (\mathcal{F}, \mathcal{R}, \mathcal{C})$, where $\mathcal{C}$ is composed of negative constraints, is *consistent* if and only if, for any BG F' that can be $\mathcal{R}$-derived from $\mathcal{F}$, $(F', \mathcal{C})$ is consistent, i.e., there is no homomorphism from a $C \in \mathcal{C}$ to F'.

Let us also point out the particular case of positive constraints with an empty condition: $(\mathcal{F}, \mathcal{R})$ is consistent with respect to to a positive constraint C^+ with $C^+_{(0)}$ empty if and only if there is a BG F' that can be $\mathcal{R}$-derived from $\mathcal{F}$ and to which $C^+_{(1)}$ can be mapped.

For both previous particular cases, the translation of $\mathcal{FRC}$ consistency to FOL is immediate and relies on the soundness and completeness of graph derivation in $\mathcal{FR}$. For the translation in general case, one may try to extend the result obtained for $\mathcal{FC}$ consistency, i.e., a KB composed of facts is inconsistent if and only if there is a BG (a FOL($\exists$, $\wedge$) formula) that can be deduced from the facts and the constraints seen as rules, but cannot be deduced from the facts alone. The next theorem shows that one direction of the above equivalence holds for a KB composed of facts and rules. And a counter-example then shows that the other direction does not hold.

Theorem 11.6 (FOL translation: a sufficient condition for $\mathcal{FRC}$ inconsistency). *Let $\mathcal{K} = (\mathcal{F}, \mathcal{R}, \mathcal{C})$ be a KB. If there exists a BG F' such that $\Phi(\mathcal{V}), \Phi(\mathcal{F}), \Phi(\mathcal{R}), \Phi(\mathcal{C}) \models \Phi(F')$ and not $\Phi(\mathcal{V}), \Phi(\mathcal{F}), \Phi(\mathcal{R}) \models \Phi(F')$, where $\Phi(\mathcal{C})$ is the translation of the constraints in $\mathcal{C}$ considered as a rules, then $\mathcal{K}$ is inconsistent.*

Proof. If there is such a BG F' then there is an $(\mathcal{R} \cup \mathcal{C})$-derivation, where the elements of $\mathcal{C}$ are considered as rules, say $F_0(= \mathcal{F}), ..., F_k$, such that F' maps to F_k. See that F_k is not deducible from $(\mathcal{F}, \mathcal{R})$ otherwise F' would also be deducible from $(\mathcal{F}, \mathcal{R})$. Let us consider the first F_i in the derivation that is not deducible from $(\mathcal{F}, \mathcal{R})$. F_i is obtained from F_{i-1} by applying a rule $C_q \in \mathcal{C}$ according to a homomorphism π. Since F_{i-1} is deducible from $(\mathcal{F}, \mathcal{R})$ there is a BG H and an $\mathcal{R}$-derivation from $\mathcal{F}$ to H such that F_{i-1} maps to H by a homomorphism π'. Consider $\pi'' = \pi' \circ \pi$ the homomorphism from the first part of C_q (its condition/hypothesis) to H. We now prove that (1) H π''-violates C_q and (2) this violation is not restorable. If (1) or (2) is false, then there is a BG H' $\mathcal{R}$-derived from H such that π'' can be extended to a homomorphism from C_q as a whole to the irredundant form of H'. This is absurd because there would be a homomorphism from F_i to a BG $\mathcal{R}$-derivable from $\mathcal{F}$, thus F_i would be deducible from $(\mathcal{F}, \mathcal{R})$. $\square$

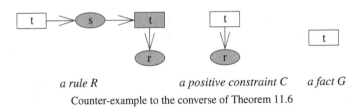

a rule R   a positive constraint C   a fact G

Counter-example to the converse of Theorem 11.6

Fig. 11.11 Counter-example

Property 11.8. The converse of theorem 11.6 is false, as shown in Fig. 11.11.

In Fig. 11.11, check that every graph that can be $\{R,C\}$-derived from G can also be $\{R\}$-derived from G. However, the KB is inconsistent: G violates constraint C and this violation cannot be restored.

We now show that when the set of rules is restricted to a finite expansion set (cf. Definition 10.12), $\mathcal{FRC}$ consistency can be translated to FOL in the same way as $\mathcal{FC}$ consistency, i.e., in this particular case the converse of Theorem 11.6 holds.

Let $\mathcal{K} = (\mathcal{F}, \mathcal{R}, \mathcal{C})$ be a KB. Let us first point out that if $\mathcal{R}$ is a finite expansion set, checking the consistency of $\mathcal{K}$ is equivalent to checking the consistency of the full graph $\mathcal{F}_{\mathcal{R}}$.

Property 11.9 (Consistency with a finite expansion set of rules). Let $\mathcal{K} = (\mathcal{F}, \mathcal{R}, \mathcal{C})$ be a KB where $\mathcal{R}$ is a finite expansion set. $\mathcal{K}$ is consistent if and only if $(\mathcal{F}_{\mathcal{R}}, \mathcal{C})$ is consistent. Thus, a BG Q can be deduced from $(\mathcal{F}, \mathcal{R}, \mathcal{C})$ if and only if it can be deduced from $(\mathcal{F}_{\mathcal{R}}, \mathcal{C})$.

This property is used to prove that the converse of Theorem 11.6 is true when $\mathcal{R}$ is a finite expansion set.

Theorem 11.7 (FOL translation of $\mathcal{FRC}$ inconsistency with finite expansion set of rules). *Let $\mathcal{K} = (\mathcal{F}, \mathcal{R}, \mathcal{C})$ be a KB where $\mathcal{R}$ is a finite expansion set. $\mathcal{K}$ is inconsistent if and only if there is a BG F' such that $\Phi(\mathcal{V}), \Phi(\mathcal{F}), \Phi(\mathcal{R}), \Phi(\mathcal{C}) \vDash \Phi(F')$ and not $\Phi(\mathcal{V}), \Phi(\mathcal{F}), \Phi(\mathcal{R}) \vDash \Phi(F')$ (where $\Phi(\mathcal{C})$ is the translation of the constraints in $\mathcal{C}$ considered as a rules).*

Proof. ($\Leftarrow$) holds as a particular case of Theorem 11.6. ($\Rightarrow$) Since $\mathcal{K}$ is inconsistent, the previous property asserts that $(\mathcal{F}_{\mathcal{R}}, \mathcal{C})$ is inconsistent. Theorem 11.2 ensures that there is a graph H such that (1) $\Phi(\mathcal{V}), \Phi(\mathcal{F}_{\mathcal{R}}), \Phi(\mathcal{C}) \vDash \Phi(H)$ and not (2) $\Phi(\mathcal{V}), \Phi(\mathcal{F}_{\mathcal{R}}) \vDash \Phi(H)$. Since $\Phi(\mathcal{V}), \Phi(\mathcal{F}), \Phi(\mathcal{R}) \vDash \Phi(\mathcal{F}_{\mathcal{R}})$ (cf. soundness of rule application in Chap. 10) we have $\Phi(\mathcal{V}), \Phi(\mathcal{F}), \Phi(\mathcal{R}), \Phi(\mathcal{C}) \vDash \Phi(H)$. Let us now assume that $\Phi(\mathcal{V}), \Phi(\mathcal{F}), \Phi(\mathcal{R}) \vDash \Phi(H)$ and prove that it is absurd. In this case, there would be a graph F' $\mathcal{R}$-derivable from $\mathcal{F}$ such that H maps to F' (cf. completeness of rule application in Chap. 10). Since H maps to F', which maps to $\mathcal{F}_\mathcal{R}$, by transitivity H maps to $\mathcal{F}_\mathcal{R}$. We should have $\Phi(\mathcal{V}), \Phi(\mathcal{F}_\mathcal{R}) \vDash \Phi(H)$ (soundness of BG homomorphism), which contradicts the hypothesis. □

11.3.2 $\mathcal{FEC}$: Constraints and Evolution Rules

Let now now consider another use of rules. The objects and application mechanisms are the same, but rules are used with another intention: Instead of representing implicit knowledge about a world, they represent possible *actions* leading from one world to another. We call them *evolution rules*. As these rules have the same form as inference rules, they only add information. It would be worthwhile to generalize these rules to true transformation rules, also allowing to delete information, but this has not yet been done. Consider, for instance, the bicolored graph in Fig. 11.12.

Fig. 11.12 An evolution rule

Taken as an inference rule, it would say that "all persons are in all offices." Taken as an evolution rule, it states that "when there are a person and an office, (try to) assign this office to that person." Consider a KB composed of facts, evolution rules and constraints. Facts describe an initial world; evolution rules represent possible transitions from worlds to other worlds; constraints define the consistency of each world; a successor of a consistent world is obtained by an evolution rule application; given a BG Q, the deduction problem asks whether there is a path of consistent worlds evolving from the initial one to a world satisfying Q.

Definition 11.4 (Deduction in $\mathcal{FEC}$). Let $\mathcal{K} = (\mathcal{F}, \mathcal{E}, \mathcal{C})$ be a KB, and let Q be a BG. Q can be deduced from $\mathcal{K}$ if there is an $\mathcal{E}$-derivation $F_0(= \mathcal{F}) \ldots, F_k$ such that, for $0 \leq i \leq k$, $(F_i, \mathcal{C})$ is consistent and Q can be deduced from F_k. We call $\mathcal{FEC}$-DEDUCTION the associated decision problem.

To summarize, in $\mathcal{FEC}$, $\mathcal{F}$ is seen as the initial world, root of a potentially infinite tree of possible worlds, and $\mathcal{E}$ describes possible evolutions from worlds to others. Each world has to be consistent. In $\mathcal{FRC}$, $\mathcal{F}$ provided with $\mathcal{R}$ is a finite description of a "potentially infinite" world, that has to be consistent.

11.3.3 $\mathcal{FREC}$: Constraints, Inference and Evolution Rules

The $\mathcal{FREC}$ model combines both derivation schemes of the $\mathcal{FRC}$ and $\mathcal{FEC}$ models. Now, $\mathcal{F}$ describes an initial world, inference rules in $\mathcal{R}$ complete the description of any world, constraints in $\mathcal{C}$ define the consistency of a world, evolution rules in $\mathcal{E}$ try to make evolve a consistent world into a new, consistent one. The deduction problem asks whether $\mathcal{F}$ can evolve into a consistent world satisfying the goal.

Definition 11.5 (Deduction in $\mathcal{FREC}$). A BG G' is an *immediate $\mathcal{RE}$-evolution* from a BG G if there is an $\mathcal{R}$-derivation from G to to a BG G'' and an *immediate $\mathcal{E}$-derivation* from G'' to G'. An *$\mathcal{RE}$-evolution* from a BG G to a BG G' is a sequence of BGs $G_0(=G),\ldots,G_k = G'$ such that, for $0 \leq i \leq k$, $(G_i, \mathcal{R}, \mathcal{C})$ is consistent and, for $1 \leq i \leq k$, G_i is an immediate $\mathcal{RE}$-evolution from G_{i-1}. Given a KB $\mathcal{K} = (\mathcal{F}, \mathcal{R}, \mathcal{E}, \mathcal{C})$, a BG Q can be *deduced* from $\mathcal{K}$ if there is an $\mathcal{RE}$-evolution $F_0(=\mathcal{F}),\ldots,F_k$ such that Q can be deduced from $(F_k, \mathcal{R})$. We call $\mathcal{FREC}$-DEDUCTION the associated deduction problem.

When $\mathcal{E} = \emptyset$ (resp. $\mathcal{R} = \emptyset$), one obtains the $\mathcal{FRC}$ model (resp. the $\mathcal{FEC}$ model).

11.3.4 Complexity of Combining Rules and Constraints

As the deduction in $\mathcal{FR}$ is not decidable, this is also the case in $\mathcal{FEC}$ and $\mathcal{FRC}$. And unsurprisingly, the deduction is more difficult in $\mathcal{FRC}$ than in $\mathcal{FEC}$.

Theorem 11.8 (Complexity in $\mathcal{FEC}/\mathcal{FRC}$). *Deduction in $\mathcal{FEC}$ is semi-decidable. Consistency and deduction in $\mathcal{FRC}$ are totally undecidable.*

Proof. $\mathcal{FEC}$ includes $\mathcal{FR}$ thus $\mathcal{FEC}$-deduction is not decidable. When Q is deducible from $\mathcal{K}$, a breadth-first search of the tree of all derivations from $\mathcal{K}$, each graph being checked for consistency, ensures that F_k is found in finite time. For $\mathcal{FRC}$, we show that checking consistency is totally undecidable. Let $\mathcal{K} = (\mathcal{F}, \mathcal{R}, \mathcal{C})$ be a KB where $\mathcal{C}$ contains a positive constraint C^+ and a negative constraint C^-, both with an empty condition. To prove consistency, one has to prove that C^- is not deducible from $(\mathcal{F}, \mathcal{R})$, and the algorithm does not necessarily stop in this case (from semi-decidability of deduction in $\mathcal{FR}$). The same holds for the complementary problem (proving inconsistency) taking C^+ instead of C^-. Indeed, since C^+ has an empty condition, $(\mathcal{F}, \mathcal{R})$ violates C^+ if and only if C^+ is not deducible from $(\mathcal{F}, \mathcal{R})$. Hence the total undecidability. □

As a generalization of $\mathcal{FRC}$, deduction in $\mathcal{FREC}$ is totally undecidable. The next section exhibits decidable fragments of $\mathcal{FREC}$ and classifies some of them in the polynomial hierarchy.

11.4 Complexity in $\mathcal{FRC}/\mathcal{FEC}/\mathcal{FREC}$ for Particular Cases of Rules and Constraints

In this section, we study how particular cases of rules and constraints influence the decidability and complexity of deduction in models combining rules and constraints, i.e., $\mathcal{FRC}$, $\mathcal{FEC}$ and $\mathcal{FREC}$. In short, we show that considering finite expansion sets of rules leads to decidability of deduction in these models. When rules are range-restricted, all deduction problems can be classified in the polynomial hierarchy. We obtain lower complexity if we furthermore restrict constraints to disconnected constraints or to more specific negative constraints.

11.4.1 Particular Cases of Rules

Property 11.10 (Complexity with constraints and finite expansion sets of rules).

- When $\mathcal{R}$ is a finite expansion set, deduction in $\mathcal{FR}$ is decidable, consistency and deduction in $\mathcal{FRC}$ are decidable, deduction in $\mathcal{FREC}$ is semi-decidable.
- When $\mathcal{E}$ is a finite expansion set, deduction in $\mathcal{FEC}$ is decidable, but remains totally undecidable in $\mathcal{FREC}$.
- When $\mathcal{R} \cup \mathcal{E}$ is a finite expansion set, deduction in $\mathcal{FREC}$ is decidable.

Proof. Assume that $\mathcal{R}$ is a finute expansion set. Decidability of deduction in $\mathcal{FR}$ is proven in Chap. 10. Decidability of deduction and consistency in $\mathcal{FRC}$ follows from Property 11.9. In $\mathcal{FREC}$, when the answer is "yes", it can be obtained in finite time; we proceed as for $\mathcal{FEC}$ (cf. proof of Theorem 11.8) but consistency checks are done on the full graph instead of the graph itself.

Now, assume that $\mathcal{E}$ is a finite expansion set. The full graph $\mathcal{F_E}$ exists, thus the derivation tree in $\mathcal{FEC}$ is finite, and consistency checks may only cut some parts of this tree. Deduction in $\mathcal{FREC}$ remains undecidable because when $\mathcal{E} = \emptyset$, one obtains the $\mathcal{FRC}$ model, in which deduction is undecidable.

Finally, if $\mathcal{R} \cup \mathcal{E}$ is a finite expansion set, the full graph $\mathcal{F}_{\mathcal{R} \cup \mathcal{E}}$ exists, thus the derivation tree is finite, and consistency checks may only cut parts of this tree. $\square$

Note that the condition "$\mathcal{R} \cup \mathcal{E}$ is a finite expansion set" is stronger than "both $\mathcal{R}$ and $\mathcal{E}$ are finite expansion sets." When both $\mathcal{R}$ and $\mathcal{E}$ are finite expansion sets, $\mathcal{FREC}$ remains non-decidable (see bibliographical notes).

Let us now consider range restricted rules (cf. Chap. 10). As previously, we assume that the arity of relation types is bounded by a constant.

Theorem 11.9 (Complexity with range restricted rules and constraints). *When $\mathcal{E}$ and $\mathcal{R}$ are range restricted rules:*

- *Deduction in $\mathcal{FR}$ is NP-complete.*
- *Consistency and deduction in $\mathcal{FRC}$ are Π_2^P-complete.*
- *Deduction in $\mathcal{FEC}$ and $\mathcal{FREC}$ is Σ_3^P-complete.*

Proof. The following results heavily rely on Property 10.7: Since all derivations involved being of polynomial length, they admit a polynomial certificate (the sequence of homomorphisms used to build the derivation).

NP-completeness of $\mathcal{FR}$-DEDUCTION. Cf. Property 10.8.

Π_2^P*-completeness of* $\mathcal{FRC}$-CONSISTENCY *and* $\mathcal{FRC}$-DEDUCTION. Recall that the consistency check involves the irredundant form of $\mathcal{F}$. In order to lighten the problem formulation, we assume here that all BGs are irredundant, but irredundancy can be integrated without increasing the consistency check complexity: See the proof of Theorem 11.3. $\mathcal{FRC}$-consistency belongs to Π_2^P since it corresponds to the language $L = \{x \mid \forall y_1 \exists y_2 \, R(x, y_1, y_2)\}$, where x encodes an instance $\mathcal{K} = (\mathcal{F}, \mathcal{R}, \mathcal{C})$ of the problem and $(x, y_1, y_2) \in R$ iff $y_1 = (d_1; \pi_0)$, with d_1 being a derivation from $\mathcal{F}$ to F', π_0 is a homomorphism from the condition of a constraint $C_{i(0)}$ to F', $y_2 = (d_2; \pi_1)$, d_2 is a derivation from F' to F'' and π_1 is a homomorphism from C_i to F'' such that $\pi_1[C_{i(0)}] = \pi_0$. R is polynomially decidable and polynomially balanced (since the lengths of d_1 and d_2 are polynomial in the size of the input). When $\mathcal{R} = \emptyset$, one obtains the problem $\mathcal{FC}$-CONSISTENCY, thus the Π_2^P-completeness. Since the $\mathcal{FRC}$-DEDUCTION consists of solving two independent problems, $\mathcal{FRC}$-CONSISTENCY (Π_2^P-complete) and $\mathcal{FR}$-DEDUCTION (NP-complete), and since NP is included in Π_2^P, $\mathcal{FRC}$-DEDUCTION is also Π_2^P-complete.

Σ_3^P*-completeness of* $\mathcal{FEC}$-DEDUCTION. As for $\mathcal{FRC}$ (see above), we assume that all BGs are irredundant. The question is "Is there a derivation from $\mathcal{F}$ to F' and a homomorphism from Q to F', such that for all F_i of this derivation, for all constraint C_j, for all homomorphism π from $C_{j(0)}$ to F_i, there is a homomorphism π' from C_j to F_i such that $\pi'[C_{j(0)}] = \pi$?" R is polynomially decidable and polynomially balanced (since the size of the derivation from $\mathcal{F}$ to F' is polynomially related to the size of the input). Thus $\mathcal{FEC}$-DEDUCTION is in Σ_3^P. In order to prove the completeness, we build a reduction from a special case of the problem B_3, where the formula is a 3-CNF (i.e., an instance of 3-SAT): Given a formula E, which is a conjunction of clauses with three literals, and a partition $\{X_1, X_2, X_3\}$ of its variables, is there a truth assignment for the variables in X_1, such that for every truth assignment for the variables of X_2, there is a truth assignment for the variables of X_3 such that is E true? This problem is Σ_3^P-complete [Sto77, Theorem 4.1]. Let us call it 3-SAT$_3$.

The transformation we use is illustrated in Fig. 11.13. The 3-SAT formula in this example is again $(a \lor b \lor \neg c) \land (\neg a \lor c \lor \neg d)$, and the partition is $X_1 = \{a, b\}, X_2 = \{c\}, X_3 = \{d\}$. The graph F obtained is the same as in the proof of theorem 11.3, Fig. 11.7, except that concept nodes [x] corresponding to variables in X_1 are not linked to the nodes [xt] and [xf] representing their possible values.

First check that in this initial world no constraint is violated, but the query Q is not satisfied. By applying the evolution rule E once, we try some valuation of variables in X_1 and obtain a world F_1, that contains an answer to Q. But this world has to satisfy the positive constraint C_+, expressing that "For every valuation of the variables in $X_1 \cup X_2$, there must be a valuation of the variables in X_3 such that the formula evaluates to *true*." If F_1 satisfies this constraint, this means that we have found (by applying E) a valuation of variables in X_1 such that for all valuations of variables in $X_1 \cup X_2$ (which can be simplified in "for all valuations of variables in

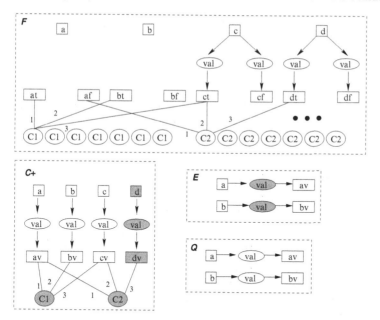

Fig. 11.13 Example of transformation from 3-SAT_3 to $\mathcal{FEC}$-DEDUCTION

X_2," since there is only one such valuation for X_1), there is a valuation of variables in X_3 such that the formula evaluates to *true*. Then there is an answer *yes* to the 3-SAT_3 problem. Conversely, suppose an answer *no* to the $\mathcal{FEC}$ problem. This means that for every world F_1 that can be obtained by applying the rule E, the constraint C^+ is violated (Otherwise Q could be mapped to F_1 and the answer would be *yes*). Thus there is no assignment of variables in X_1 satisfying the constraint, *i.e.*, the answer to the 3-SAT_3 problem is *no*.

Σ_3^P-*completeness of* $\mathcal{FREC}$-DEDUCTION. $\mathcal{FREC}$-deduction stays in the same complexity class as $\mathcal{FEC}$-deduction. Indeed, the question is "Is there an $\mathcal{RE}$-derivation from $\mathcal{F}$ to F' and a homomorphism from Q to F', such that for all F_i either equal to $\mathcal{F}$ or obtained by an $\mathcal{E}$-evolution, for all F_i' derived from F_i by an $\mathcal{R}$-derivation, for all constraint C_j, for all homomorphism π from $C_{j(0)}$ to F_i', there is an $\mathcal{R}$-derivation from F_i' to F_i'' and a homomorphism π' from C_j to F_i'' such that $\pi'[C_{j(0)}] = \pi$?" and the lengths of all derivations are polynomial in the size of the input. When $\mathcal{R} = \emptyset$, one obtains $\mathcal{FEC}$-DEDUCTION, thus the Σ_3^P completeness. □

Let us point out that in the general case deduction is more difficult in $\mathcal{FRC}$ (totally undecidable) than in $\mathcal{FEC}$ (semi-decidable), but the converse holds for the particular case of range-restricted rules.

11.4.2 Particular Cases of Constraints

One may consider the case where not only rules but also constraints are restricted. Let us first consider the meaningful category of negative constraints.

Theorem 11.10 (Complexity with negative constraints). *Without any assumption on the rules in $\mathcal{E}$ or $\mathcal{R}$, but using only negative constraints:*

- *Consistency in $\mathcal{FC}$ is co-NP-complete.*
- *Deduction in $\mathcal{FC}$ is DP-complete.*
- *Inconsistency in $\mathcal{FRC}$ is semi-decidable.*
- *Deduction in $\mathcal{FEC}$ remains semi-decidable.*
- *Deduction in $\mathcal{FRC}$ and $\mathcal{FREC}$ remains totally undecidable.*

Proof. Co-NP-completeness of $\mathcal{FC}$-CONSISTENCY and DP-completeness of $\mathcal{FC}$-DEDUCTION are proven in Theorem 11.4.

Semi-decidability of $\mathcal{FRC}$-INCONSISTENCY: To prove the inconsistency of a KB, we must find some violation of a constraint that will never be restored. But no violation of a negative constraint can ever be restored (further rule applications can only add information, thus more possible homomorphisms, and cannot remove the culprit). So we only have to prove that one constraint of $\mathcal{C}$ can be deduced from $(\mathcal{F}, \mathcal{R})$: It is a semi-decidable problem. *Undecidability of $\mathcal{FRC}$-DEDUCTION* follows: We must prove that Q can be deduced from $(\mathcal{F}, \mathcal{R})$, but that no constraint of $\mathcal{C}$ can.

The arguments proving semi-decidability of deduction in $\mathcal{FEC}$ and undecidability of deduction in $\mathcal{FREC}$ are the same as those used in the proof of Theorem 11.8. □

The restriction to negative constraints decreases the problem complexity in the $\mathcal{FC}$ model, but it does not help much as soon as rules are involved, since problems remain undecidable. When range restricted rules and negative constraints are combined, more interesting complexity results are obtained:

Theorem 11.11 (Complexity with range-restricted rules and negative constraints). *If only range-restricted rules and negative constraints are present in the knowledge base:*

- *Consistency in $\mathcal{FRC}$ is co-NP-complete.*
- *Deduction in $\mathcal{FRC}$ is DP-complete.*
- *Deduction in $\mathcal{FEC}$ and $\mathcal{FREC}$ is Σ_2^P-complete.*

Proof. Inconsistency in $\mathcal{FRC}$ admits a polynomial certificate: A derivation (of polynomial length) from $\mathcal{F}$ leading to a graph to which a constraint of $\mathcal{C}$ can be mapped, and this homomorphism. Inconsistency is thus in NP, and completeness follows from the particular case when $\mathcal{R}$ is empty. For deduction, we must prove that no constraint can be deduced from $(\mathcal{F}, \mathcal{R})$, but that Q can. So the problem is in DP. Completeness follows from the particular case when $\mathcal{R}$ is empty.

To prove that $\mathcal{FEC}$-DEDUCTION with range restricted rules and negative constraints is Σ_2^P-complete, we will first show that it belongs to Σ_2^P, and then exhibit a reduction from a Π_2^P-complete problem to its co-problem $\mathcal{FEC}$-NON-DEDUCTION (since co-$\Sigma_i^P = \Pi_i^P$).

$\mathcal{FEC}$-DEDUCTION corresponds to the language $L = \{x \mid \exists y_1 \forall y_2 \; R(x, y_1, y_2)\}$, where x encodes an instance $(Q; (\mathcal{F}, \mathcal{E}, \mathcal{C}))$ of the problem, and $(x, y_1, y_2) \in R$ if y_1 encodes an $\mathcal{E}$-derivation from $\mathcal{F}$ to F' and a homomorphism from Q to F', and y_2 encodes a mapping from some constraint of $\mathcal{C}$ to F' that is not a homomorphism (note that if F' does not violate any constraint, no graph in the derivation from $\mathcal{F}$ to F' neither does).

We now exhibit a reduction from the general $\mathcal{FC}$-CONSISTENCY problem to $\mathcal{FEC}$-NON-DEDUCTION with range restricted rules and negative constraints. Let $(\mathcal{F}, \mathcal{C} = \{C\})$ be an instance of $\mathcal{FC}$-CONSISTENCY (without loss of generality, we restrict the problem to consider only one positive constraint). The transformation we consider builds an instance of $\mathcal{FEC}$-NON-DEDUCTION $(Q(C); (\mathcal{F}, \mathcal{E}(C), \mathcal{C}^-(C)))$ as follows. As for a rule, we call the *frontier* of the positive constraint C the set of nodes in the condition (*i.e.*, colored by 0) having at least one neighbor in the obligation. The definition of colored graphs implies that frontier nodes are concept nodes (their neighbors are thus relation nodes). Let us denote these frontier nodes by $1, \ldots, k$. The evolution rule $\mathcal{E}(C)$ has for hypothesis the condition of C, and for conclusion a relation node typed $found$, where $found$ is a new k-ary relation type that is incomparable with all other types. The ith neighbor of this node is the frontier node i. Check that $\mathcal{E}(C)$ is a range restricted rule. The negative constraint $C^-(C)$ is the subgraph of C composed of its obligation $C_{(1)}$ and the relation node typed $found$, linked to the frontier nodes in the same way as above. Finally, the query $Q(C)$ is made of one relation node typed $found$ and its neighbor frontier nodes. This transformation is illustrated in Fig. 11.14.

Without loss of generality we can assume that $\mathcal{F}$ is irredundant: In this case, $\mathcal{FC}$-CONSISTENCY is still Π_2^P-complete (See that the transformation used in the proof of Theorem 11.3 produces an irredundant graph G). Now suppose that $(\mathcal{F}, \mathcal{C})$ is consistent: This means that either the condition of C does not map to $\mathcal{F}$, and in this case the rule $\mathcal{E}(C)$ will never produce the needed $found$ node, or every (existing) homomorphism from the condition of C to $\mathcal{F} = irr(\mathcal{F})$ can be extended to a homomorphism from C as a whole. So every application of $\mathcal{E}(C)$ produces a violation of $C^-(C)$. In both cases, $Q(C)$ cannot be deduced from the knowledge base. Conversely, suppose that $\mathcal{F}$ π-violates C, then the application of $\mathcal{E}(C)$ according to π produces a graph that does not violate $C^-(C)$, and we can deduce $Q(C)$. $\Box$

The above theorem shows a complexity decrease when general positive constraints are restricted to negative ones. $\mathcal{FC}$-CONSISTENCY falls from Π_2^P to co-NP and, when also considering range restricted rules, $\mathcal{FRC}$-CONSISTENCY falls from Π_2^P to DP, and $\mathcal{FEC}$-DEDUCTION falls from Σ_3^P to Σ_2^P. It would be interesting to exhibit particular cases of constraints that are more general than negative ones and that make this complexity fall into intermediary classes. We considered *disconnected constraints* and *range-restricted constraints* in Sect. 11.2.3.

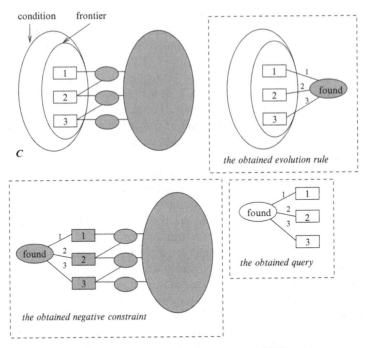

Fig. 11.14 Transformation from $\mathcal{FC}$-CONSISTENCY to a restricted $\mathcal{FEC}$-NON-DEDUCTION

Theorem 11.12 (Complexity with disconnected constraints). *When $\mathcal{C}$ contains only disconnected constraints:*

- $\mathcal{FC}$-CONSISTENCY *is co-DP-complete.*
- $\mathcal{FRC}$-CONSISTENCY *and* $\mathcal{FRC}$-DEDUCTION *remain undecidable, but* $\mathcal{FRC}$-CONSISTENCY *is co-DP-complete when rules are range-restricted.*
- $\mathcal{FEC}$-DEDUCTION *remains semi-decidable, but is* Σ_2^P-*complete when rules are range-restricted.*
- $\mathcal{FREC}$-DEDUCTION *remains undecidable, but* Σ_2^P-*complete when rules are range-restricted.*

Proof. $\mathcal{FC}$-INCONSISTENCY belongs to DP, since we must prove that for one constraint there is a homomorphism from its condition and no homomorphism from its obligation. DP-completeness is proven with a reduction from SAT/UNSAT (as in proof of Theorem 11.10). $\mathcal{FC}$-CONSISTENCY is thus co-DP-complete.

Arguments for undecidability of $\mathcal{FRC}$-CONSISTENCY, $\mathcal{FRC}$-DEDUCTION and $\mathcal{FREC}$-DEDUCTION, as well as semi-decidability of $\mathcal{FEC}$-DEDUCTION, are the same as in the proof of Theorem 11.8: The constraints we used were already disconnected.

When rules are range-restricted, $\mathcal{FRC}$-INCONSISTENCY belongs to DP: We must prove that the condition of the constraint can be deduced from $(\mathcal{F}, \mathcal{R})$, but not its obligation. These problems belong to NP and co-NP. DP-completeness comes from the particular case where $\mathcal{R}$ is empty. $\mathcal{FRC}$-CONSISTENCY is thus co-DP-complete.

$\mathcal{FEC}$-DEDUCTION belongs to Σ_2^P when rules involved are range-restricted. Although this property does not appear with an immediate formulation of the problem, the problem can be stated as follows: "Is there a sequence of graphs $F_0 (= \mathcal{F})$, ..., F_p, F_{p+1}, where $F_0 (= \mathcal{F})$, ..., F_p is an $\mathcal{E}$-derivation and F_{p+1} is the disjoint union of F_p and $C_{(1)}$, a homomorphism from Q to F_p and a homomorphism from $C_{(1)}$ to F_k, $0 \le k \le p+1$ such that for every graph $F_i, 0 \le i < k$, for every mapping π from $C_{(0)}$ to F_i, π is not a homomorphism ?" Note that no F_i before F_k in such a sequence triggers the constraint ($C_{(0)}$ does not map to F_i) and that all F_i from F_k satisfy it (since $C_{(1)}$ maps to F_i). Thus all F_i of the sequence are consistent. F_{p+1} ensures that $C_{(1)}$ maps to at least one graph of the sequence, which allows the above formulation of the problem. Completeness follows from the particular case of negative constraints.

Proof for $\mathcal{FREC}$-DEDUCTION is similar: In the expression of the above problem, the derivation is now an $(\mathcal{E} \cup \mathcal{R})$-derivation, the F_i considered are only those obtained after the application of a rule in $\mathcal{E}$, and instead "every mapping π from $C_{(0)}$ to F_i", we consider "every mapping π from $C_{(0)}$ to a graph that can be $\mathcal{R}$-derived from F_i". $\square$

One may expect consistency checking to be easier with range-restricted constraints than with general constraints, but as far as we know there are currently no results on the complexity of this particular case.

Table. 11.1 summarizes the decidability and complexity results obtained. The following notations are used: *fes* and *rr* for respectively *finite expansion set* and *range restricted* sets of rules, C^- for a set of negative constraints and C^d for a set of disconnected constraints. The sign / indicates a particular case that is not applicable to the problem considered.

	General case	$\mathcal{R}$ fes	$\mathcal{E}$ fes	$\mathcal{R} \cup \mathcal{E}$ fes	$\mathcal{R} \& \mathcal{E}$ rr	C^-	$\mathcal{R} \& \mathcal{E}$ rr, C^-	$\mathcal{R} \& \mathcal{E}$ rr, C^d
$\mathcal{F}$-DED.	NP-C	/	/	/	/	/	/	/
$\mathcal{FC}$-CONS.	Π_2^P-C	/	/	/	/	co-NP-C	co-NP-C	co-DP-C
$\mathcal{FC}$-DED.	Π_2^P-C	/	/	/	/	DP-C	DP-C	?
$\mathcal{FR}$-DED.	semi-dec.	dec.	/	dec.	NP-C	/	NP-C	NP-C
$\mathcal{FRC}$-CONS.	undec.	dec.	/	dec.	Π_2^P-C	co-semi-dec.	co-NP-C	co-DP-C
$\mathcal{FRC}$-DED.	undec.	dec.	/	dec.	Π_2^P-C	undec.	DP-C	?
$\mathcal{FEC}$-DED.	semi-dec.	/	dec.	dec.	Σ_3^P-C	semi-dec.	Σ_2^P-C	Σ_2^P-C
$\mathcal{FREC}$-DED.	undec.	semi-dec.	undec.	dec.	Σ_3^P-C	undec.	Σ_2^P-C	Σ_2^P-C

Table 11.1 Summary of Complexity Results

11.5 Bibliographic Notes

This chapter is essentially based on [BM02], which introduces the BG family and studies the complexity of problems with rules and contraints in detail. The model names have been changed for consistency with the notations used in the present book: BG corresponds to SG in the paper, and the $\mathcal{F}$, $\mathcal{FC}$, $\mathcal{FR}$, $\mathcal{FRC}$, $\mathcal{FEC}$ and $\mathcal{FREC}$ models correspond to the $\mathcal{SG}$, $\mathcal{SGC}$, $\mathcal{SR}$, $\mathcal{SRC}$, $\mathcal{SEC}$ and $\mathcal{SREC}$ models in the paper.

Let us point out two results of [BM02] that have not been included in this chapter. First, the tight connection between $\mathcal{FC}$-CONSISTENCY and a problem on constraint networks called MIXED-SAT in [FLS96]. This latter paper considers a generalization of constraint networks, called a mixed-network, in which the set of variables is decomposed into controllable and uncontrollable variables, say X and Δ. The MIXED-SAT problem takes a binary mixed-network as input and asks whether it is consistent, that is: Is it true that every solution to the subnetwork induced by Δ can be extended to a solution of the whole network? MIXED-SAT is shown to be Π_2^P-complete. This result provides us with another proof of Π_2^P-completeness for $\mathcal{FC}$-consistency. Secondly, we mention in this chapter that $\mathcal{FREC}$-DEDUCTION is not decidable when $\mathcal{R}$ and $\mathcal{E}$ are both finite expansion sets (but their union is not a finite expansion set). The proof of this result is in [BM02].

Examples throughout this chapter are inspired from the modelling of the SYSI-PHUS-I case-study presented in [BGM99b] and [BGM99a].

Relations to other constraints.

As already noted for rules, our BG constraints have the same shape as the TGDs (Tuple Generating Dependencies) in databases. Different forms of constraints can be found in the conceptual graph literature. To the best of our knowledge, the most general ones are the minimal/maximal descriptive constraints defined in [DHL98]. BG-constraints are a particular case of these minimal descriptive constraints. A minimal descriptive constraint can be seen as a set of BG-constraints with the same condition A; its intuitive semantics is "If A holds so must B_1 or B_2 or ... B_k." A BG satisfies a minimal descriptive constraint if it satisfies at least one element of the set. Note that the "disjunction" does not increase complexity with respect to BG-constraints. The proof of Theorem 11.3 (complexity of $\mathcal{FC}$-CONSISTENCY) can be used to show that consistency of minimal descriptive constraints is also Π_2^P-complete. Particular cases of BG-constraints found in the CG literature are the topological constraints used by [MM97], which are indeed disconnected constraints. In this paper non-validity intervals in the BG space (pre)-ordered by specialization are defined. A non-validity interval is a pair of BGs u and v, such that $v \leq u$. A graph G is said to fall within this interval, if $G \leq u$ and not $G \leq v$. Then it is said to be invalid. This can be expressed in terms of BG-constraints in the following way: G is invalid if it violates the disconnected constraint "If u holds, so must v." Let us add that in these CG works constraints are used to solely check the consistency of facts (as in $\mathcal{FC}$) and not of richer knowledge bases composed of rules (as in $\mathcal{FRC}$) and they are not integrated in more complex reasoning schemes (as in $\mathcal{FEC}$ or in $\mathcal{FREC}$).

Concerning the combined use of rules and constraints, there are related works about verification of knowledge bases composed of logical rules (e.g., Horn rules). Let us mention in particular [LR96] [LR98], in which constraints are TGDs, so they have the same form as ours. However, the consistency notion is not defined in the same way. One major difference is that facts are not part of the knowledge base, and the verification notion for a knowledge base is defined with respect to any set of facts (satisfying some validity conditions). Finally, evolution rules could be generalized into transformation rules, i.e., rules that add knowledge but may also delete knowledge. As part of this programme, it would be interesting to consider the work on graph transformation rules carried out in the graph grammar domain (cf. for instance [Roz97]).

Chapter 12
Conceptual Graphs with Negation

Overview

Basic or simple graphs constitute the kernel of conceptual graphs. They can be used as such, to represent facts or queries. They are also basic bricks for more complex constructs, corresponding to more expressive conceptual graphs, for instance rules (cf. Chap. 10). In this chapter, we consider the addition of negation to basic graphs.

Full conceptual graphs (FCGs) are obtained when negation is added without restriction. FCGs are inspired from Peirce's existential graphs, which form a diagrammatical system for logics. In the first section, we first briefly present the main characteristics of existential graphs, then define Sowa's full conceptual graphs and their translation to FOL with an extension of the mapping Φ defined on BGs. By adding negation to existential quantification and conjunction, one obtains the full expressive power of FOL. A translation Ψ from FOL formulas (without functions) to FCGs is given. It is shown that for every closed FOL formula f, $\Phi(\Psi(f)) \equiv f$. We mention a variant of FCGs, introduced by Dau, which in our opinion yields simpler diagrams. Dau's FCGs come with a model semantics and sound and complete calculus with respect to FOL. The calculus for these two FCG models are both based on the existential graph calculus. Peirce's diagrammatical rules, thus FCG rules, are not adapted to automated reasoning. Thus, we briefly present a third FCG version introduced by Kerdiles, who proposes an automated reasoning method that mixes a tableau method with graph homomorphism checks.

FCGs extend BGs with negation, but in a way that does not suit the approach to knowledge representation developed throughout this book. Sect. 12.1 thus only briefly outlines FCGs. In our opinion, full FOL is too complicated at the end-user level, for building knowledge-based systems and understanding how they work, also from a computational viewpoint. Most conceptual graphs applications are based on BGs and extensions that keep their intuitive graphical appeal, as nested graphs, rules and constraints. One advantage of these extensions is to help to distinguish between different kinds of knowledge—a key issue in knowledge-based systems building. Note however that these extensions only provide an implicit and very specific form of negation.

In the second section, we study a limited form of negation, namely atomic negation. We will see that the graphical aspect is respected but further work is required to obtain practical computational efficiency. Basic graphs plus atomic negation yield polarized graphs (PGs), which are equivalent to the FOL fragment of existentially closed conjunctions of positive and negative literals. We also study $PG^{\neq}s$ (resp. $BG^{\neq}s$), which are PGs (resp. BGs) added with inequality (also called difference). We point out the equivalence of PG-DEDUCTION with the containment of conjunctive queries with negation. It is proven that PG-DEDUCTION is Π_p^2-complete, as well as $BG^{\neq}$-DEDUCTION (hence $PG^{\neq}$-DEDUCTION). Special cases for PG-DEDUCTION, based on the notion of *exchangeable* literals, are exhibited. Algorithmic improvements of the brute-force algorithm for PG-DEDUCTION are detailed. We then shift from PG-DEDUCTION to querying PGs. We review both closed-world and open-world assumptions. With open-world assumption, there may be no answer to a query in a fact base, even if the query can be deduced from the base. We discuss the role of the excluded-middle law in this distinction, and propose intuitionistic logic as a way to translate the notion of answer. Furthermore, with this logic, PG-deduction and query answering are NP-complete, thus negation is processed without overhead complexity compared to BGs. We end with a note on new sources of complexity when PG rules are considered.

12.1 Full Conceptual Graphs

We briefly glance at existential graphs to highlight their main characteristics, and then get to full conceptual graphs and their equivalence with FOL.

12.1.1 Existential Graphs: a Diagrammatical System for Logics

At the end of the 19th century, the logician Peirce introduced a diagrammatical system for logics, i.e., *existential graphs*. Existential graphs are not graphs in the classical graph theoretical meaning. They are simply diagrams. Thus, to avoid confusion, the existential graphs are called existential diagrams. There are two levels of existential diagrams: alpha diagrams, corresponding to propositional logic and beta diagrams corresponding to FOL. There is also an unfinished third level, i.e., gamma diagrams, which introduces various kinds of modalities.

Let us begin with alpha diagrams. An alpha diagram is built from two syntactical constructs: proposition and cut representing negation. Asserting a diagram consists of writing it down in an area, called the sheet of assertions. Juxtaposing several propositions in this area corresponds to asserting them together, i.e., asserting their conjunction. Encircling a proposition by an oval corresponds to cutting it from the sheet of assertions, i.e., negating it. The line used to draw the oval is thus called a cut. The space within a cut is the area of the cut. An area is said to be oddly (resp.

evenly) enclosed if it is enclosed by an odd (resp. even) number of cuts. For instance, the first alpha diagram D in Fig. 12.1 asserts a proposition p and the negation of "p and the negation of q". Translated into propositional logic it asserts the formula $(p \land \neg(p \land \neg q))$, i.e., $(p \land (p \to q))$.

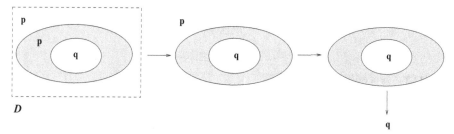

Fig. 12.1 A proof with alpha rules

Existential graph calculus is composed of rules defining diagrammatical transformations. The alpha system is composed of five rules:

- Erasure: every evenly enclosed diagram can be erased;
- insertion: any diagram can be written in an oddly enclosed area;
- iteration: if a diagram occurs in an area, then a copy of this diagram can be written in the same area or in any nested area which does not belong to the diagram;
- deiteration: any diagram whose occurrence could be the result of iteration may be erased: any diagram may be erased if a copy of this diagram occurs in the same area or in an enclosing area;
- double cut: any double cut may be inserted around or removed from any diagram of any area (including the empty diagram).

Figure 12.1 shows a proof in the alpha system. It illustrates the *modus ponens* reasoning: If p and $p \to q$ is true, then q is true. First, the diagram corresponding to $p \land (p \to q)$ is asserted. Secondly, by the deiteration rule, the oddly enclosed proposition p is erased and the second diagram corresponding to $p \land \neg\neg q$ is obtained. Thirdly, the erasure rule deletes p. Forthly, q is obtained by the double cut rule.

In the first-order version of diagrams, i.e., beta diagrams, symbols of predicate with any arity are introduced and lines, called lines of identity, are used to represent objects, to connect a predicate to objects, as well as to express the identity between objects. Several connected identity lines form a network, called a ligature, and represent a single object. There are no variable names. The only—implicit—quantifier is the existential quantifier, and the universal quantifier is obtained as the negation of the existential one. For instance, in Fig. 12.2, the diagram has five predicate occurrences (let us say five literals) and several identity lines. Human is a unary predicate, and hasParent is a binary predicate. A first ligature connects the argument of the literal with predicate Human, and the first argument of both literals with predicate hasParent (arguments are here implicitly ordered from left to right).

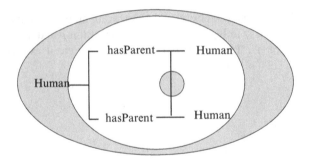

Fig. 12.2 A beta existential diagram

It thus represents one object. Two other identity lines connect, respectively, the second argument of a literal with predicate hasParent and the argument of a literal with predicate Human. These lines are themselves connected by a line that crosses a cut and thus denotes the non-identity of its extremities. Intuitively, this diagram asserts that there is no Human who does not have as parents two distinct Human, i.e., every Human has (at least) two parents who are Human. The associated formula could be
$\neg(\exists x(Human(x) \land \neg(\exists y \exists z(hasParent(x,y) \land hasParent(x,z) \land \neg(y=z)))))$.

Beta rules extend the alpha rules with the treatment of identity lines.

12.1.2 Full Conceptual Graphs (FCGs)

Inspired by Peirce's existential diagrams, Sowa defines full CGs (FCGs) from BGs by adding *boxes* representing negation, and *lines* (the coreference links) representing equality of terms. In drawings, negation is represented by a rectangular box preceded by a ¬ symbol. For instance, the FCG G in Fig. 12.3 says that the relation r is transitive on entities of type t, i.e., for all x, y, z of type t, if $r(x, y)$ and $r(y, z)$, then $r(x, z)$. Its logical translation is as follows (cf. the next section for details):
$$\Phi(G) = \neg(\exists x\, \exists y\, \exists z\, (t(x) \land t(y) \land t(z) \land r(x,y) \land r(y,z) \land \neg r(x,z)))$$
which is equivalent to: $\forall x\, \forall y\, \forall z\, (t(x) \land t(y) \land t(z) \land r(x,y) \land r(y,z) \rightarrow (r(x,z)))$.

G is equivalent to a bicolored BG rule (cf. Chap. 10). More generally, a λ-rule $R = (c_1...c_n)H \Rightarrow C$ is equivalent to an FCG of form $\neg[\,H\ \neg C\,]$ with a coreference link between c_{1_i} and c_{2_i} for all $i = 1...n$.

Figure 12.4 shows how inequality is represented. Like the existential diagram in Fig. 12.2, it says that "Every Human as (at least) two parents", i.e., for every Human x there is a Human y and there is a Human z, such that y and z are different and are both parents of x. To express that $y \neq z$, we have to say "No entity exists that is equal to both y and z." Note that Dau's FCGs (cf. Sect. 12.1.6) provide a more readable way of expressing difference.

Let us now formally define FCGs. BGs are first combined within more complex structures called graph propositions.

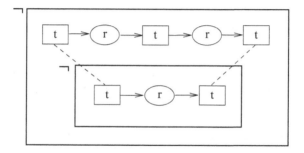

G

Fig. 12.3 A full CG

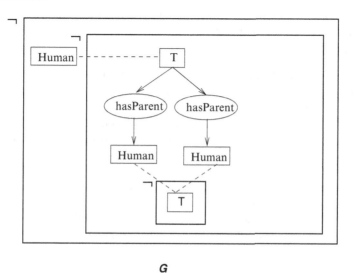

G

Fig. 12.4 A full CG with inequality

Definition 12.1 (Graph proposition). A *graph proposition* on a vocabulary $\mathcal{V}$ is:

- either a box $[G\ p_1\ ...\ p_n]$, which contains a BG G on $\mathcal{V}$ (possibly empty) and finitely many graph propositions $p_1\ ...\ p_n$, $n \geq 0$. This box is also called *the context* of G.
- or a negated graph proposition $\neg p$. A *negated context* is a box preceded by a negation.

All BGs composing a graph proposition are disjoint.

A box $[G]$, where G is a BG, can be identified with G. A box $[G_0\ [G_1],\ldots,[G_n]]$, where each G_i is a BG can be identified with the disjoint sum of the G_i. More generally, a box in which no negation occurs can be identified with the disjoint sum of all BGs appearing in this box.

Note that these contexts should not be confused with the contexts of nested conceptual graphs (cf. Chap. 9). Both kinds of contexts have radically different logical translations.

Definition 12.2 (Domination). A context p *dominates* a context q if q is contained in p or $p = q$. By extension, we say that a node of a BG G_1 *dominates* a node of a BG G_2 if the context of G_1 dominates the context of G_2.

Definition 12.3 (FCG). A full conceptual graph (FCG) on a vocabulary $\mathcal{V}$ is a pair $(p, coref)$, where:

1. p is a graph proposition on $\mathcal{V}$.

2. Let $C(p)$, called the set of concept nodes of p, be the union of the concept node set of all BGs composing p. $coref$ is a binary symmetric relation over $C(p)$, such that $(c_1, c_2) \in coref$ implies that c_1 dominates c_2 or c_2 dominates c_1. (c_1, c_2) is called a coreference link.

This definition calls for a few comments. Since FCGs are meant to represent any formula of FOL, coreference links are allowed between concept nodes of any label: Coreferent nodes are not necessarily compatible (as for coreference defined on SGs); in particular, there may be two individuals with different markers. Indeed, in classical FOL, if a and b are two constants, they may be interpreted by the same element of a domain, thus $(a = b)$ is a well-formed formula; in other words, the unique name assumption usually made in knowledge representation does not hold here. Coreference links can link two nodes of the same context (since a context dominates itself), or of nested contexts.

We now give a precise definition of the meaning of an FCG by way of a translation to FOL, which extends the mapping defined for BGs.

12.1.3 Logical Semantics of FCGs

Let us first point out that the universal type $\top$ is translated in a special way. The atom assigned to a concept node $c = [\top : m]$ is not $\top(term(c))$, where $term(c)$ is the term assigned to c, but the trivial equality $term(c) = term(c)$. This ensures that the graphs $[\top : *]$ or $[\top : a]$ are valid (i.e., the associated formulas are valid). By convention, the empty conceptual graph is translated to *true*.

Briefly explained, basic graphs occurring in the FCG are first translated. Distinct variables are associated with distinct generic concept nodes. Each basic graph G is interpreted as a conjunction of atoms $\alpha(G)$ (with the special processing of $\top$), without quantification for the moment. Coreference links are processed in the dominated context: Let (c_1, c_2) be a coreference link, with c_1 dominates c_2, and c_2 belongs to G; the atom $(term(c_1) = term(c_2))$ is added to $\alpha(G)$. Let $\alpha'(G)$ be the (free) formula obtained.

Now, given an FCG $G = (p, coref)$, $\Phi(G) = \Phi(p)$, and $\Phi(p)$ is defined according to the inductive definition of p. If p is a box $[K\ p_1\ ...\ p_n]$ then $\Phi(p)$ is the existential closure of the conjunction of $\alpha'(K)$ and the $\Phi(p_i)$. If p is a negated proposition $[\neg p']$, then $\Phi(p) = \neg\Phi(p')$. Let us consider the FCG G of Fig. 12.3. Let us call G_1 and G_2 the basic graphs composing G, with G_1 being the graph in the outermost context. We first assign distinct variables to generic nodes and free formulas to G_1 and G_2 (cf. Fig. 12.5). $\alpha'(G_1) = t(x_1) \wedge t(y) \wedge t(z_1) \wedge r(x_1, y) \wedge r(y, z_1)$

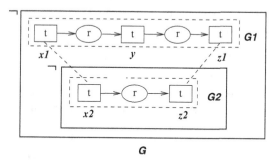

Fig. 12.5 Variables assigned to the FCG of Fig. 12.3

$\alpha'(G_2) = t(x_2) \wedge t(z_2) \wedge r(x_2, z_2) \wedge (x_1 = x_2) \wedge (z_1 = z_2)$

The graph proposition underlying G has the following structure: $\neg[G_1\neg[G_2]]$. The closed formula $\Phi([G_2])$ is $\exists x_2 \exists z_2 \alpha'(G_2)$. The closed formula $\Phi([G_1\neg[G_2]])$ is $\exists x_1 \exists y \exists z_1 (\alpha'(G_1) \wedge \neg(\exists x_2 \exists z_2 \alpha'(G_2)))$. Finally, one obtains

$\Phi(G) = \neg(\exists x_1 \exists y \exists z_1 (t(x_1) \wedge t(y) \wedge t(z_1) \wedge r(x_1, y) \wedge r(y, z_1)) \wedge \neg(\exists x_2 \exists z_2 (t(x_2) \wedge t(z_2) \wedge r(x_2, z_2) \wedge (x_1 = x_2) \wedge (z_1 = z_2))))$.

This formula can be rewritten in a more classical way as:

$\Phi(G) = \forall x_1\ \forall y\ \forall z_1 ((t(x_1) \wedge t(y) \wedge t(z_1) \wedge r(x_1, y) \wedge r(y, z_1)) \rightarrow \exists x_2 \exists z_2 (t(x_2) \wedge t(z_2) \wedge r(x_2, z_2) \wedge (x_1 = x_2) \wedge (z_1 = z_2)))$.

It is equivalent to the following formula without equality:

$\Phi(G) = \forall x\ \forall y\ \forall z ((t(x) \wedge t(y) \wedge t(z) \wedge r(x, y) \wedge r(y, z)) \rightarrow r(x, z))$.

The FCG in Fig. 12.4 is composed of three BGs, say G_1, G_2 and G_3. G has the structure $\neg[G_1\neg[G_2\neg G_3]]$. Let $\alpha'(G_1) = Human(x_1)$, $\alpha'(G_2) = (x_2 = x_2) \wedge Human(y) \wedge Human(z) \wedge hasParent(x_2, y) \wedge hasParent(x_2, z)$, $\alpha'(G_3) = (v = v) \wedge (y = v) \wedge (z = v)$. The formula associated with the negated proposition $\neg G_3$ is $\Phi(\neg G_3) = \neg \exists v((v = v) \wedge (y = v) \wedge (z = v))$. $\Phi(\neg G_3)$ is equivalent to $(y \neq z)$. After removal of trivial equalities, we obtain $\Phi(\neg[G_2\neg G_3]) = \neg(\exists x_2 \exists y \exists z(Human(y) \wedge Human(z) \wedge hasParent(x_2, y) \wedge hasParent(x_2, z) \wedge (x_1 = x_2)))$. Finally, $\Phi(G) = \neg \exists x_1 (Human(x_1) \wedge \neg(\exists x_2 \exists y \exists z(Human(y) \wedge Human(z) \wedge hasParent(x_2, y) \wedge hasParent(x_2, z) \wedge (x_1 = x_2))))$. It is equivalent to the following formula:

$\forall x(Human(x) \rightarrow \exists y \exists z(Human(y) \wedge Human(z) \wedge hasParent(x, y) \wedge hasParent(x, z) \wedge (y \neq z)))$.

Conversely, every first-order formula f (without function) can be translated into a CG G, such that $\Phi(G) \equiv f$, as detailed in the next section.

12.1.4 Equivalence of CGs and FOL

FCGs are equivalent to FOL (without functions) in the sense that every FOL formula can be translated into an FCG, say by a mapping Ψ, and, reciprocally, every FCG can be translated to a FOL formula, by Φ, such that the composition of the two translations keeps logical equivalence: For every FOL formula f, $\Phi(\Psi(f)) \equiv f$. This kind of correspondence allows one to consider FCGs as a diagrammatical representation of FOL formulas. For instance FCGs could be used at an interface level: The user manipulates CGs, which are translated into FOL formulas, on which logical provers can be used to do reasoning, and the results are translated back into CGs. Note however that for a formula f we only have equivalence between $\Phi(\Psi(f))$ and f and not equality. The question of whether using FCGs at an interface level only is interesting at all is another issue we will not discuss.

Before defining Ψ, we introduce another representation of FCGs. Indeed, as in positive nested conceptual graphs, coreference links destroy the recursive structure of conceptual graphs. They are not convenient for proofs of correspondence with logical formulas. They can be replaced by special variables. Let V be a set of variables. Let T denote the set $V \cup \mathcal{I}$. An *FCG with variables* over a vocabulary $\mathcal{V}$ and a set of variables V is a triple $(p, id, links)$ where p is a graph proposition, id is a mapping which assigns to each node in $C(p)$ its marker if it is individual. Otherwise a new variable (similarly to Φ), *links* is a mapping which assigns to each node c in $C(p)$ a subset of T such that each variable in this set is associated with a concept node that dominates c.

There are canonical translations between FCGs (with coreference links) and FCGs with variables. Let $G = (p, coref)$ be an FCG. To obtain an FCG with variables, define an arbitrary id (satisfying the condition in the definition of a CG with variables) and for each c take $links(c) = \{id(c') | c'$ dominates $c\}$.

Conversely, let $G = (p, id, links)$ be a CG with variables. Define coref as follows: $coref = \{(c, c') | c \in C(p), c' \in C(p), id(c') \in links(c)$ or $id(c) \in links(c')\}$.

We are led to consider conceptual graphs with free variables. In such graphs, the condition on *links* is weakened: If a variable x in $links(c)$ is assigned to a concept c' then c' dominates c. Thus, there are also variables which are not assigned to any node. These variables are said to be free. Indeed, they are free in $\Phi(G)$. And if Ψ is applied to a formula with free variables, these variables will stay free in the obtained CG.

Let us now define Ψ. A conceptual graph vocabulary is obtained from a logical language as follows. Concept types come from unary predicates and relation types come from the other predicates. In addition, there is a new universal concept type, denoted $\top$, which is the maximal concept type. All other concept types are pairwise incomparable. The same holds for relation types. The individual marker set comes from the constant set. In addition, there is an infinite set of variables, including those of the logical language.

Without loss of generality, we can assume that the FOL formulas considered are built only with $\exists$, $\wedge$, $\neg$ and equality, that they are closed, and that each variable is bound exactly once. Ψ maps such a formula to a CG with variables.

The overall ideas are as follows. Each occurrence of variable or constant e is translated into a concept node c of type t for an occurrence in an atom $t(e)$; otherwise (in an atom with an arity greater than one, in an equality or in a quantification $\exists e$) of type $\top$. If e is a variable x, then $id(c) = x$ only if the occurrence of x considered is $\exists x$, otherwise $id(c)$ is a fresh variable, and $links$ is used to relate c to x. If e is a constant a, then $id(c) = a$ and $links$ is used only if the occurrence of a considered is in an equality.

Given a formula f in appropriate form, $\Psi(f)$ is recursively defined as follows:

1. $\Psi(\exists x f') = [g_\top \Psi(f')]$ where $g_\top$ is restricted to a concept node $c = [\top : *]$, $id(c) = x$ and $links(c) = \emptyset$;
2. $\Psi(f_1 \wedge f_2) = [\emptyset \ \Psi(f_1) \ \Psi(f_2)]$
3. $\Psi(\neg f') = \neg \Psi(f')$
4. $\Psi(t(x)) = [g_t]$ where g_t is restricted to a concept node $c = [t : *]$, $id(c) = z$ and $links(c) = \{x\}$; z is a fresh variable that does not occur in the formula nor in the conceptual graph;
5. $\Psi(t(a)) = [g_t]$ where g_t is restricted to a concept node $c = [t : a]$, $id(c) = a$ and $links(c) = \emptyset$;
6. $\Psi(x = y) = [g]$ where g is restricted to a concept node $c = [\top : *]$, $id(c) = z$ and $links(c) = \{x, y\}$; z is a fresh variable that does not occur in the formula nor in the conceptual graph;
7. $\Psi(x = a) = [g]$ where g is restricted to a concept node $c = [\top : a]$, $id(c) = a$ and $links(c) = \{x\}$;
8. $\Psi(a = b) = [g]$ where g is restricted to two concept nodes $c_a = [\top : a]$ and $c_b = [\top : b]$, $id(c_a) = a, id(c_b) = b$, $links(c_a) = \{b\}$ and $links(c_b) = \{a\}$;
9. $\Psi(r(e_1 ... e_n))$ is the elementary star graph $r(c_1, ..., c_n)$ with a relation node of type r and n concept nodes; each c_i is of form $[\top : *]$ or $[\top : a]$, depending of whether e_i is a variable or a constant. If e_i is a variable x, then $c_i = [\top : *]$, $id(c_i) = z_j$ and $links(c_i) = \{x\}$; if e_i is a constant a, then $c_i = [\top : a]$, $id(c_i) = a$ and $links(c_i) = \emptyset$; the variables z_j are fresh variables that do not occur in the formula nor in the conceptual graph;

See for example Fig. 12.6, where $f = \Phi(G)$, with G is the FCG in Fig.12.3. In the picture, variables have been replaced by coreference links.

If formula f is closed, then $\Psi(f)$ is a well-formed CG with variables. We are now able to state the equivalence between FCGs and FOL at a descriptive level:

Theorem 12.1 (Equivalence FCGs and FOL). *Let $\mathcal{V}$ be a vocabulary and $\mathcal{L}_\mathcal{V}$ the corresponding first-order language (including equality). For each FOL formula f on $\mathcal{L}_\mathcal{V}$ (possibly with free variables), one has:* $\Phi(\Psi(f)) \equiv f$.

Hints for a proof. It can be checked that for each basic case in the definition of Ψ one has $\Phi(\Psi(f)) \equiv f$. By induction on the depth of a formula f, it is proven that $\Phi(\Psi(f)) \equiv f$ for any formula f in appropriate form. $\square$

This result shows that FCGs can represent any formula of first-order logic with equality (and without functions). Note however that the graph obtained by Ψ is completely split into trivial BGs, which are star graphs or isolated concept nodes

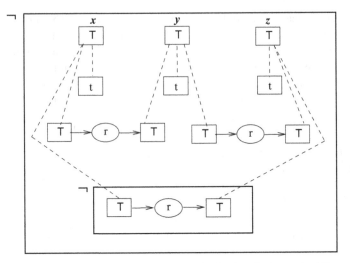

Fig. 12.6 $\Psi(f)$ with $f = \neg(\exists x\, \exists y\, \exists z(t(x) \wedge t(y) \wedge t(z) \wedge r(x,y) \wedge r(y,z) \wedge \neg r(x,z)))$

(compare for instance the FCGs in Fig. 12.3 and 12.6). It contains one concept node per occurrence of a term in the formula. To obtain a "true" graph form, one has to merge coreferent concept nodes (that is individual nodes with the same individual marker or nodes related by *links* or *coref*) belonging to BGs in the same box. Up to this transformation, CGs can be used as a graph representation for FOL.

12.1.5 FCG Calculus

FCGs are provided with a calculus inspired from Peirce's alpha and beta systems. A first version of this calculus was proposed by Sowa, and some minor flaws corrected by Wermelinger. Wermelinger proved the soundness and completeness of his set of rules (however the proof is not available in English, see bibliographical notes). Soundness and completeness of the FCG calculus states that FCGs and FOL are equivalent not only at a descriptive level but also at a *reasoning* level, i.e., given two FCGs G and H, G can be derived from H by the FCG calculus if and only if $\Phi(V), \Phi(H) \models \Phi(G)$ (equivalently: given two FOL formulas f and g in appropriate form, $f \models g$ if and only if $\Psi(g)$ can be derived from $\Psi(f)$ by the FCG calculus).

To the best of our knowledge, there is no automated reasoning tool that implements these rules. The main reason is certainly that Peirce's rules, thus FCG rules, are not suited to an automatic proof procedure. Indeed, the construction of a proof with these rules strongly relies on human intuition. The *rule of insertion*, which can be seen as the deduction of $\neg(A \wedge B)$ from $\neg(A)$, leads especially to an infinity of possibilities, by allowing insertion of any graph at a place surrounded by an odd number of negative contexts. Peirce's (and FCG) proof system is not analytical, in the sense it does not allow to build proofs by analyzing/ decomposing the input into its successive components.

12.1.6 Dau's FCGs

A notable variant of Sowa's FCGs was proposed by Dau, who called them *concept graphs*. Dau's FCGs are based on Peirce's cut notion rather than negative contexts. Differing from negative contexts, cuts can be drawn around arbitrary subgraphs, and not necessarily basic graphs. Furthermore, identity is represented by a binary equality relation, and this relation can be surrounded by a cut, like any other relation. In a sense, concept graphs are thus more directly related to existential graphs. Figures 12.7 and 12.8 show concept graphs equivalent to the CGs in Fig. 12.3 and 12.4, respectively. Cuts are drawn as rectangles with round corners. Non-numbered edges are implicitly directed from left to right.

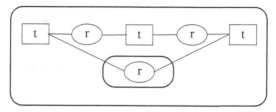

Fig. 12.7 A concept graph

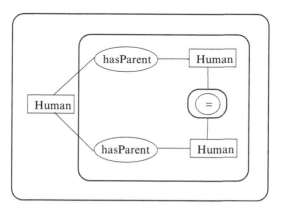

Fig. 12.8 A concept graph with inequality

Dau's FCGs are provided with inference rules based on Peirce's alpha and beta rules. They have a set semantics, called "contextual semantics" since graphs are interpreted in structures that are called power context families coming from Formal Concept Analysis. They have also a translation Φ to first-order logic. The rules are proven to be sound and complete with respect to these semantics. An inverse mapping Ψ from FOL formulas to concept graphs is given. The logical systems of concept graphs and FOL are shown to be equivalent via these mappings.

As they are based on Peirce's rules, concept graph rules are not analytical and, like FCG rules, not adapted to automated reasoning.

12.1.7 Kerdiles' FCGs

Kerdiles defined FCGs with a slightly different definition of coreference links (extending the one we gave for SGs), and another extension of Φ, which maps (his) FCGs to FOL without equality. He provided his FCGs with a sound and complete tableau calculus.

In Kerdiles' FCGs, coreference is not translated by equality but by multiple occurrences of the same term. The coreference relation is not only reflexive and symmetric, it is also transitive. It is actually an equivalence relation over generic nodes occurring in the graph proposition. Moreover, in each coreference class, there is at least one node that dominates all the others (there are possibly several such nodes). In other words, among the contexts of these nodes, there is at least a dominating context. These coreference conditions prevent some configurations of coreference links such as the one in the FCG of Fig. 12.4, for instance, where coreference is not transitive. And consequently, these full conceptual graphs can be mapped to FOL without equality. Roughly said, the translation is done in three steps. First, associate a term to each coreference class. The term assigned to a generic concept node is that assigned to its coreference class. Then, for each BG K composing the FCG, let $x_1 \ldots x_q$ be variables associated with its concept nodes coreferent with strictly dominating nodes, and let $x_{q+1}\ldots x_s$ be variables associated with the other generic nodes. Let $\alpha(K)$ be the conjunction of atoms associated with K. Finally, construct the whole formula in the following inductive way: If the graph proposition of G is a box $p = [K\ p_1\ \ldots\ p_n]$, then $\Phi(G) = \exists x_{q+1}\ldots x_s(\alpha(K) \wedge \Phi(p_1)\ \ldots\ \wedge \Phi(p_n))$ (note that the variables, if any, corresponding to nodes strictly dominated are not yet quantified). If the graph proposition of G is a negated proposition $[\neg p]$, $\Phi(G)$ is the negation of $\Phi(p)$.

Example. The formula associated with the FCG G in Fig. 12.3 is constructed as follows. First let x, y and z be the variables respectively assigned to each coreference class.

$\alpha(G_1) = t(x) \wedge t(y) \wedge t(z) \wedge r(x,y) \wedge r(y,z)$
$\alpha(G_2) = t(x) \wedge t(z) \wedge r(x,z)$
$\Phi([G_2]) = \alpha(G_2)$ since all nodes of G_2 are coreferent with strictly dominating nodes.

$\Phi([G_1 \neg [G_2]]) = \exists x \exists y \exists z(\alpha(G_1) \wedge \neg \alpha(G_2))$.
$\Phi(G) = \forall x \forall y \forall z(\alpha(G_1) \rightarrow \alpha(G_2))$.

An interesting point is that Kerdiles' FCGs are provided with a tableau calculus, thus a potentially efficient procedure, which is sound and complete. Moreover, the unique aspect relative to other logical tableau methods is that BGs are considered as basic constructs, thus are never decomposed into more elementary elements. A branch of a tableau is said to be closed if it contains a node ¬[G] where G is a BG, and there is a BG H that is possibly obtained by summing several graphs of the branch, such that G can be mapped into the normal form of H. We just give an example illustrating the role of BG homomorphism in this method.

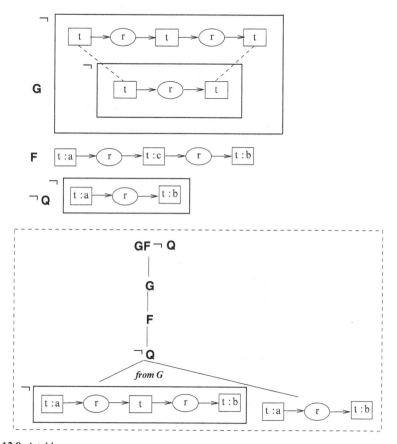

Fig. 12.9 A tableau

Let G be the BG of Fig. 12.3. Let F be the BG $[t : a] \rightarrow (r) \rightarrow [t : c] \rightarrow (r) \rightarrow [t : b]$. And Q is the BG $[t : a] \rightarrow (r) \rightarrow [t : b]$. One wants to prove that Q is deducible from G (a rule) and F (a fact), i.e., prove that $((G \wedge F) \rightarrow Q)$ is valid. Figure 12.9 shows a closed tableau for its negation, $(G \wedge F \wedge \neg Q)$. First, this formula is decomposed into G, F and ¬Q, on the same branch. Then, from G, which is of form $\neg[g_1 g_2]$, one deduces $\neg g_1$ OR $\neg g_2$, thus splits the branch. Coreferent nodes are instantiated by the same constant, which in this case may be any constant (new or occurring in the tableau). The left branch is closed because the leaf graph is the negation of a BG which maps to F. The right branch is also closed because graph $\neg Q$ is the negation of a BG that maps to the leaf graph (actually $\neg Q$ is the negation of the leaf graph). Thus the tableau is closed, which proves that $(G \wedge F \wedge \neg Q)$ is unsatisfiable (thus $(G \wedge F \rightarrow Q)$ is valid).

12.2 Conceptual Graphs with Atomic Negation

Basic or simple conceptual graphs express conjunctions of positive knowledge. As explained in the preceding sections, FCGs extend them with negation, but in a way that does not comply with our approach. We therefore consider here a basic but fundamental form of negation, namely *atomic negation*. The scope of atomic negation is an atom. Instead of considering conjunctions of atoms, we consider now conjunctions of literals, i.e., atoms or negation of atoms, and keep the existential quantifier. Atomic negation allows us to express knowledge of form "this relation does not hold between these entities," "this entity does not have that property" or "this entity is not of that type."

12.2.1 Polarized Graphs

Beside positive relation nodes, we now have negative relation nodes. Such graphs are said to be polarized. Below we define polarized graphs as extensions of BGs. Coreference will be considered with its opposite, i.e., difference, in Sect. 12.2.2. Negation is defined on relation types only, but as discussed hereafter, the definitions and results can be easily extended to concept types.

Definition 12.4 (Polarized Graph (PG)). A *polarized graph* (PG) is defined similarly to a BG (cf. Definition 2.2) except that relation nodes are labeled not only by a type but also by a polarity (denoted $+$ or $-$). A *positive* (resp. *negative*) relation node is labeled by $+r$ (resp. $-r$), where r is a relation type. $+r$ can also be noted r.

A negative relation node with label $-r$ and arguments $(c_1...c_k)$ expresses that "there is no relation of type r between $c_1...c_k$" (or if $k = 1$, "c_1 does not possess the property r"); it is logically translated by Φ into the literal $\neg r(e_1...e_k)$, where e_i is the term assigned to c_i.

We denote by $+r(c_1...c_k)$ (resp. $-r(c_1...c_k)$) a positive (resp. negative) relation node with type r and $c_1...c_k$ being the list of its arguments. FOL$\{\exists, \wedge, \neg_a\}$ denotes the extension of FOL$\{\exists, \wedge\}$ to negative literals. Translations between BGs and FOL $\{\exists, \wedge\}$ presented in Sect. 4.3 are naturally extended to translations between PGs and FOL$\{\exists, \wedge, \neg_a\}$. PGs are thus equivalent to the FOL$\{\exists, \wedge, \neg_a\}$ fragment.

Negation on Concept Types

Although we will not deal with it in this chapter, negation in concept labels can be defined in a similar way. If conjunctive concept types are considered, a concept node is labeled by a set $\{\sim t_1 \ldots \sim t_k\}$, where $\sim$ can be $+$ or $-$ and t_i is a primitive concept type. A concept node labeled by $-t$ is interpreted as *"there is an entity that is not of type t,"* and not as *"there is not an entity of type t,"* i.e., we keep an existential interpretation. Since the universal concept type is supposed to represent all entities, it cannot be negated. Let us point out that, although negation on concept types is interesting from a modeling viewpoint, it does not add expressiveness. Indeed, concept types can be processed as unary relation types with a transformation similar to the mapping from bicolored rules to λ-rules (cf. Sect. 10.1.2). More precisely, consider BGs on a vocabulary $\mathcal{V}$. Let $\mathcal{V}'$ be the vocabulary built by translating all concept types, except the universal type $\top$, into unary relation types keeping the same partial order. The concept type set of $\mathcal{V}'$ is composed of the single type $\top$. Then, BGs on $\mathcal{V}$ can be transformed into BGs on $\mathcal{V}'$, while preserving homomorphisms: Each concept node with label $l = (\{\sim t_1 \ldots \sim t_k\}, m)$ is translated into a concept node with label $(\top, m)$ and, for each $\sim t_i$ in l, with $t_i \neq \top$, one incident relation node with label $\sim t_i$. A simple and uniform way of processing negation on concepts and relations thus involves applying the transformation sketched above, processing the obtained graphs with the algorithms of this chapter and, if needed, applying the reverse transformation to present the results. Another solution is to adapt the definitions and algorithms presented hereafter, which is straightforward.

Since negation is introduced, a PG can be inconsistent.

Definition 12.5 (inconsistent PG). A PG is said to be *inconsistent* if its normal form contains two relation nodes $+r(c_1 \ldots c_k)$ and $-s(c_1 \ldots c_k)$ with contradictory labels, i.e., with $r \leq s$. Otherwise it is said to be *consistent*.

The following property is immediate:

Property 12.1. For any PG G on a vocabulary $\mathcal{V}$, G is inconsistent iff $\Phi(\mathcal{V}) \cup \{\Phi(G)\}$ is inconsistent.

Proof. ($\Rightarrow$): trivial. ($\Leftarrow$): see that $\Phi(\mathcal{V})$ cannot be inconsistent since it only contains positive information. Now consider the clausal form of $\Phi(\mathcal{V}) \cup \{\Phi(G)\}$: The only way to deduce the empty clause from it is to have two clauses of form $r(e)$ and $\neg s(e)$ with same argument list e and $r \leq s$ (which allows us to obtain $\neg(r(e))$ from $\neg s(e)$ and $\Phi(\mathcal{V})$). $\square$

The order on relation labels is extended as follows: We set $-r_1 \leq -r_2$ if $r_2 \leq r_1$.

Definition 12.6 (Extended order on relation labels). Given two relation labels l_1 and l_2, $l_1 \leq l_2$ if, either l_1 and l_2 are both positive labels, say $l_1 = r_1$ and $l_2 = r_2$, and $r_1 \leq r_2$, or l_1 and l_2 are both negative labels, say $l_1 = -r_1$ and $l_2 = -r_2$, and $r_2 \leq r_1$.

Given this extended order on relation labels, homomorphism can be used without changing its definition. It is still logically sound, as expressed by the next property (which can be seen as a corollary of the forthcoming Theorem 12.2):

Property 12.2. Given two PGs G and H on a vocabulary $\mathcal{V}$, if there is a homomorphism from G to H then $\Phi(\mathcal{V}), \Phi(H) \models \Phi(G)$.

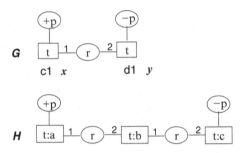

Fig. 12.10 Atomic negation and homomorphism (1)

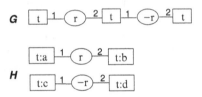

Fig. 12.11 Atomic negation and homomorphism (2)

The bad news is that homomorphism is no longer complete (even if the target graph is in normal form), as illustrated by Figs. 12.10 and 12.11. In Fig. 12.10, the formulas assigned to G and H by Φ (here we ignore the atoms associated with concept nodes) are respectively $\Phi(G) = \exists x \exists y (p(x) \land \neg p(y) \land r(x, y)))$ and $\Phi(H) = p(a) \land r(a,b) \land r(b,c) \land \neg p(c)$. $\Phi(G)$ can be deduced from $\Phi(H)$ using the tautology $p(b) \lor \neg p(b)$ (indeed, every model of $\Phi(H)$ satisfies either $p(b)$ or $\neg p(b)$; if it satisfies $p(b)$, then x and y are interpreted as b and c; in the opposite case, x and y are interpreted as a and b; thus every model of $\Phi(H)$ is a model of $\Phi(G)$). In the second example, illustrated by Fig. 12.11, the target graph is composed of two connected components, one containing positive information only, the other negative information only. There is no homomorphism from G to H, even if $\Phi(G)$ can be deduced from $\Phi(H)$, using the tautology $(r(b,c) \lor \neg r(b,c))$.

More generally, negation introduces disguised disjunctive information that cannot be taken into account by homomorphism. This disjunctive information is related to the law of the excluded-middle which holds in classical logic: Given a proposition P, either P is true, or $\neg P$ is true. This leads to reasoning by cases: If a relation is not asserted in a fact, either it is true or its negation is true. We thus have to consider all ways of *completing* the knowledge asserted by a PG. Let us look again at the example in Fig. 12.10. H does not say whether the unary relation p holds for b. We thus have to consider two cases: Either a relation node with label $+p$ or a relation node with label $-p$ can be attached to b. Let H_1 and H_2 be the graphs respectively obtained from H (see Fig. 12.12). There is a homomorphism from G to H_1 and there is a homomorphism from G to H_2. We conclude that G can be deduced from H.

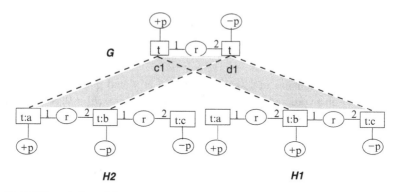

Fig. 12.12 When the law of the excluded-middle intervenes

The next definition specifies the notion of completion of a PG relative to a vocabulary $\mathcal{V}$.

Definition 12.7 (Complete PG). A *complete PG* on a vocabulary $\mathcal{V}$ with relation type T_R is a consistent (normal) PG satisfying the following completion condition: For each relation type r of arity k in T_R, for each k-tuple of concept nodes $(c_1, \ldots, c_k)$, where $c_1, \ldots, c_k$ are not necessarily distinct nodes, there is a relation $+s(c_1, \ldots, c_k)$ with $s \leq r$ or (exclusive) there is a relation $-s(c_1, \ldots, c_k)$ with $r \leq s$. A PG is complete with respect to a subset of relation types $T \subseteq T_R$ if the completion condition considers only elements of T. If a PG G^c that is complete with respect to T is obtained by adding relations to a graph G, it is called a T-completion of G (or simply a *completion* of G if T is implicit).

Property 12.3. If a relation node is added to a complete PG, either this relation node is redundant (There is already a relation node with the same argument list and a label less or equal to it) or it makes the PG inconsistent.

A complete PG is obtained from a consistent PG G by repeatedly adding positive and negative relations as long as adding a relation brings new information and does not yield an inconsistency. Since a PG is a finite graph defined over a finite vocabulary, the number of different complete PGs that can be obtained from it is finite. We can now define the deduction problem on PGs in terms of completions.

Definition 12.8 (PG-DEDUCTION). PG-DEDUCTION takes two PGs G and H as input, with H being consistent, and asks whether G can be PG-deduced from H, i.e., whether each complete PG H^c obtained from H is such that $G \succeq H^c$.

The following theorem expresses that PG-DEDUCTION is sound and complete with respect to the deduction in FOL.

Theorem 12.2. *Let G and H be two PGs defined on a vocabulary $\mathcal{V}$. H is a consistent PG. Then G can be PG-deduced from normal(H) if and only if $\Phi(\mathcal{V}), \Phi(H) \models \Phi(G)$.*

To prove this theorem, we first need to extend to PGs some definitions concerning BGs. The notion of $\mathcal{L}$-substitution (cf. Definition 4.8) is extended in a straightforward way:

Definition 12.9 (Extension of $\mathcal{L}$-substitution). Let f and g be two formulas of FOL$\{\exists, \wedge, \neg_a\}$ on an ordered language $\mathcal{L}$, with $vars(g) \cap vars(h) = \emptyset$. A $\mathcal{L}$-substitution σ from f to g is a substitution such that:

- for all positive literals $p(e_1, \ldots, e_k)$ in f, there is a positive literal $q(\sigma(e_1, \ldots, e_k))$ in g such that $q \leq_{\mathcal{L}} p$,
- for all negative literals $\neg\, p(e_1, \ldots, e_k)$ in f, there is a negative literal $\neg\, q(\sigma(e_1, \ldots, e_k))$ in g such that $q \geq_{\mathcal{L}} p$.

A good assignment is defined as follows.

Definition 12.10 (Good assignment). Given a model $M = (D, \delta)$ of a logical language $\mathcal{L}$, a *good assignment* α of a FOL$\{\exists, \wedge, \neg_a\}$ formula f on $\mathcal{L}$ to M is a mapping α from the terms of f to D, which maps each constant c to $\delta(c)$, such that for all positive literal $p(e_1, \ldots, e_k)$, $\alpha(e_1, \ldots, e_k) \in \delta(p)$ and for all negative literal $\neg p(e)$, $\alpha(e) \notin \delta(p)$.

It is immediately checked that the equivalence between $\mathcal{L}$-substitution and homomorphism is preserved (extension of Property 4.3).

Property 12.4 ($\mathcal{L}$-substitution and PG Homomorphism Equivalence). Let G and H be two PGs on a vocabulary $\mathcal{V}$. There is a homomorphism from G to $normal(H)$ iff there is a $\mathcal{L}_{\mathcal{V}}$-substitution from $\Phi(G)$ to $\Phi(H)$, where $\mathcal{L}_{\mathcal{V}}$ is the ordered language associated with $\mathcal{V}$. Let f and g be two FOL$\{\exists, \wedge \neg_a\}$ formulas on an ordered language $\mathcal{L}$. There is a $\mathcal{L}$-substitution from f to g if and only if there is a homomorphism from $f2g(f)$ to $f2g(g)$ (where $f2g$ is the natural extension of the mapping defined in Sect. 4.3.3.1).

We can now prove Theorem 12.2.

Proof. $\Rightarrow$ Assume that G can be PG-deduced from $normal(H)$ (actually, that the target graph is *normal* is not needed in this direction). Let M be a model of $\Phi(\mathcal{V}), \Phi(H)$. By definition of complete graphs, there is a complete graph, say H', obtained from $normal(H)$, such that M is a model of $\Phi(H')$. By hypothesis there is a homomorphism from G to H', thus a $\mathcal{L}_{\mathcal{V}}$-substitution from $\Phi(G)$ to $\Phi(H')$. The composition of the assignment from $\Phi(H')$ to M and this $\mathcal{L}_{\mathcal{V}}$-substitution defines a good assignment from $\Phi(G)$ to M. M is thus a model of $\Phi(G)$.

$\Leftarrow$ Let H' be any complete PG obtained from $normal(H)$.

Consider M, the canonical model (D, δ) of $\Phi(H')$, in which

- D is the set of terms in $\Phi(H')$,
- δ is the identity over constants and for each k-ary predicate q in $\mathcal{L}$, $(e_1, \dots, e_k) \in \delta(q)$ if and only if there is $p(e_1, \dots, e_k)$ in $\Phi(H')$ with $p \leq q$.

Add the appropriate literals to M such that: For any predicate p of arity k, if $(e_1, \dots, e_k) \in \delta(p)$ then for all $q \geq p$, $(e_1, \dots, e_k) \in \delta(q)$.

M is a model of $\Phi(H)$ and $\Phi(\mathcal{V})$. Thus it is a model of $\Phi(G)$ by hypothesis. There is thus a good assignment from $\Phi(G)$ to M, which is here a mapping from $\Phi(G)$ to $\Phi(H')$. This mapping is a $\mathcal{L}_{\mathcal{V}}$-substitution from $\Phi(G)$ to $\Phi(H')$. There is thus a homomorphism from G to $normal(H') = H'$ (since H' is built from $normal(H)$, it is in normal form). $\quad\square$

From now on, we can thus make no distinction between PG-deduction and logical deduction on the associated formulas, and we will sometimes simply say "deduction."

Algorithm 27 presents a brute-force algorithm scheme for PG-DEDUCTION. An immediate observation for generating the completions H^c is that we do not need to consider all relation types but only those appearing in G. The algorithm generates all complete PGs relative to this set of types and for each of them checks whether G can be mapped to it. A complete graph to which G cannot be mapped can be seen as a counter-example to the assertion that G is deducible from H. Algorithmic improvements are studied in Sect. 12.2.6.

Algorithm 27: PG-Deduction(G,H)

Input: PGs G and H, such that H is consistent and normal
Output: true if G can be deduced from H, false otherwise
begin
    Compute $\mathcal{H}$ the set of complete PGs obtained from H with respect to relation types in G
    forall $H^c \in \mathcal{H}$ **do**
        if *there is no homomorphism from G to H^c* **then**
            return *false*;   `// `H^c` is a counter-example`
    return *true*
end

12.2.2 Handling Coreference and Difference

In this section, we introduce inequality, also called *difference*, as a special element of the CG syntax, called a *difference link*. A difference link between two nodes c_1 and c_2 expresses that c_1 and c_2 represent distinct entities. In Fig. 12.13, difference links are represented by crossed lines. Due to the unique name assumption, there is an implicit difference link between nodes having distinct individual markers. Formally, difference is added to PGs (or only to BGs or SGs) as a symmetrical and antireflexive relation on concept nodes, called *dif*. Let us recall that coreference is represented in SGs as an equivalence relation on concept nodes added to BGs. Similar to difference links, we can distinguish between implicit coreference links that relate concept nodes with the same individual marker, and other links, which are explicit coreference links. In the next definitions, we distinguish between the set of explicit coreference and difference links (E_{coref} and E_{dif}) and the relations (*coref* and *dif*) obtained from explicit *and* implicit links. Finally, let us point out that, since we consider coreference, we also consider conjunctive concept types.

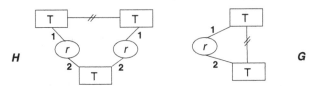

Fig. 12.13 $PG^{\neq}$: G is deducible from H

A $PG^{\neq}$ is a PG added with coreference and difference. A $BG^{\neq}$ is a BG added with coreference and difference. It is a particular $PG^{\neq}$ in which all relation labels are positive. It can also be seen as an SG added with difference.

Definition 12.11 ($PG^{\neq}$, $BG^{\neq}$). A PG (resp. BG) with inequality, notation $PG^{\neq}$ (resp. $BG^{\neq}$) is a 6-tuple (C, R, E, l, E_{coref}, E_{dif}), where:

- (C, R, E, l) is a PG (resp. BG),
- E_{coref} and E_{dif} are sets of specific edges between *distinct* nodes of C.

Definition 12.12 (*coref* relation). The relation *coref* on a $PG^{\neq}$ $G = (C, R, E, l, E_{coref}, E_{dif})$ is the equivalence relation over C defined as the reflexive and transitive closure of the union of:

- TShe symmetrical relation induced by E_{coref} over C;
- implicit links due to multiple occurrences of individual markers: $\{\{c, c'\} \mid c, c' \in C, marker(c) \neq * \text{ and } marker(c) = marker(c')\}$;

Before defining *dif*, we introduce a relation *Dif* on the *equivalence classes* of *coref*.

Definition 12.13 (*Dif* relation). The relation *Dif* on a $PG^{\neq}$ $G = (C, R, E, l, E_{coref}, E_{dif})$ is the symmetrical relation over equivalence classes of *coref* defined as the union of:

1. The symmetrical relation induced by E_{dif}: $\{\{C_1, C_2\} \mid \text{there are } c_1 \in C_1, c_2 \in C_2 \text{ with } \{c_1, c_2\} \in E_{dif}\}$,
2. implicit links due to unique name assumption: $\{\{C_1, C_2\} \mid \text{there are } c_1 \in C_1, c_2 \in C_2 \text{ with } marker(c_1) \neq *, marker(c_2) \neq * \text{ and } marker(c_1) \neq marker(c_2)\}$,
3. implicit links due to incompatible types: $\{\{C_1, C_2\} \mid \text{there are } c_1 \in C_1, c_2 \in C_2 \text{ such that } type(c_1) \text{ and } type(c_2) \text{ are incompatible }\}$,

4. implicit links due to contradictory relations: $\{(C_1, C_2) \mid$ there are $c_1 \in C_1$, $c_2 \in C_2$, $(c_1, c_2) \notin coref$ and $+r(d_1, \ldots, d_i, c_1, d_{i+1}, \ldots, d_q), -s(e_1, \ldots, e_i, c_2, e_{i+1}, \ldots, e_q) \in G$ such that $r \leq s$ and for all $i \in \{1..q\}$, one has $\{d_i, e_i\} \in coref\}$.

Set 4 is illustrated by Fig. 12.14: The relation nodes have opposite labels and coreferent arguments, except for c and c'; making c and c' coreferent would lead to an inconsistent graph. Set 4 can be removed when only $BG^{\neq}$ are considered.

Definition 12.14 (*dif* relation). The relation dif on a $PG^{\neq}$ $G = (C, R, E, l, E_{coref}, E_{dif})$ is the symmetrical relation over C defined as the cartesian product of all pairs of coref classes belonging to Dif (i.e. if $\{C_1, C_2\} \in Dif$ then for all $c_1 \in C_1$ and $c_2 \in C_2$, $\{c_1, c_2\} \in dif$).

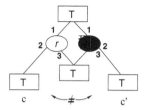

Fig. 12.14 Implicit *Dif* link

For a $PG^{\neq}$ in normal form, one has $dif = Dif$. A $PG^{\neq}$ is consistent if $coref$ and dif are not contradictory (i.e. $coref \cap dif = \emptyset$). Note that dif and Dif are then antireflexive.

The FOL formula assigned to a $PG^{\neq}$ translates coreference by assigning the same term (variable or constant) to coreferent nodes, or equivalently, by adding an atom $e_1 = e_2$ for each pair of coreferent nodes with assigned terms e_1 and e_2. dif is naturally translated by $\neq$ (i.e. $\neg =$). Every consistent $PG^{\neq}$ possesses a normal form which is obtained by merging all concept nodes of the same $coref$ class and is logically equivalent to it.

Let us recall how coreference is processed by homomorphism. A homomorphism π from an SG G to an SG H has to respect coreference: For all nodes c_1 and c_2 of G, if $\{c_1, c_2\} \in coref_G$ then $\{\pi(c_1), \pi(c_2)\} \in coref_H$ (where $\pi(c_1)$ and $\pi(c_2)$ can be the same node). Completeness is obtained only if the target SG is in normal form. Recall that homomorphism can be replaced by coref-homomorphism to ensure completeness without this condition. If G and H are $BG^{\neq}$s (or $PG^{\neq}$s), homomorphism has to respect the dif relation as well: For all nodes c_1 and c_2 in G, if $\{c_1, c_2\} \in dif_G$ then $\{\pi(c_1), \pi(c_2)\} \in dif_H$. Concerning completeness, the same discussion as that put forward for negative relations on the use of the law of excluded middle can apply, as illustrated by Fig. 12.13. Formulas assigned to H and G are respectively $\Phi(H) = \exists x \exists y \exists z \ (r(x, z) \wedge r(y, z) \wedge \neg(x = y)))$ and $\Phi(G) = \exists x \exists y \ (r(x, y) \wedge \neg(x = y))$. $\Phi(G)$ can be deduced from $\Phi(H)$, using the law of excluded middle for $x = z$ (i.e., either $x = z$ or $x \neq z$) and/or $y = z$, while there is no homomorphism from G to H. Difference thus introduces reasoning by cases as negative relations. Let us say that a consistent $BG^{\neq}$ H is complete if for all c, c' distinct concept nodes in H, either $\{c, c'\} \in dif_H$ or $\{c, c'\} \in coref_H$. We will also say that it is dif-complete. A $PG^{\neq}$ is complete if it is complete as a PG (which is relative to relation types) and complete as a $BG^{\neq}$, i.e., dif-complete.

Algorithm 28 is a brute-force algorithm computing all dif-completions obtainable from H. Computing completions incrementally during deduction checking would of course be more efficient (see Sect. 12.2.6 about algorithmic improvements). Note that, in the subalgorithm CompleteRec, case 1 updates $coref$ but may also involve updating dif (due to potential contradictory relations, as illustrated in Fig. 12.14), while case 2 updates dif only.

Algorithm 28: AllCompleteGraphsForDif(H)

Input: a $BG^{\neq}$ or a $PG^{\neq}$ H
Output: the set of all dif-completions from H
begin
    $CompleteSet \leftarrow \emptyset$
    CompleteRec(H) // see Algorithm 29
    return $CompleteSet$
end

Algorithm 29: CompleteRec(H) *subalgorithm of Algorithm 28*

Input: a $BG^{\neq}$ or a $PG^{\neq}$ H
Output: computes recursively the dif-completions of H and adds them to CompleteSet
begin
    if $dif \cup coref$ *is complete* **then**
        // all pair of distinct concept nodes are in dif or $coref$
        $CompleteSet \leftarrow CompleteSet \cup \{H\}$
    else
        Choose two (distinct) nodes c, c' in H such that $\{c, c'\} \notin dif \cup coref$
        // case 1: make them coreferent
        let H_1 *be obtained from H by adding* $\{c, c'\}$ *to* E_{coref}
            CompleteRec(H_1)
        // case 2: make them ``different''
        let H_2 *be obtained from H by adding* $\{c, c'\}$ *to* E_{dif}
            CompleteRec(H_2)
end

A brute-force algorithm for $BG^{\neq}$-DEDUCTION is obtained from the brute-force algorithm for PG-DEDUCTION (Algorithm 27), by replacing $\mathcal{H}$ by the set of all dif-completions of H. A brute-force algorithm for $PG^{\neq}$-DEDUCTION is obtained by combining the two kinds of completion.

12.2.3 PG-DEDUCTION *and Equivalent Problems*

As pointed out above, PGs are equivalent to the FOL$\{\exists, \wedge, \neg_a\}$ fragment of first-order-logic. Without loss of generality, we can consider formulas in prenex form, i.e., such that all (existential) quantifiers are in front of the formula. PG-DEDUCTION is thus equivalent to the following problem:

FOL$\{\exists, \wedge, \neg_a\}$-DEDUCTION

Input: existentially closed conjunctions of (positive and negative) literals f_1 and f_2.
Question: Does $f_1 \models f_2$ hold, i.e. is f_2 a consequence of f_1?

In Chap. 5, we show that the containment problem for positive conjunctive queries (CQs) is equivalent to BG-DEDUCTION. Let us now consider conjunctive queries with atomic negation (*CQNs*), i.e., queries with the following form:

$$q = ans(u) \leftarrow r_1(u_1), \ldots r_n(u_n), \neg s_1(y_1), \ldots \neg s_m(y_m) \quad n \geq 1, m \geq 0$$

Given a CQN q defined as above and a database instance D, an answer to q in D is a tuple $\mu(u)$, where μ is a substitution of variables in q by constants in D such that for any i in $\{1, ..., n\}$, $\mu(u_i) \in D(r_i)$ and for any j in $\{1, ..., m\}$, $\mu(y_j) \notin D(s_j)$ (in other words, the query evaluation makes the closed-world assumption, see Sect. 12.2.7). The containment problem for conjunctive queries with atomic negation is defined as follows:

CONTAINMENT OF CONJUNCTIVE QUERIES WITH NEGATION (CQNC)

Input: two conjunctive queries with negation q_1 and q_2.
Question: Is q_1 contained in q_2 ($q_1 \sqsubseteq q_2$), i.e., is the set of answers to q_1 included in the set of answers to q_2 for any database D?

A CQN on a schema can be naturally seen as a PG on a flat vocabulary, and reciprocally, by a simple extension of the transformations $q'2b$ and $b2q'$ detailed in Chap. 5. Similarly to the extension of BG homomorphism to PG homomorphism, we can extend the notion of positive query homomorphism to CQN homomorphism and define the notions of a consistent CQN and a complete CQN. The following property expresses the containment of CQNs in a form similar to PG deduction. The equivalence of CQNC and PG-DEDUCTION follows immediately.

Property 12.5. Given two CQNs q_1 and q_2 (with q_1 being consistent), $q_1 \sqsubseteq q_2$ if and only if for each complete query q_1^c generated from q_1, there is a CQN homomorphism from q_2 to q_1^c.

Proof. ($\Rightarrow$) Each complete query q_1^c generated from q_1 corresponds to a database (one just has to remove negative literals and the literal $ans(u)$) that contains a canonical answer to q_1: tuple u. By hypothesis, u is also an answer to q_2; thus there is a CQN homomorphism h from q_2 to q_1^c.

($\Leftarrow$) By contradiction. Suppose that $q_1 \not\sqsubseteq q_2$, then there is a database D and a tuple w such as $w \in q_1(D)$ but $w \notin q_2(D)$; there is thus a CQN homomorphism h_1 from q_1 to $ans(w) \leftarrow D^{c-}$, where D^{c-} is the negative completion of D obtained by adding solely negative literals, but there is no CQN homomorphism from q_2 to $ans(w) \leftarrow D^{c-}$. Let q_1^c be a complete query built from q_1 in the following way: For each relation r in $\mathcal{R}$ of arity k and each k-tuple $(t_1, \ldots, t_k)$ of terms in q_1, add a literal with predicate r, arguments $(t_1, \ldots, t_k)$ and the same polarity as the literal l in D^{c-} with predicate r and arguments $(h_1(t_1), \ldots, h_1(t_k))$. Since D^{c-} is complete and consistent, l exists and is unique. By construction, there is a homomorphism h_c from q_1^c to $ans(w) \leftarrow D^{c-}$. By hypothesis, there is a CQN homomorphism h_2 from q_2 to q_1^c. Hence, $h_c \circ h_2$ is a CQN homomorphism from q_2 to $ans(w) \leftarrow D^{c-}$, which leads to a contradiction.   □

One can identify the notions of substitution on $FOL\{\exists, \wedge, \neg_a\}$ formulas (in prenex form), PG homomorphism and CQN homomorphism. Similarly, one can identify the notions of canonical model of a $FOL\{\exists, \wedge, \neg_a\}$ formula, completion of a consistent PG and completion of a consistent CQN.

12.2.4 Complexity of PG-DEDUCTION

Theorem 12.3 (PG-DEDUCTION **Complexity**). PG-DEDUCTION *is* Π_2^P*-complete.*

Proof. The way PG-DEDUCTION is stated makes it clear that it is in Π_2^P. Indeed, it corresponds to the language $L = \{x \mid \forall y_1 \, \exists y_2 \, R(x, y_1, y_2)\}$, where x encodes an instance (G, H) of the problem and $(x, y_1, y_2) \in R$ if and only if y_1 is a completion of H and y_2 encodes a homomorphism π from G to this completion. Note that a completion has a size exponential in the maximum arity of a relation, but without loss of generality we can assume that this size is bounded by a constant and even by 2, as n-ary relations can be polynomially transformed into binary relations.

The Π_2^P-completeness is proven by reduction from the following Π_2^P-complete problem:

GENERALIZED RAMSEY NUMBER [SU02]
Input: an undirected graph $K = (V, E)$, where V is the set of vertices and E the set of edges; a partial two-coloring σ of E (i.e., a partial function $\sigma : E \mapsto \{0, 1\}$); and an integer k.
Question: Does every total extension of σ produce a one-color k-clique[1] (i.e., a k-clique with all edges colored 0 or with all edges colored 1)?

Let $(K = (V, E), \sigma, k)$ be an instance of GENERALIZED RAMSEY NUMBER. We build two (consistent) PGs G and H on a very flat vocabulary with set of relation types $\{r, d, p\}$, where r and d are binary relations and p is a unary relation. All concept nodes are labeled by $(\top, *)$.

The set of concept nodes of G is $C_1 \cup C_2$ where C_1 and C_2 are copies of the set of edges of a k-clique (thus in each C_i there is one concept node x for each edge ab of a k-clique). There is a $r(x, y)$ if $\{x, y\} \subseteq C_1$ or $\{x, y\} \subseteq C_2$, and x and y represent distinct edges with a common endpoint in the k-clique. There is a $d(x, y)$ between all distinct concept nodes x and y in G. There is a $p(x)$ if $x \in C_1$ and there is a $-p(x)$ if $x \in C_2$.

---

[1] A clique is an undirected graph with all edges between distinct vertices. A k-clique is a clique with k vertices.

Intuitively: G is composed of two components, C_1 and C_2; for each component, the concept nodes and the relation nodes labeled by r represent the intersection graph of the edges of a k-clique; distinct concept nodes anywhere in G are related by d; the concept nodes in C_1 are marked by p, those in C_2 are marked by $-p$. p can be seen as denoting the color 0 and $-p$ the color 1.

The set of concept nodes of H is $C_1' \cup E \cup C_2'$ where, as for G, C_1' and C_2' are copies of the set of edges of a k-clique; E is a copy of the set of edges in G. There is a $r(x,y)$ if $\{x,y\} \subseteq C_1'$ or $\{x,y\} \subseteq C_2'$ or $\{x,y\} \subseteq E$, and x and y represent distinct edges with a common endpoint in the k-clique. There is a $d(x,y)$ between distinct nodes x and y in $C_1' \cup E$ and between distinct nodes x and y in $E \cup C_2'$ (actually, it is not necessary to have d in both directions between C_1 and C_2 in G, and between C_1' and E and E and $C'2$ in H: Instead, d could lead from C_1 to C_2 in G, and from C_1' to E and from E to $C'2$ in H). There is a $p(x)$ if $x \in C_1'$, or $x \in E$ and $\sigma(x)$ is defined and equal to 0; there is a $-p(x)$ if $x \in C_2'$, or $x \in E$ and $\sigma(x)$ is defined and equal to 1.

Intuitively: H has the same components C_1 and C_2 as G but, in addition, there is a component E representing the intersection graph of the edges of G. The elements in E are marked by p, $-p$ or nothing, depending of the color of the corresponding edge in G. In G, C_1 and C_2 are "directly connected" by d; in H they are not: E is between them.

The construction of G and H is illustrated by Fig. 12.15; in this figure, all edges labeled d stand for two symmetrical relation nodes with label d.

Let us now prove that all extensions of σ produce a one-color k-clique if and only if, for all completions $H*$ of H, there is a homomorphism from G to $H*$.

Let us call subgraph "induced by" a set of concept nodes S the subgraph with set of concept nodes S plus all relation nodes having all their neighbors in S.

First see that there is a *bijection*, say b, between an edge-coloring of G extending σ and a $\{p\}$-completion of the subgraph induced by E in H. As the subgraphs induced by C_1' and C_2' are already $\{p\}$-complete, such a completion is also a $\{p\}$-completion of H itself. Also note that r and d occur only positively, thus they are not needed in completions: There is a homomorphism from G to each $\{r,d,p\}$-completion of H if and only if there is a homomorphism from G to each $\{p\}$-completion of H (see Sect. 12.2.6.1).

Let σ' be an extension of σ that produces a k-clique of color 0 (resp. 1). In $b(\sigma')$, i.e., the completion of H assigned to σ' by b, the subgraph induced by E contains concept nodes corresponding to the edges of a k-clique S, with all these nodes being attached to a p relation node (resp. a $-p$ relation node). Let $edges(S)$ denote this set of nodes. There is a homomorphism from G to the subgraph of $b(\sigma')$ induced by $C_1' \cup edges(S)$ (resp. $C_2' \cup edges(S)$).

Reciprocally, let $H*$ be a $\{p\}$-completion of H and let h be a homomorphism from G to $H*$. See that h is necessarily injective on the set of concept nodes in G due to d relation nodes between all distinct concept nodes in G^2. h maps G either

---

[2] A homomorphism from a clique C to a graph G is necessarily injective, but it is no longer true when C and G are replaced by the intersection graphs of their edges. The role of d relation nodes in the subgraphs induced by C_1 and by C_2 is to enforce injectivity.

to the subgraph of $H*$ induced by C'_1 and the nodes corresponding to edges of a
k-clique that are all attached to a $-p$, or to the subgraph induced by C'_2 and edges of
a k-clique that are all attached to a p. Thus, there is an extension of σ that produces
a k-clique of color 0 or of color 1.

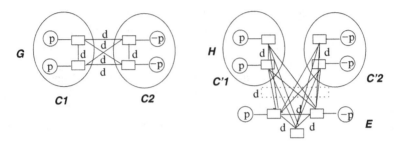

Fig. 12.15 Schema of the reduction in theorem 12.3 proof

The preceding reduction is easily adapted into a reduction to Deduction in
FOL$\{\exists, \wedge, \neg_a\}$ or CQNC. Note that PGs obtained in the reduction are built on a
very flat vocabulary and only unary relation nodes are negated. Hence the property:

Property 12.6. PG-DEDUCTION (thus FOL$\{\exists, \wedge, \neg_a\}$-DEDUCTION) remains Π^P_2-
complete if the vocabulary is not ordered and negation concerns only unary relation
nodes. Deduction on BGs augmented with negation on concept types only is Π^P_2-
complete.

The above reduction can be adapted to provide a proof of Π^P_2-completeness of
BG$^{\neq}$-DEDUCTION (hence PG$^{\neq}$-DEDUCTION since PG$^{\neq}$-DEDUCTION is clearly in
Π^P_2).

Theorem 12.4 (BG$^{\neq}$-DEDUCTION **Complexity**). BG$^{\neq}$-DEDUCTION *is Π^P_2-complete.*

Sketch of proof. The previous transformation is adapted as follows: The p relation
disappears and is replaced by dif links (whose role is to relate nodes of different
color). In G there is a dif link between all nodes x and y, where x is a concept node
in C_1 and y is a concept node in C_2. In H there is a dif link between all nodes x and
y such that x is in E, and, if x is of color 0 (i.e., was previously marked by p) then y
is in C'_2, and if x is of color 1 (i.e., was previously marked by $-p$) then y is in C'_1.

12.2.5 Special Cases with Lower Complexity for PG-DEDUCTION

The existence of a PG homomorphism from G to H is a sufficient condition for
G to be deducible from H. However, it is not a necessary condition, as we have
seen before. In this section, we study the question "when is a homomorphism from

G to H a necessary condition for G to be deducible from H?". Answers to this question yield particular cases where the theoretical complexity of PG-DEDUCTION decreases. We shall also identify special subgraphs of G for which there must be a homomorphism to H when G is deducible from H. These subgraphs can be used as filters or guides during the completion algorithm (see Sect. 12.2.6).

Let us first identify relation nodes in G that may play a role in the problem complexity, in the sense that they may lead to using the excluded-middle law.

Definition 12.15 (Opposite relation labels and nodes). Two relation labels are said to be *opposite* if they have opposite polarities and if the type r of the positive label is greater than the type s of the negative label ($r \geq s$). By extension, two relation nodes are said to be opposite if they have opposite labels $+r$ and $-s$.

Let us say that two opposite relation nodes of G are "exchangeable" if they can have the same list of images for their arguments by homomorphisms to (necessarily distinct) completions of H.

Definition 12.16 (Exchangeable relations). Two relations $+r(c_1,\ldots,c_k)$ and $-s(d_1,\ldots,d_k)$ in G are *exchangeable* with respect to H if (1) they are opposite, (2) there are two completions of H, say H_1 and H_2, and two homomorphisms π_1 and π_2, respectively from G to H_1 and from G to H_2, such that for all $i : 1\ldots k$, $\pi_1(c_i) = \pi_2(d_i)$.

See, for instance, the PG G in Fig. 12.10. Let us consider the opposite relation nodes $r_1 = p(c_1)$ and $r_2 = -p(d_1)$. These nodes are exchangeable, as can be seen in Fig. 12.12: There is a homomorphism π_1 from G to a completion H_1 of H and there is a homomorphism π_2 from G to another completion H_2 of H, such that $\pi_1(c_1) = \pi_2(d_1)$ (and is the concept node in H with marker b).

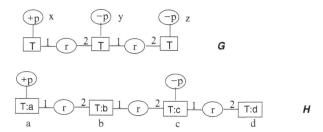

Fig. 12.16 Exchangeable and opposite relation nodes

The definition of exchangeable relations is strictly more restrictive than the definition of opposite relations. See Fig. 12.16 for instance, where x and y are opposite nodes, as well as x and z. x and y are exchangeable, which can be seen with the following two completions of H: In one completion, say H_1, $-p(b)$ is added (and a homomorphism from G to H_1 maps the neighbor of y to b); in another completion, say H_2, $p(b)$ and $-p(d)$ are added (and a homomorphism from G to H_2 maps the neighbor of x to b). It can be checked that x and z are not exchangeable: There are no two completions such that their argument can be mapped to the same node.

Property 12.7. Let G and H be two PGs, with G having no pair of exchangeable relations with respect to H, and H being consistent and normal. If G is deducible from H, then there is a homomorphism from G to H.

Proof. Let H^{c+} be a completion of H with solely positive relations, and furthermore with all possible relations: For all n-ary type p and for all n-tuple u of concept nodes in H, if there is no relation $-q(u)$ with $q \geq p$, then $+p(u)$ is added if it is not already present in H. We call it the *maximal positive completion* of H. If there is no homomorphism from G to H but G is deducible from H, then for each homomorphism from G to H^{c+}, there is at least one added relation in H^{c+}, say $p(u)$, such that a relation $p'(v)$ ($p' \geq p$) in G is mapped to $p(u)$. Let us replace *all* such $p(u)$ by $-p(u)$. Note that it remains a completion and does not lead to an inconsistency: Indeed, for all $+p'(u)$ with $p' \leq p$, $p'(u)$ is also reversed. Let $H^{c'}$ be this completion. Let h be a homomorphism from G to $H^{c'}$ (there is such a homomorphism since G is deducible from H). h maps a relation $-p''(w)$ in G to a relation $-p(u)$ ($p'' \leq p$); otherwise there would be a homomorphism from G to H. By construction, there is a relation $p'(v)$ mapped to $p(u)$ by a homomorphism from G to H^{c+}; thus $p'(v)$ and $-p''(w)$ (one has $p' \geq p''$) are exchangeable relations, which contradicts the hypothesis on G. $\square$

We thus obtain a case for which PG-DEDUCTION has the same complexity as homomorphism checking (and is thus NP-complete):

Property 12.8. Let G and H be two PGs, with G having no pair of exchangeable relations with respect to H, and H being consistent and normal. G is deducible from H if and only if there is a homomorphism from G to H.

Note also that G is deducible from H if and only if each connected component of G is deducible from H. Moreover, splitting an individual concept node into several nodes does not change the logical semantics of the graph and preserves a homomorphism from this graph to any other graph. Thus, one can consider each piece of G (cf. definition 8.22) with respect to its individual concept nodes, i.e. each connected subBG of G induced by all concept nodes such that there is a path between them which does not go through an individual node (however, individual nodes can be extermities of such a path). Let us call *detached form* of G the set of its pieces with respect to its individual concept nodes. Hence the property:

Property 12.9. Let G and H be two PGs, such that each piece of the detached form of G has no pair of exchangeable relations with respect to H, and H is consistent and normal. G is deducible from H if and only if there is a homomorphism from G to H (i.e., there is a homomorphism from each of these connected components to H).

If the detached form of G is acyclic (more generally has bounded treewidth or hypertreewidth when seen as a hypergraph) then homomorphism checking is polynomial (see Chap. 7), hence PG-DEDUCTION.

A desirable property is that recognizing exchangeable relations is not difficult compared to PG-DEDUCTION complexity, which is indeed the case:

Property 12.10. Let EXCHANGEABLE be the problem that takes two PGs G and H as input and asks if G possess a pair of exchangeable relations with respect to H. EXCHANGEABLE is NP-complete.

Proof. EXCHANGEABLE is in NP: A polynomial certificate is given by two relation nodes of G and two applications from concept nodes of G to concept nodes of H (assumed to provide the two wanted homomorphisms from G to two completions of H). For NP-completeness, a reduction is built from BG-HOMOMORPHISM. Let (G_1, G_2) be an instance of BG-HOMOMORPHISM. "Gadgets" are added to G_1 and G_2, yielding G'_1 and G'_2 respectively, such that there is a homomorphism from G_1 to G_2 if and only if G'_1 possess exchangeable relations with respect to G'_2. Take, for instance, the graphs G and H in Fig. 12.10, and choose the types t, r and p such that they do not occur in G_1 and G_2. G'_1 (resp. G'_2) is obtained by making the disjoint sum of G_1 and G (resp. of G_2 and H). The only candidate exchangeable relation nodes in G'_1 are the nodes labeled p and $-p$. This construction is easily adapted to yield connected graphs if connected graphs are preferred.   □

The following property will be used to prove other properties:

Property 12.11. Let G and H be two PGs, where H is consistent and normal, and G is deducible from H. Let G' be a subgraph of G having no pair of exchangeable relation nodes with respect to H. Then there is a completion H^c of H and a homomorphism from G to H^c that maps G' entirely to H.

Proof. Consider any maximal completion of H, say H^c (for each n-ary type p and each n-tuple u of concept nodes in H, either one has $+p(u)$ or one has $-p(u)$). Assume that there is no homomorphism from G to H^c that maps G' to H. For each homomorphism from G to H^c, there is at least one added relation $\sim p(u)$ in H^c (with $\sim$ stands for $+$ or $-$) which is the image of a relation in G' such that H does not contain a node $\sim q(u)$ with $\sim p \geq \sim q$. Let R be the set of *all* such relation nodes in $H^c \setminus H$ for all homomorphisms from G to H^c. Let us inverse the polarity of the nodes in R. The graph obtained cannot be inconsistent[3] and it is of maximum size: Thus it is again a maximal completion of H. Let $H^{c'}$ be this maximal completion. As G' does not possess exchangeable relations, there is no homomorphism from G to $H^{c'}$ that maps a relation node in G' to a node in R. If there is no homomorphism from G to $H^{c'}$ that maps G' entirely to H, let R' be the set of all relation nodes $\sim p(u)$ in $H^{c'} \setminus H$ which are images of a node in G' and such that H does not contain a node $\sim q(u)$ with $\sim p \geq \sim q$. As previously, reverse the polarity of all nodes in R', which yields the graph $H^{c''}$. Add the nodes of R' to R. We thus build a sequence of maximal completions of H and a set R of relation nodes of these completions not belonging to H (nor redundant with nodes of H). As R grows strictly from one

---

[3] Indeed, assume we obtain two contradictory relation nodes $-q(u)$ and $p(u)$, with $q \geq p$. One of these nodes does not belong to R, otherwise G would have exchangeable nodes. Let x be this node and y be the node that belongs to R. The label of x in H^c is necessarily more general than the label of y in H^c (since both nodes were comparable and had the same polarity in H^c). Thus, by inverting the label of y, it is impossible to obtain an inconsistency.

completion to another, this sequence is finite. The last graph of this sequence is a completion satisfying the property.   □

If G is PG-deducible from H, for each subgraph of G without exchangeable relations, there must be a homomorphism from this subgraph to H. Moreover, there must be such a homomorphism that is potentially extensible to a homomorphism from the entire G to a completion of H. We call it a *compatible* homomorphism.

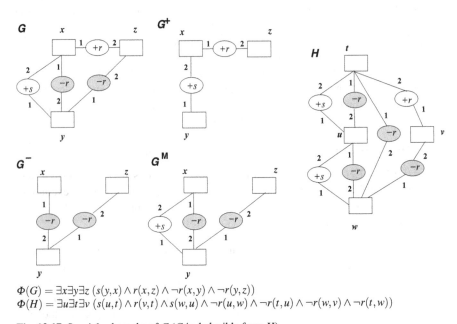

$$\Phi(G) = \exists x \exists y \exists z \; (s(y,x) \wedge r(x,z) \wedge \neg r(x,y) \wedge \neg r(y,z))$$
$$\Phi(H) = \exists u \exists t \exists v \; (s(u,t) \wedge r(v,t) \wedge s(w,u) \wedge \neg r(u,w) \wedge \neg r(t,u) \wedge \neg r(w,v) \wedge \neg r(t,w))$$

Fig. 12.17 Special subgraphs of G (G is deducible from H)

Definition 12.17 (compatible homomorphism). Given two PGs G and H, and G' any subgraph of G, a homomorphism h from G' to H is said to be *compatible* (with respect to G) if for each relation node x of G that does not belong to G' but has all its neighbors in G', say $c_1...c_k$, there is no relation node y with neighbor list $h(c_1)...h(c_k)$ in H and with a label contradictory to that of x (i.e. with label $-r$ if the label of x is $+s$, or with label $+s$ if the label of x is $-r$, such that $s \leq r$).

Example. See Fig. 12.17, where all concept nodes are assumed to have the same label $(\top, *)$ and relation types are incomparable. Let us consider G^-, a subgraph of G. There are three homomorphisms from G^- to H: $h_1 = \{x \rightarrow t, y \rightarrow u, z \rightarrow w\}$, $h_2 = \{x \rightarrow t, y \rightarrow w, z \rightarrow v\}$, $h_3 = \{x \rightarrow u, y \rightarrow w, z \rightarrow v\}$. To check the compatibility of these homomorphisms, we have to consider $+s(y,x)$ and $+r(x,z)$. h_1 is not compatible because it cannot be extended to $+r(x,z)$ due to the presence of $-r(t,w)$ in H.

Property 12.12. If G is deducible from H, then there is a compatible homomorphism from every subgraph of G without exchangeable relations to H.

Proof. Let G' be any subgraph of G without exchangeable relations. From Property 12.11, there is a homomorphism from G to a completion of H which maps G' entirely to H. By restricting the domain of this homomorphism to G', we have a homomorphism from G' to H, which is compatible with respect to G. $\square$

Let us point out some easily identifiable subgraphs without exchangeable relations. The *positive subgraph* of G, denoted G^+, is the subgraph obtained from G by selecting all concept nodes and only the positive relation nodes. The *negative subgraph* G^- of G is the dual notion, i.e., the subgraph obtained from G by selecting all concept nodes in G and only the negative relation nodes. Negative and positive subgraphs are particular cases of subgraphs without opposite relations. A subgraph of G without opposite relations and maximal for the inclusion is easily built by selecting, for each relation type appearing in G, either all its positive occurrences or all its negative occurrences, while satisfying the following constraint: If one selects the positive (resp. negative) occurrences of a relation type r, then the same choice must be done for all subtypes (resp. supertypes) of r.

Example. In Fig. 12.17, several subgraphs without opposite relations of G are pictured. G^+ and G^- are respectively the positive and negative subgraphs of G. G has two subgraphs without opposite nodes maximal for inclusion: G^+ and G^M.

12.2.6 Algorithmic Improvements

Let us say that a concept or relation label l_x occurring in G has a support in H if there is a label l_y in H with $l_y \leq l_x$ (and l_y is said to support l_x). By extension, we say that a node x in G has a support in H if there is a node y in H such that the label of x is supported by the label of y (and y is said to support x).

A first observation is that if a node in G has no support in H then G is not deducible from H. This is trivial for concept nodes. For relation nodes, if this node is negative (resp. positive), consider a completion of H with solely positive (resp. negative) relation nodes. There is no homomorphism from G to this completion.

12.2.6.1 Limitation of the Completion Vocabulary

Let us call "completion vocabulary" the set of relation types used to build completions of H. The size of the completion vocabulary determines the number of completions of H. The number of completions of H is itself a key element in the complexity of PG-deduction checking. It is thus essential to decrease the number of relation types involved in completions as much as possible. One can observe that the completion vocabulary can be restricted to the relation types occurring in G, and furthermore to relation types occurring in opposite relations of G:

Property 12.13. Let G and H be two PGs, where H is consistent and normal. G is deducible from H if and only if G can be mapped to each completion of H with respect to relation types occurring in opposite relations of G (i.e., r and s such that there are nodes in G with labels $+r$ and $-s$ and $s \leq r$).

Proof. Let T_R be the set of relation types in the vocabulary, and let T be the set of relation types occurring in opposite relations in G. ($\Leftarrow$) We prove that if G can be mapped to each T-completion of H, then it can be mapped to each T_R-completion of H. Indeed, let H^c be any T_R-completion of H. Let $H^{c'}$ be the graph obtained from H^c by replacing all relations with types outside T with a set of relations built as follows: Let r be a node labeled by $+t$ (resp. $-t$) such that $t \notin T$. Let $\{t_1...t_n\}$ be the types in T greater than t (resp. less than t). If r is positive, consider the minimal elements of this set, otherwise consider the maximal elements of this set. Let S be the obtained set. Replace r with $|S|$ relation nodes, each labeled by a type in S, with the same polarity and the same arguments as r. $H^{c'}$ is a T-completion of H. By construction, there is a homomorphism, say h_1, from $H^{c'}$ to H^c (which is the identity on concept nodes). By hypothesis, there is a homomorphism from G to $H^{c'}$, say h. The composition of these homomorphisms $h_1 \circ h$ is a homomorphism from G to H^c.

($\Rightarrow$) Let G be deducible from H and assume that H^c is a T-completion of H such that there is no homomorphism from G to H^c. We show that this assumption leads to a contradiction. From H^c, we build the following T_R-completion of H, say $H^{c'}$. For all type t in $T_R \setminus T$, let us add only $(+t)$ nodes if $+t$ does not support the label of any node in G; otherwise, add only $(-t)$ nodes if $-t$ does not support the label of any node in G (if neither $+t$ nor $(-t)$ support node labels in G, relation nodes typed t can be added with any polarity); if both $+t$ and $-t$ support node labels in G, there are opposite nodes in G with label $+r$ and $-s$ and $r \geq t \geq s$, thus r and s belong to T, and nodes with type t would be redundant in $H^{c'}$ thus are not needed to obtain a completion. Since G is deducible from H, there is a homomorphism from G to $H^{c'}$. By construction, no node in G can be mapped to an added node. Thus this homomorphism is a homomorphism from G to H^c, which contradicts the hypothesis on H^c. □

We can even restrict the vocabulary completion to the relation types of *exchangeable* relations in G.

Theorem 12.5. *Let G and H be two PGs, where H is consistent and normal. G is deducible from H if and only if G can be mapped to each completion of H with respect to relation types occurring in exchangeable relations of G with respect to H (i.e., relation types r such that there is a pair of exchangeable relations in G with one of the two labeled by $+r$ or $-r$).*

Proof. Let *exchangeable*(G) denote the types occurring in exchangeable relations in G.
($\Leftarrow$) Same as in the proof of Property 12.13, where T is replaced by *exchangeable*(G).
($\Rightarrow$) Let H^c be a completion of H with respect to exchangeable(G) such that there

is no homomorphism from G to it. As in the proof of Property 12.13, we build a completion of H, say $H^{c'}$, as follows: For any type t occurring in G but not in exchangeable(G), let us add only $(-t)$ nodes if $-t$ does not support any label in G, $(+t)$ nodes if $+t$ does not support any label in G. Let us add it positively if both $-t$ and $+t$ support labels in G. A homomorphism from G to $H^{c'}$ is a homomorphism to H^c plus the nodes added positively for types supporting both forms. Let us now inverse the polarity of these latter nodes if they are images of nodes in G. No node in G can be mapped to these nodes, otherwise their type would be in exchangeable(G). Thus, we have a homomorphism from G to H^c, which contradicts the hypothesis on H^c. □

12.2.6.2 Space Algorithm

Consider the space of graphs leading from H to its completions. All graphs in this space have the same set of concept nodes. The space is ordered as follows: Given two graphs H_1 and H_2 in this space, $H_2 \le H_1$ if for each relation x in H_1, there is a relation with the same list of neighbors in H_2 and a label less or equal to the label of x. The question "Is there a homomorphism from G to each completion H^c" can be reformulated as follows "Is there a *covering set* of completions, that is a subset of incomparable graphs of this space $\{H_1, ..., H_k\}$ such that (1) there is a homomorphism from G to each H_i ; (2) for each H^c there is a H_i with $H^c \le H_i$."

The brute-force algorithm (Algorithm 27) takes the set of all completions of H as covering set.

The next algorithm (Algorithm 30 and recursive Subalgorithm 31) searches the space in a top-down way starting from H and tries to build a covering set with partial completions of H. Reasoning by cases is applied at each step: For a given relation type r with arity k and a tuple $(t_1...t_k)$ of concept nodes, such that neither $+r$ nor $-r$ is supported by the label of a relation node on $(t_1...t_k)$ in the current partial completion, two graphs are generated according to each case. Note that if $+r$ or $-r$ is supported by a $\sim s$ in the current completion, then adding $+r(t_1...t_k)$ or $-r(t_1...t_k)$ to it would lead to a redundancy or an inconsistency.

The algorithm is justified by the following property:

Theorem 12.6. *Let G and H be two PGs, where H is consistent and normal. G is deducible from H if and only if:*
*1. There is a homomorphism π from G to H **or***
2. G is deducible from H' and H'' where H' (resp. H'') is obtained from H by adding the positive relation $r(t_1...t_k)$ (resp. the negative relation $-r(t_1...t_k)$) where r is a relation type of arity k occurring in G (and more specifically r belongs to the completion vocabulary) and $t_1...t_k$ are concept nodes of H such that neither $+r$ nor $-r$ is supported by the label of a relation node on $(t_1...t_k)$ in H.

Proof. (sketch) ($\Rightarrow$) Any completion of H' or H'' is a completion of H. ($\Leftarrow$) Condition 1 corresponds to Property 12.2. For condition 2, check that $\{H', H''\}$ is a covering set (of completions of H). □

Subalgorithm 31 is supposed to have direct access to data available in the main Algorithm 30. The choice of r and $t_1...t_k$, in Algorithm 31, can be guided by a compatible homomorphism from a special subgraph of G.

Algorithm 30: Check by space exploration

Input: Consistent PGs G and H, with H being consistent and normal
Output: true if G is deducible from H, false otherwise
begin
    Result ← **Filtering**()
    if *(Result ≠ undetermined)* **then**
        ∟ **return** *Result*
    Let $\mathcal{R}$ be the completion vocabulary
    return RecCheck(*H*); // see algorithm 31
end

The filtering subalgorithm performs "simple" tests corresponding to necessary or sufficient conditions of deduction that would allow us to conclude without entering the completion steps:

- If a concept or relation node of G has no support in H, then return false.
- If there is a homomorphism from G to H, then return true.
- Compute some subgraphs of G without exchangeable relations (for instance a subgraph without opposite relations maximal for the inclusion). If one of these subgraphs does not map to H by a compatible homomorphism, then return false.

Algorithm 31: RecCheck(*H*) *subalgorithm of Algorithm 30*

Input: Consistent and normal PG H **Access**: $G, \mathcal{R}$
Output: true if G is deducible from H, false otherwise
begin
    if *there is a homomorphism from G to H* **then**
        ∟ **return** *true*
    if *H is complete with respect to $\mathcal{R}$* **then**
        ∟ **return** *false*
    $(r, t_1...t_k)$ ← **ChooseRelationTypeToAdd**()
    /* r is a relation type of $\mathcal{R}$, $t_1...t_k$ are concept nodes in H
       and neither +r nor −r is supported by a the label of a
       relation on $(t_1...t_k)$ in H   */
    Let H' be obtained from H by adding the relation node $r(t_1...t_k)$
    Let H'' be obtained from H by adding the relation node $-r(t_1...t_k)$
    return (**RecCheck**(*H'*) AND **RecCheck**(*H''*))
end

The following property ensures that Algorithm 31 does not generate the same graph several times, which is a crucial point for complexity. Otherwise the algorithm could be worse than the brute-force algorithm in the worst-case.

Property 12.14. The subspace explored by Algorithm 31 is a (binary) tree.

Indeed, at each recursive call, $\{H', H''\}$ is a covering set inducing a bipartition of the covered space: Each graph in this space is below exactly one of these two graphs.

Property 12.15. The time complexity of Algorithm 30 is in $O(2^{(n_G)^k \times |\mathcal{R}|} \times hom(G, H^c))$, where n_G is the number of concept nodes in G, k is the maximum arity of a relation, $\mathcal{R}$ is the completion vocabulary and $hom(G,H^c)$ is the complexity of checking the existence of a homomorphism from G to H^c. Its space complexity is in $O(max(size(G), size(H), (n_G)^k \times |\mathcal{R}|))$.

Proof. The number of completions of H is bounded by $2^{(n_G)^k \times |\mathcal{R}|}$. Property 12.14 ensures that the number of graphs generated is at most twice the number of completions of H (in the worst case, all leaves of the generated tree of graphs are complete graphs). If the relation types are not ordered, all completions have the same number of relation nodes; the number of their relation nodes typed by an element of $\mathcal{R}$ is $\sum_{r \in \mathcal{R}} (n_G)^{arity(r)}$; checking whether a graph is complete can then be done in constant time if the number of relation nodes typed by an element of $\mathcal{R}$ is incrementally maintained. When relation types are ordered, the size of completions varies according to the order in which relation types are considered. One solution is to count the addition of a relation node $\sim r(t_1...t_k)$ not for one, but for n, where n is the number of types s in $\mathcal{R}$, such that $\sim s$ is supported by $\sim r$ and was not before. Computing n at each node addition can be roughly bound by $|\mathcal{R}|^2$, which can be reasonably considered as less than $hom(G,H^c)$. For space complexity, see that the tree is explored in depth-first way. □

12.2.7 Querying Polarized Graphs

The fundamental problem on BGs, namely *deduction*, takes two BGs as input and asks whether there is a homomorphism from the first one to the second. Querying BGs highlights another fundamental problem, namely *query answering*. The query answering problem takes as input a knowledge base (KB) composed of BGs representing facts and a BG Q, which represents a query, and asks for all answers to Q in the KB. Each homomorphism from Q to a fact defines an *answer*. If we consider the query answering problem in its decision form ("Is there an answer to Q in the KB?") we obtain the deduction problem ("Is Q deducible from the KB?")

Let us now consider query answering on PGs. Classically, there are two ways of understanding negation in a query. Briefly, when a query asks "find the x and y such that $p(x,y)$ and *not* $r(x,y)$," "not" can be understood in several ways. It might mean "the knowledge $r(x,y)$ cannot be proven" or "the knowledge *not* $r(x,y)$ can be proven." The first view is consistent with the closed-world assumption, the second one with the open-world assumption. Both are of interest in real-world applications. The closed-world assumption assumes complete knowledge about the world. It follows that only positive information needs to be coded in the fact base, with negative information being obtained by difference with the content of the base. *not $r(x,y)$* is then understood as "$r(x,y)$ is not present in the base," or more generally "$r(x,y)$ cannot be deduced from the base" (note that the *negation by failure* used in logic programming can be seen as an implementation of the closed-world assumption, where deduction, generally undecidable, is replaced by a decidable proof notion: *not $r(x,y)$* holds if $r(x,y)$ cannot be obtained by a finite proof). The closed-world assumption is commonly made in databases because it eases the representation of knowledge and improves performance (e.g., see the survey [Bid91]). To define a property, it is easier to list the individuals satisfying this property than to enumerate all individuals and indicate, for each of them, whether it satisfies the property or not. It may even be impossible to list all individuals. Consider, for instance, the property "being a registered user": Assuming complete knowledge about this property seems reasonable; any individual who does not appear in the list of registered users should be considered as not registered. Finally, let us recall that, if this assumption comes naturally in databases, it is impossible in FOL to infer negative facts (for instance *not registered(a)*) from positive ones. Various non-monotonic logics have been proposed for translating the closed-world assumption (e.g., see the survey in [AB94]).

Closed-world negation can be easily integrated into queries represented as PGs: By definition, a query Q is closed-world deducible from a BG or a PG G if it is deducible from its negative completion, i.e., if there is a homomorphism from Q to the completion of G with negative relation nodes only. Of course, the negative completion does not need to be computed effectively: A query Q is closed-world deducible from a BG or a PG G if there is a homomorphism π from the positive subPG of Q (i.e., the subPG obtained by deleting negative relation nodes) to G, which does not contradict any negative relation in Q: For all $-r(c_1 \ldots c_k)$ in Q, G does not contain a positive relation node $+s(\pi(c_1) \ldots \pi(c_k))$ with $s \leq r$. Consider, for instance, the example of Fig. 12.18. G describes a situation where there is a pile of three cubes A, B and C; A is blue and C is not blue. Q can be seen as a yes/no question asking *whether* there is a blue cube on top of a non-blue cube. It can also be seen as a query, that asks for *exhibiting* objects having these properties. Whether B is blue or not is not specified. Thus, with closed-world assumption, B is considered as not blue. There is a homomorphism from the positive part of Q to G, which maps x to A and y to B. Hence, the answer to Q as a boolean question is "yes," and the answer to Q as a query is $\{(x,A),(y,B)\}$.

Open-world assumption assumes incomplete knowledge about the world. Consequently, missing information, or more generally information not deducible from the base, is simply unknown. *not $r(x,y)$* is true if it can be deduced from the base. The value of the property "being a parent," for instance, might not be known for all individuals, so an individual not known as being a parent should not be considered as without children. In turn, the notion of deduction can have several meanings depending on the logics. In particular, the law of the excluded middle in classical logic has important consequences for query answering. Let us come back to the pile of cubes example (Fig. 12.18) with an open-world assumption: Nothing is known about the color of cube B. What should be answered to Q (as a boolean question or a query)?

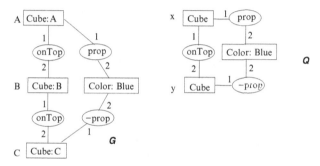

Fig. 12.18 The pile of cubes example

Let us first point out that spontaneously a non-logician (a typical end-user for instance) would say that the answer to the boolean question is *no*; this intuition corresponds to the observation that there is no answer to the query. However, in classical FOL, the answer to the boolean question is *yes*. Indeed the logical formulas assigned to Q and G by Φ are respectively of form $\Phi(Q) = \exists x \exists y \, (p(x, Blue) \land \neg p(y, Blue) \land r(x, y))$ and $\Phi(G) = p(A, Blue) \land r(A, B) \land r(B, C) \land \neg p(C, Blue)$ (where $p = prop$, $r = onTop$ and atoms assigned to concept nodes are ignored). $\Phi(Q)$ can be deduced from $\Phi(G)$ using the valid formula $p(B, Blue) \lor \neg p(B, Blue)$ ($\Phi(Q)$ is obtained by interpreting x and y as B and C if $p(B, blue)$ holds, and as A and B in the opposite case). Classical logical deduction thus ensures that there is a "solution" to Q, but it is not able to *construct* it. Hence there is no answer to Q as a query. This example leads to the following observations:

- The assertions "Q is (classically) deducible from G" and "the set of answers to Q in G is not empty" might disagree. In other words, deduction and the decision problem associated with query answering are different problems (which is not the case for BGs).
- The difference between the notions of deduction and the existence of an answer is due to the use of the law of excluded middle, which states here that "either B is blue or it is not blue."

This observation shifts the attention to logics in which the law of excluded middle does not hold. Let us consider one of these logics, i.e., intuitionistic logic [Fit69], a well-established logic which differs from classical FOL by admitting only *constructive* proofs. For instance, a proof of $(A \lor B)$ is given by a proof of A or a proof of B; a proof that the falsity of $(A \lor B)$ leads to a contradiction (i.e., a *reductio-ad-absurdum* proof) does not yield a proof of $(A \lor B)$ since it does not determine which of A or B is true. Each intuitionistic logic theorem is classical logic theorem, but not conversely. In particular, the theorem $(A \lor \neg A)$, i.e., the law of excluded middle, does not hold.

Let us come back to the example in Fig. 12.18. According to intuitionistic logic, the formula $p(B, Blue) \lor \neg p(B, Blue)$ can be considered as true only if it can be shown that $p(B, Blue)$ is true, or that $\neg p(B, Blue)$ is true. Since none of these two statements can be proven, Q cannot be deduced; hence the answer to Q as a yes/no question is *no*, which corresponds to the fact that there is no answer to Q as a query.

More generally, this logic appears to be completely in line with the notion of answer: Q is intuitionistically deducible from G if and only if the set of answers to G is not empty. It has been proven that PG homomorphism is sound and complete with respect to intuitionistic deduction in the FOL$\{\exists, \wedge, \neg_a\}$ logical fragment (cf. bibliographical notes). More precisely:

Theorem 12.7. *Let Q and G be two PGs defined on a vocabulary $\mathcal{V}$, with G being consistent. $Q \succeq normal(G)$ if and only if $\Phi(\mathcal{V}), \Phi(G) \Vdash \Phi(Q)$, where $\Vdash$ denotes the intuitionistic deduction .*

It follows that atomic negation can be introduced in BGs with no overhead cost for the query answering problem.

Note that another candidate for translating the existence of an answer would have been 3-value logic, in which three truth values are considered instead of two: *true*, *false* and *undetermined*.

Finally, let us point out that, independently of the answer notion, there are real world situations in which the law of excluded middle is not desirable, or not desirable for all properties or relations. Properties or relations might be neither true nor false because they cannot be determined with certainty (for instance, it might not be true that a cube is either blue or not blue because its color might be not "really" blue ...) or they intrinsically admit "truth value gaps" (this is the case, for instance, for a property like "being a sportsperson": Some people occasionally practice sports, and cannot be considered as sportspersons nor as non-sportspersons).

12.2.8 Note on Negation and Rules

When we want to integrate atomic negation and rules, i.e., to process rules where the hypothesis and the conclusion are no longer basic graphs but rather polarized graphs, there are two intrinsic sources of difficulty. First, forward chaining is not complete in the presence of negation (thus neither is backward chaining). This incompleteness occurs even if negation occurs only in facts and even in propositional logic. Consider, for instance, the rule $R : A \rightarrow B$ and the fact $F = \neg B$. Proposition $\neg A$ cannot be derived by forward chaining from R and F, but it should be. It could be derived from F and the contrapositive rule $R' : \neg B \rightarrow \neg A$ equivalent to R. A technique for overcoming this incompleteness consists of compiling the rule base before its use for deduction or query answering [RM00]. The compilation step adds all rules necessary to achieve completeness. For instance, let $\mathcal{R} = \{r_1, r_2, r_3\}$, with $r_1 = A \rightarrow B, r_2 = A \rightarrow C$ and $r_3 = B \wedge C \rightarrow D$. Check that $\neg A$ is deducible from $\mathcal{R}$ and $F = \neg D$, but it is not derivable from $\mathcal{R}$ and F. The compilation adds rules obtained by rewriting r_1, r_2 and r_3 in all forms logically equivalent to their clausal form (i.e., $\neg B \rightarrow \neg A$ for r_1, $\neg C \rightarrow \neg A$ for r_2, and $B \wedge \neg D \rightarrow \neg C$ and $C \wedge \neg D \rightarrow \neg B$), but also induced rules like $A \rightarrow D$ (which is actually not necessary as it can be simulated by

r_1 and r_2, then r_3) and $\neg D \rightarrow \neg A$. The size of the new base can be exponential in the size of the original base. Another source of complexity when passing to first-order is that rule application itself is much more complex: indeed, to check whether a rule can be applied, i.e., whether its hypothesis can be deduced from the facts, finding a homomorphism is not sufficient if we stick to classical deduction.

12.3 Bibliographic Notes

Full Conceptual Graphs

FCGs, and their logical semantics, as well as associated diagrammatical rules of inference, are introduced in [Sow84]. It was claimed that the full FOL was obtained, with arguments relying on the fact that Peirce's calculus was sound and complete. However, no formal proof of soundness and completeness of the calculus was provided.

Concerning Peirce's work, see for instance the historical survey [Rob03]. The version of FCG called concept graphs was introduced by Dau in his PhD thesis [Dau03b]. Besides the development of concept graphs, Dau's thesis provides an in-depth presentation of existential graphs. Concept graphs are also related to Formal Concept Analysis, cf. [GW99] for an introduction to this area. The version of FCGs equivalent to FOL without equality, and for which a sound and complete tableau calculus is defined, is studied in Kerdiles' Ph.D. thesis [Ker01].

Sowa presents negative contexts as a special case of conceptual graph contexts (see Chap. 9 about nested graphs). A negative context is seen as a concept with a graph referent, and a unary relation "NEG" (for negation) attached to it. We disagree with this view because the semantics of both kinds of concepts is different, and the NEG relation has to be translated as the logical negation operator.

In [Wer95a], Wermelinger corrected minor mistakes in Sowa's original definitions. In particular, he pointed out that the universal type $\top$ has to be translated in such a way that the graphs $[\top : *]$ or $[\top : a]$ are valid. He proposed their translation into a trivial equality ($x = x$ or $a = a$). A proof of completeness was given in his Master's thesis [Wer95b], but it is not available in English. Wermelinger's definitions took into account higher order concept and relation types that we do not want to consider. Baader, Molitor and Tobies stripped higher order features from Werlinger's definitions, and proved the equivalence of full CGs with FOL [BMT99] [BMS98]. Since no calculus over full CGs was considered, equivalence was shown at a descriptive level. The definition of full CGs given in this chapter and the associated Φ translation to logics rely on [BMT99] and the Theorem 12.1 is from [BMS98].

The *(loosely) guarded fragment* [vB97], noted lgFOL, is a decidable fragment of FOL with negation. This fragment has interesting properties, such as the finite model property and others [AvBN96]. It contains the FOL translation of many modal logics and description logics. In [BMT99] the authors identify a restricted form of full conceptual graphs corresponding to lgFOL. One can polynomially check if a conceptual graph is loosely guarded (the graph has to satisfy some syntactic restrictions), and compute the lgFOL formula equivalent to G, and reciprocally. Note that deduction in the loosely guarded fragment does not include deduction in the $\mathcal{FR}$ model, even when rules are restricted to range restricted rules. For instance, the following formula that asserts the transitivity of a binary relation R is not a (loosely) guarded formula and has no equivalent formula in the loosely guarded fragment [Grä99]: $\forall x \forall y \forall z (R(x, y) \land R(y, z) \to R(x, z))$.

Atomic Negation.

Few works have considered BGs with atomic negation. The name "polarized graphs" is from Kerdiles [Ker01]. Simonet exhibited examples showing that homomorphism is no longer complete and proposed an algorithm based on an adaptation of the resolution method (unpublished note, 1998; see also [Mug00]). Kerdiles exhibited simpler examples (cf. Figs. 12.10 and 12.11) and showed that homomorphism remains complete in a very particular case (briefly when positive and negative relations are separated into distinct connected components) [Ker01]. Klinger gave examples of problems related to the introduction of negation on relations (including equality) in the framework of protoconcept graphs (which can be translated into PGs) [Kli05]. Leclère and Mugnier discussed different kinds of negation (as well as difference) in the context of querying BGs and gave sound and complete mechanisms based on homomorphism for each of them [LM06][ML07]. The reduction in the proof of Π_p^2-completeness of PG-DEDUCTION is essentially due to Bagan, who built it when he was a Master's student at Montpellier University in November 2004 (see also [Mug07]). Properties concerning exchangeable relations and algorithmic improvements were published in [LM07] in the context of the containment problem of conjunctive queries with negation (CQNC). In this context, the vocabulary is restricted to a flat set of relation types and a set of individual markers. In this chapter, we have extended the results of this paper to PGs on general vocabularies and presented some additional results about exchangeable relations (namely Properties 12.10 and 12.11). The idea of Property 12.10 is from Thomazo [Tho07] (in French). In the same work, it is proven that PG-DEDUCTION is NP-complete if the source graph contains at most one pair of exchangeable relations. The question of PG-DEDUCTION complexity with a source graph containing two pairs of exchangeable relations, and more generally a bounded number of such pairs, is open.

Finally, let us mention the proof of Π_p^2-completeness of CQNC in [FNTU07], which relies on a reduction from the validity problem of a quantified boolean formula of the form $\forall^* \exists^* C$, where C is a conjunction of 3-clauses. Since CQNC and PG-DEDUCTION are equivalent, it can be seen as another proof of PG-DEDUCTION Π_p^2-completeness.

Chapter 13
An Application of Nested Typed Graphs: Semantic Annotation Bases

Overview

This chapter is devoted to the problem of resource retrieval using resource annotations. This problem is a promising application domain for nested typed graphs. In the first section, semantic annotations are defined as a kind of metadata. Two applications, information retrieval and editing, are also briefly described, and the annotation-based resource retrieval (*ARR*) problem is stated. The main components of an *ARR* system are also presented. In the second section, it is shown how modular vocabularies as well as exact and plausible knowledge can be defined and used in an *ARR* system. Ways to answer queries to an annotation base are discussed in the third section. Finally, relationships between an *ARR* system and the semantic web are briefly discussed in the fourth section.

13.1 Annotation

13.1.1 Annotations, Metadata and Resources

An annotation always annotates something. It does not exist in isolation. For instance, an annotation of a text is usually defined as a note, in natural language, giving explanations or comments about this text. In this chapter, an annotation can annotate any kind of object. The historical interest of a castle, the rhetoric study of a political discourse, a critique on a movie, or a description of a painting in modern art, are annotations about non-textual resources. Such resources can be electronic, e.g., an html file, jpeg image or database, or non-electronic, e.g., a book, person or tourist resort. Thus, in a way, we extend the usual meaning of an annotation, which is generally restricted to a text annotation. An annotation can annotate anything, not only texts. But, in another way, we restrict annotations from natural language notes to formal annotations. In this chapter, formal annotations are well-defined computational objects, particularly nested typed graphs. In a computational environment, an annotation is thus simply a data (sometimes called secondary data) about a data (sometimes called primary data), i.e., metadata.

Metadata can be roughly classified into two classes: *objective* metadata and *subjective* metadata. The name of a file, its type, its address or its date of creation are examples of objective metadata. A comment on a movie, an explanation of the plot of a detective story, or the description of the content of a book in the bibliographical system of a library (i.e., indexation of a book), are examples of subjective metadata. These metadata can be called subjective because they largely depend on the author of the metadata.

This sort of metadata, i.e., subjective metadata, are usually called *semantic annotations*. Subjective metadata are generally more complex than objective metadata. Therefore, any model able to manage subjective metadata should also be able, at least theoretically, to manage objective metadata, but probably in a less efficient way. Subjective metadata are simply knowledge, in the Artifical Intelligence sense, about resources. Processing such subjective metadata concerns knowledge-based systems while processing objective metadata is within the scope of database management systems. Hence, hereafter we focus on subjective metadata.

A resource can be structured, e.g., a book can be divided into chapters, or a multimedia document can be divided in texts, videos, etc. A set of resources can also be structured. In classical libraries, documents are classified by domain (e.g., humanities, sciences) or genus (e.g., books, reviews, theses), etc. Even if a document is not structured, for instance a picture, one can possibly want to decompose it and annotate each part of it. In what follows, for the sake of simplicity, we consider that each part of a resource is a resource and that the annotator cannot access the description of the structure of the resource (or the resource set), which is outside the annotation base. In other words, if the resource structure must be taken into account (for instance when querying an annotation base), this structure has to be represented in the annotation base; a part R' of a resource R is a resource and in order to consider this information in the query/answering mechanism one has to represent the fact that R' is a part of R in an annotation associated with R or/and with R'.

13.1.2 Examples of Annotation Base Uses

13.1.2.1 Information Retrieval

An Information Retrieval System (IRS) is usually defined as a software system that helps a user to find some required information. However, long ago it was noted that an IRS generally does not directly give the information needed, but only documents that should contain that information, and the user has to find the sought information within these documents. Generally, during a search task, an IRS cannot directly access the documents, it can only access them through metadata about documents (e.g., document content descriptions, data about the author, title of the document, etc), and these metadata contain references to the document base.

Usually, the relevance of a document is determined solely by a match between a query and the representation of the document content, so the way the content of a document is represented is crucial. Simple content representations, called indexes, are used in many IRSs. Basically, an index consists of a set of weighted words (or terms, or concepts) chosen within a finite set. In these systems, queries are often represented by a boolean expression of terms. Such descriptions by a set of terms are generally incomplete and imprecise (even if full text indexing is considered). It is commonly acknowledged that content representation and content-based access are crucial for improving the precision (the ratio of the number of relevant documents retrieved to the total number of documents retrieved) and the recall (the ratio of the number of relevant documents retrieved to the total number of relevant documents in the document base) of a user's search.

More precise document content representations and more precise queries can be obtained by using structures composed of terms enriched by relations between these terms, i.e., by labeled graphs. We see further how nested typed graphs can be used to represent document content as well as elementary queries. Graph homomorphisms are used for computing answers to a query and this approach to the search process is in line with the logical view of information retrieval. The logical view of information retrieval, as stated by van Rijsbergen [vR86], consists of considering that a document, whose content is represented by a formula d, is relevant to a formula query q if q can be inferred from d via some logic.

Information retrieval is intrinsically vague because documents are imprecisely and incompletely represented and also because queries are themselves vague. Thus, exclusively computing exact answers often leads to too much silence (or to a too low recall since the silence is the ratio of the number of relevant documents not retrieved to the total number of relevant documents). Different approaches have been developed to compute approximate answers and thus account for this intrinsic vagueness. Contrary to the situation where only exact answers are sought, approximate answers have to be ordered. Otherwise, any resource could be considered as answering a query and there would be too much noise. Thus, computing approximate answers induces the construction of a ranking function. Whenever queries and indexations are built on the same language, ranking functions are usually based on a measure of the distance between a query and an indexation. Numerical approaches are generally used, especially vectorial methods (There are also probabilistic approaches (cf. [vR79] or [Fuh00]) and fuzzy approaches (cf. [TBH03b])). Nevertheless, combinatorial approaches using graph transformations and van Rijsbergen's uncertainty principle can also be used (cf. [GC05]).

The main drawback of a labeled graph approach is that it requires (at least partial) manual indexing, since it is hard to automatically build faithful complex indexations. This point is discussed in the conclusion. Information retrieval in its present acceptation described above, i.e., document retrieval, is a specific resource localization problem. A *resource localization problem* can be described as follows: Given a set of resources described by a metadata base, and given a query, find the resources answering the query. More precisely, find in a metadata base the metadata satisfying the properties expressed by a query. We focus here on content-search retrieval or more generally on annotation-based retrieval, i.e., find in an annotation base the annotations (therefore the resources) which satisfy a query. Let us call this the *ARR* problem.

13.1.2.2 Editing

The *ARR* problem is itself important but it can also be a subtask of a very general editing problem. Let us assume that one wants to edit a new document using existing materials rather than from scratch. Materials are then annotated (existing) documents or document parts. Document annotations are used for retrieving documents relevant to a new publication specification. Thus, in this situation, annotations are information added to resources in order to transform them into bricks useful for building new documents. One goal of annotations is then to allow the "recontextualization," the "repurposing," of the resources. If queries are able to express publication specifications then the editing process can be aided by an *ARR* system. Finally, an annotation base can be considered as a knowledge base associated with a set of resources, designed to facilitate editing problems. In this framework, information retrieval can be considered as the simplest editing problem. Indeed, it can be considered like the construction of a new document composed of a list of (existing) documents responding to a query.

13.1.3 Components of an Annotation System

The main components of an annotation system based on conceptual graphs are briefly described as follows.

- *The data language for representing annotations* is composed of nested typed graphs.
  The vocabulary is decomposed into modules and graph types represent sorts of annotations. Signatures and constraints are used to prevent the construction of absurd annotations. Rules are used to represent common implicit (and exact) knowledge. Plausible knowledge, defined by schema graphs, are used to help annotators. For instance, a schema graph associated with a concept type t gathers typical or plausible information commonly accompanying the occurrence of an entity of type t.
- The *query language*, used for searching annotation bases, is also based on nested typed graphs.
  Elementary queries are graphs with the same form as annotations, i.e., nested typed graphs on the same vocabulary as the annotations. Elementary queries can be combined with the usual boolean connectors (and, or, not).

- *Answers* are computed by graph operations.
  Graph homomorphisms are used for computing exact answers. An exact answer to a query is composed of all annotations which are specializations of the query. The soundess and completeness of NTG homomorphism with respect to deduction in FOL guarantees that this approach complies with the logical view of Information Retrieval. Versatile matching algorithms for computing approximate answers can be based on graph transformations. A general framework for building approximate answers is proposed hereafter.
- A *graphical user interface* is needed because the intended users are not necessarily computer scientists nor engineers. More generally, *friendly tools* for building and interrogating annotation bases have to be developed, as well as easily explainable answering mechanisms. Graphs are especially useful to explain to end-users how the system works because knowledge and reasonings can be visualized in a way that is easily interpretable in natural language. Systems aiming at helping the solution of vague problems such as information retrieval should allow users to have maximal control over each step of the process:
  - entering and understanding the data (labeled graphs),
  - understanding results given by the system (labeled graphs),
  - understanding how the system computes results (easily vizualizable graph operations).
- When the resources are electronic resources, an important task of an annotation system involves presenting the resources to the annotator, giving him/her functionalities for decomposing a resource and attaching annotations to the resources. This task markedly depends on the resource types (e.g., video, image, text documents, etc.), and this issue is not addressed here.

13.2 Annotation Base

An annotation base is basically composed of a set of annotations that are nested typed graphs built over a given vocabulary $\mathcal{V}$ gathering the notions needed for adequately representing a set of resources. Different kinds of knowledge can be added to a vocabulary in order to help and to control the work of an annotator. We distinguish between exact knowledge, which can be automatically used by the system in the search process or to complete annotations, and plausible knowledge, which must be validated by the user before being taken into account by the system.

13.2.1 Exact Knowledge

Exact knowledge consist of rules, relation signatures, constraints, and individual graphs.

- Rules can represent knowledge of the form "if information is present (within an annotation) then further information can be added (to this annotation)." This kind of rule can be represented by the rules studied in Chap. 10. Using rules lighten the annotation work since they can represent implicit knowledge (e.g., if x is the mother of y, then y is a child of x).
  Rules can be used in two ways. They can be used to automatically complete an annotation by implicit and general knowledge. They can also be used by the search process which simultaneously queries an annotation base and a set of rules without changing the annotations.
- Positive constraints ("if information is present—within an annotation– then further information has to be present—in this annotation") are used to help an annotator to not forget to represent important information. When an annotator validates an annotation, if a positive constraint is violated then the missing part is shown, and the annotation will be accepted only when it will be supplemented in order to respect the positive constraint. Such positive constraints can be represented by the constraints studied in Chap. 11.
- Relation signatures and negative constraints ("if information is present—within an annotation—then further information cannot be present—in this annotation") are used to avoid the construction of absurd annotations.
- An *individual graph* for the individual marker m is an SG with a specific concept node having m for marker and called the head of the individual graph. Such a graph represents general and exact knowledge concerning m. Individual graphs are considered facts and are added to the annotation base when querying this base. An individual marker m can have several individual graphs that represent different ways of describing the entity denoted m. As rules, representing general information in individual graphs reduces the tedious task of entering the same information several times.

Definition 13.1 (Annotation base). An *ontology* $\mathcal{O}$ is composed of a vocabulary $\mathcal{V} = (T_C(T, \mathcal{B}), T_R, T_G, \mathcal{I})$, a set $\mathcal{C}$ of relation signatures and of constraints over $\mathcal{V}$, and a set of rules $\mathcal{R}$ over $\mathcal{V}$. An *annotation base* over $\mathcal{O}$ is composed of a set $\mathcal{A}$ of nested typed graphs over $\mathcal{V}$ and of a set $\mathcal{G}$ of individual graphs over $\mathcal{V}$. Furthermore, the graphs in $\mathcal{A}$ and $\mathcal{G}$ respect $\mathcal{C}$.

Before presenting prototypical knowledge, we present the module notion, which is another way to help and control the annotator's work.

13.2.2 Modules

Graph types are used to represent different sorts of annotation. Each sort of annotation may use specific concept and relation types, and these types can be gathered within a part of $\mathcal{V}$ called a *module*. With the module notion the vocabulary can thus be restricted to a subset dedicated to representing a specific kind of annotation. For instance, let us assume that we have to annotate a collection of 18th century paintings. A part of the annotation can concern the painting technique (e.g., oil on canvas or watercolors, the palette, the distribution of forms, etc.) and another part can concern the theme (e.g., a battle or peasants in the fields). In such a situation, terms of the vocabulary needed to describe painting techniques can be pooled within part of the vocabulary, while another part can contain the terms needed to describe what is painted. Such a vocabulary part is called a *module*. More precisely,

Definition 13.2 (Module). A *module* of a vocabulary $\mathcal{V} = (T_C(T, \mathcal{B}), T_R, T_G)$, where $T_C(T, \mathcal{B})$ is the set of acceptable conjunctive types generated by the primitive type set T and the banned type set $\mathcal{B}$, is a triple $m = (T'_C(T', \mathcal{B}'), T'_R, T'_G)$, where:

- T' is a subset of T, and $\mathcal{B}'$ is the set of maximal elements of $\mathcal{B}^{\sqcap} \cap T'$
- T'_R is a subset of T_R
- T'_G is a subset of T_G

T'_C, T'_R, T'_G are ordered by the partial orders on T_C, T_R, T_G respectively. A module denoted $m(g)$ is associated with any graph type g.

The whole vocabulary $\mathcal{V}$ can be considered as a module, so if a graph type g does not correspond to an annotation sort then $\mathcal{V}$ can be associated with g, i.e. $m(g) = \mathcal{V}$. Otherwise, $\mathcal{V}$ is the default module associated with a graph type. Note that if a vocabulary is equipped with a signature function, then a module is not necessarily a (sub)vocabulary because the concept types of the signature of a relation in T'_R can be outside T'_C.

The partial order on graph types induces constraints on the associated modules. Let g and g' be two graph types with $g' \leq g$. An annotation of type g' is an annotation of type g, thus a type used in module $m(g')$ must be interpretable in module $m(g)$. The simplest way to fulfill this condition is to assume that any type in $m(g')$ is also in $m(g)$. Let us assume that $m(g) = (T_C, T_R, T_G)$ and $m(g') = (T'_C, T'_R, T'_G)$, then $m(g')$ is *included* in $m(g)$ if $T'_C \subseteq T_C$, $T'_R \subseteq T_R$, and $T'_G \subseteq T_G$.

Definition 13.3 (Modular vocabulary). A *modular vocabulary* is a triple $(\mathcal{V}, \mathcal{M}, m)$, where $\mathcal{V}$ is a vocabulary, $\mathcal{M}$ is a set of modules of $\mathcal{V}$ and m is a mapping from the graph type set of $\mathcal{V}$ to $\mathcal{M}$. Moreover, the mapping m satisfies the condition: For any graph types g and g' if $g' \leq g$ then $m(g') \subseteq m(g)$, i.e., m is monotonous for the inclusion order.

An NTG G built over $\mathcal{V}$ *respects* a modular vocabulary $(\mathcal{V}, \mathcal{M}, m)$ if for each typed graph (g_i, G_i) in a typed box of G, the graph G_i is built over $m(g_i)$, i.e., the module which is associated with the graph type g_i.

Thus, a modular vocabulary allows the restriction of the terms used when constructing part of an annotation corresponding to a given viewpoint. It is easy to check if an annotation respects a modular vocabulary. An annotator tool may also dynamically control the construction of annotations so that they respect a modular vocabulary.

13.2.3 Plausible Knowledge

An annotation can be constructed from scratch, i.e., from an empty graph. If there are several annotators, there may be a great variety among annotations of similar resources. The schema graph notion aims at directing the construction of an annotation by proposing annotation templates.

A *schema graph* is associated with a concept type or with a relation type or with a graph type. A schema defines a usual or plausible context of a concept, or relation or graph of some type. Hence, schema graphs are conceptual graphs whose specificity is to be used by annotation tools in a particular way. Contrary to exact knowledge, schemas are not used for reasoning, at least in the system we are describing, but only to suggest pieces of annotation to a user (i.e., an annotator).

Let us first explain how a graph type schema is used. Let g be a graph type corresponding to a kind of annotation. A schema for g is an NTG which respects $m(g)$, i.e., the module associated with g. Such a schema is an annotation template, it consists of general and frequent notions appearing when annotating a resource with respect to g. Several schemas can be associated with a given graph type g.

When a user begins to annotate a resource with respect to the annotation viewpoint g, the system proposes a schema associated with g as a starting point. Note that this is only a starting point, and a proposal, but the user is free to modify the proposal at will. Otherwise, if an annotation should specialize any schema of a type appearing in this annotation, then a schema would be a kind of constraint, i.e., a kind of exact knowledge.

Example. In a project aimed at annotating film interviews of researchers, the way a language is used by a researcher is described in a *rhetoric* part of an annotation. Figure 13.1 is the schema associated with this rhetoric graph type.

In a schema graph for a concept or a relation type t, there is a specific node, called the *head* of the graph, with the type t. Nodes of an annotation whose types have schemas are called *extension nodes*. Such a node c of type t is used to propose the user an extension of the annotation in construction by merging the head of a schema for t with c. Like graph type schemas, the user is free to transform the obtained annotation at will. After merging of a relation schema, the head does not respect the arity condition since it has twice the required number of neighbors. The user has to modify the thus-obtained graph in order to transform it into a graph respecting the vocabulary.

Like graph types, a concept type or a relation type can have several schemas, each schema represents a way of describing an entity or a relation.

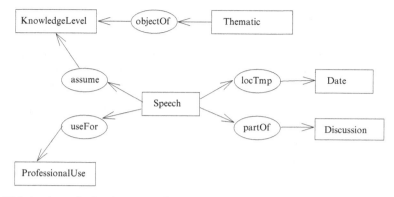

Fig. 13.1 A schema for the *rhetoric* graph type

Example. Figure 13.2 shows an example of a schema for the concept type *Movi-eTrip* corresponding to a children's viewpoint.

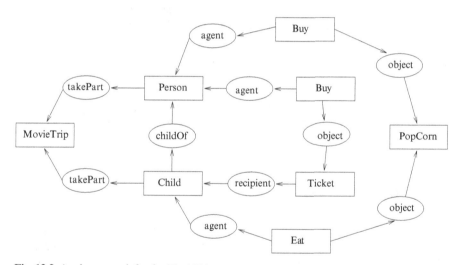

Fig. 13.2 A schema graph for the *MovieTrip* concept type

13.3 Querying an Annotation Base

First we present how exact answers to a query can be computed using graph ho-momorphisms. Then a method for computing approximate answers, based on graph transformations and graph homomorphisms and implementing van Risjbergen's un-certainty principle, is outlined.

13.3.1 Exact Search

Let $\mathcal{A}$ be an annotation base over $\mathcal{O}$. The simplest form of a query is a nested typed graph over the vocabulary of $\mathcal{O}$. Let Q be such a query. Different answers of Q over $\mathcal{A}$ can be defined as follows:

- The set $spec(Q, \mathcal{A})$ of elements in $\mathcal{A}$, i.e., annotations, which are specializations of Q, i.e., $spec(Q, \mathcal{A}) = \{A \in \mathcal{A} \mid \exists \pi,$ homomorphism from Q to $A\}$,
- for any element A in $spec(Q, \mathcal{A})$, the set of homomorphisms from Q to A or the set of images of Q by these homomorphisms,
- the set of resources $res(Q, \mathcal{A})$ which are referenced by the elements in $spec$ $(Q, \mathcal{A})$.

Q can be decorated in order to filter the answers (e.g., consider only homomorphisms that do not change the labels of some nodes specified in the query) or to restrict the answers (e.g., return only images of some nodes specified in the query).

Simple queries can be combined by boolean operators implemented as set operators on the answer sets. Let Q be a query graph, $\mathcal{A}$ an annotation base, and $spec(Q, \mathcal{A})$ the answer set. The elementary exact search can be simply extended by the classical boolean operators AND, OR, and NOT as follows.

- $spec(Q \text{ AND } Q', \mathcal{A}) = spec(Q, \mathcal{A}) \cap spec(Q', \mathcal{A})$,
- $spec(Q \text{ OR } Q', \mathcal{A}) = spec(Q, \mathcal{A}) \cup spec(Q', \mathcal{A})$,
- $spec(Q \text{ AND NOT } Q', \mathcal{A}) = spec(Q, \mathcal{A}) \setminus spec(Q', \mathcal{A})$.

To sum up, the fundamental problem to solve for computing answers in this framework is to compute the set of homomorphisms from a nested graph to a nested graph.

13.3.2 Approximate Search

The general approach described in this section has two specificities. First, it is a combinatorial approach based on labeled graphs and not a numerical approach as is usually the case for approximate search, and, secondly, it is a specialization of the van Rijsbergen's uncertainty principle for information retrieval and thus it still complies with the logical view of information retrieval.

Let us first recall van Rijsbergen's Uncertainty Principle, which can be stated as follows: *"Let f and g be two formulas in a logic $\mathcal{L}$, a measure of the uncertainty of $f \models g$ relative to a knowledge base is determined by the minimal transformation from g to g', such that f entails g' in $\mathcal{L}$."* (cf. [JN98]). Instead of transforming the conclusion, one can also consider a transformation of the hypothesis, i.e., f.

A way of implementing this very general principle in our graph-based framework is to define a set of graph transformations $\mathcal{T}$ equipped with a total preorder $\preceq$. As already explained in Sect. 13.1.2, a ranking function, i.e., a total (pre)order, of the answers is mandatory when approximate answers are searched.

The total preorder $\preceq$ defined on $\mathcal{T}$ is used to define a ranking function as follows. Let Q be a query graph over an annotation base $\mathcal{A}$. Let us assume that the answer set is defined by $spec_{\mathcal{T}}(Q,\mathcal{A}) = \{A \in \mathcal{A} \mid \exists t \in \mathcal{T}, t(A) \models Q\}$. Let t_1 (resp. t_2) be a minimal transformation such that $t_1(A_1) \models Q$ (resp. $t_2(A_2) \models Q$). If $t_1 \preceq t_2$, then A_1 is ranked before A_2. Note that if the identity transformation is the only minimal element in $\mathcal{T}$, then the exact answers are ranked before the approximate answers.

Instead of transforming the annotations, we can also transform the query. Transforming the query is frequently used in the Information Retrieval domain (it is called "query reformulation"). In an annotation-based system, it can be interesting to transform the annotations instead of the queries. Indeed, if a user is satisfied by an approximate answer when querying an annotation base, this answer has been obtained from a modified annotation. Then this modified annotation can be included (or proposed to be included) in the annotation base. Hence, transforming annotations instead of queries in a way makes "living" the annotation base.

With the transformations proposed hereafter, the two approaches can be considered as equivalent and this is basically due to the fact that annotations and queries are elements of the same language. If there is a transformation $t \in \mathcal{T}$ such that $t(A) \models Q$, then there is a transformation t' in a set of graph transformations dual to $\mathcal{T}$ such that $A \models t'(Q)$ and conversely.

The transformations are defined as sequences of elementary transformations. An *elementary transformation* can be:

- a node label substitution,
- the identification of two concept nodes,
- a node adjunction.

For any query Q, by applying such a transformation to an annotation A, the number of homomorphisms from Q to A may increase. For any query Q and any annotation A, A can be transformed in such a way that A specializes Q. Thus, if all the sequences of elementary transformations are considered, it is always possible to build transformations t such that the number of homomorphisms from Q to the transformed annotations $t(A)$ is ≥ 1, i.e., any A (approximately) answers Q. Therefore, in such an approach, a fundamental problem is to restrict the transformation set to *admissible transformations*.

A transformation t is admissible for an annotation A if, whenever A is a relevant annotation for a resource, then $t(A)$ is also relevant for this resource. $t(A)$ can be less relevant than A but nevertheless $t(A)$ should still be relevant for the resource annotated by A. Admissible transformations and ranking functions are difficult to abstractly define because these notions are strongly context-dependent. They depend on the ontology qualities (e.g., precision, completeness), on the annotation base qualities, on the user competence, etc. We only give here the main ideas in a general way which have to be specified for any application.

1. *Admissible elementary transformations* Let us begin with three sorts of elementary transformations.

- *Label substitution.* A way of defining an admissible node label substitution is to consider the ordered structure of the vocabulary. Note first that one only has to consider the substitution of a label by a non-smaller one since the substitution of a label by a smaller one is taken into account in the homomorphism notion. Then the precision of the vocabulary as well as that of the annotations have to be taken into account. Let us consider the (concept or relation) type substitution. Many definitions can be proposed on the basis of some "distance" between types. For instance, substituting y for x can be admissible if there is a path from x to y (e.g., if x and y are concept types, then this is a path in T_C) having no more than p ascents and q descents (with p and q being two non-negative integer parameters).
- *Identification of concept nodes.* In the same way, an identification of two concept nodes can also be said to be admissible if some distance between the labels is not too large (a restrictive way is to only allow identification of two concept nodes with the same label).
- *Node adjunction.* It is harder to propose a general definition of a natural admissible adjunction. Therefore, if node adjunction is an annotation transformation worth considering (for instance if it is known that the annotations are usually incomplete), a solution is to consider admissible the adjunction of a concept node labeled ⊤ or the adjunction of a maximal k-ary relation between k concepts which are not already linked.

2. *Admissible transformations* After defining admissible elementary transformations, admissible transformations have to be defined, i.e., what is an admissible sequence of elementary transformations. A sequence of elementary transformations is admissible if it consists of a limited number of admissible elementary transformations, e.g., no more than α label substitutions, no more than β concept node identifications, and no more than γ node adjunctions. All of these numbers can be ratios relative to the node numbers of A. In this way, the labeling function and structure of the annotation are only slightly modified.

3. *Total preorder.* A total preorder $\preceq$ on the set of admissible transformations now has to be defined. Functions of the numbers of elementary transformations can be used, the simplest one is the length of the sequence, but one can also weight the importance of the elementary transformation (e.g., count a for a label substitution, b for a node identification and c for a node adjunction, with $a \leq b \leq c$). The different numerical parameters can be defined by the user or by an annotation base administrator.

13.4 Annotation and the Semantic Web

The semantic web is the domain in which the term semantic annotation is the most frequently used. So, let us make a little digression into the semantic web even though the approach here described is not specially dedicated to web documents (cf. [HS03a]).

Usually a semantic annotation of a web document, i.e., a web annotation, is a RDF (Resource Description Framework) graph associated with the document. RDF is dedicated to the web (cf.http://www.w3.org/TR/rdf-primer): "RDF is intended to provide a simple way to make statements about Web resources, e.g., Web pages ... RDF is based on the idea of identifying things using Web identifiers (called Uniform Resource Identifiers, or URIs), and describing resources in terms of simple properties and property values. This enables RDF to represent simple statements about resources as a graph of nodes and arcs representing the resources, and their properties and values." An RDF graph is a set of RDF statements, and an RDF statement is a 3-tuple (resource, property, value).

The feasibility of this approach (i.e., associating RDF graphs to a large number of web documents) has been criticized. The main critiques are as follows: The RDF model is too complicated for most of users, there is no methodology (no strict template structure even if one considers ontology-based semantic annotations) so the graphs are too author-dependent, it is difficult to assess if an RDF graph is a sufficient and relevant annotation, it is a labor-intensive and time-consuming task even for users fluent with RDF, one does not know how to build a common vocabulary for a large part of the web, the size of the RDF graph can be important (e.g., complexity of an RDF graph describing a photo). Despite these criticisms, web annotation platforms have been constructed (cf. for instance [Ann01], [HS03b]).

Note that besides this mainstream approach there are other approaches to the semantic web which may seem less ambitious but may be more effective in bringing some "semantics" into the web (cf. [QV07]).

Naturally, almost all criticisms on RDF can also apply to CGs. Indeed, an RDF graph is a labeled multigraph and can be easily translated into an SG.

Building vocabularies, rules or constraints for large parts of the web seems unfeasible in the near future. Thus, we do not recommend replacing RDF graphs by CGs but rather to use CGs for specific applications (not "The Web" even though web resources can be considered) in which the expensive process of construction of a knowledge base is not too costly relative to the hoped benefit, e.g., aided publication and editing.

Tim Berners-Lee in [BL01] states: "I think the tendency of the CG examples to relate to natural language rather than hard logic made it more difficult for someone of my own leanings toward rigid logical processing for the Sweb to understand what the CG folks were driving at." And at the end of the same note: "All in all, there is a huge overlap, making the two technologies very comparable and hopefully easily interworkable."

Since then, studies have compared RDF graphs and CGs (e.g., [ME99], [CDH00]). An important result was published in 2005 that relates RDF reasonings and graph homomorphisms (cf. [Bag05]). Simple RDF graphs are provided with a model theoretic semantics [Hay04]. This semantic is used to define an entailment relation: An RDF graph G entails an RDF graph H iff H is true whenever G is. There is a finite procedure characterizing entailments (the Interpolation Lemma [Hay04]), which is sound and complete with respect to entailment. It has been extended to account for more expressive languages (RDF, RDFS [Hay04], and other languages, e.g., [tH04]). All of these extensions rely on a polynomial-time initial treatment of graphs, with the hard kernel being the basic simple entailment, which is an NP-hard problem. Baget [Bag05] has reformulated this fundamental entailment relation as a graph homomorphism.

The benefits of such a reformulation are frequently mentioned throughout this book. However, the huge size of the web and its intrinsic unstructured nature mean that a complete procedure for exact search would likely be unfeasible in the near future.

13.5 Conclusion

Applications using annotation bases generally require an interactive system to be built in which a user and the search process communicate. The intrinsic vagueness of annotations and sometimes of the user's goals lead to searches for exact answers but also approximate answers. In such situations, it is important for the user to understand the way the system functions. This is hard to achieve for numerical approaches, which are generally used for computing and ranking approximate answers, but it is one of the main motivations for using graphs. As very often stated in this book, (simple) graph transformations on (rather) small graphs are easily visualizable and understandable by non-specialist users.

Conceptual graph systems have been developed (e.g., [CH93], [CH94], [Mar97], [CDH00]), and the nested conceptual graph framework presented in this chapter was implemented and used in different applications (e.g., [Gen00],[GC05], [MLCG07]).

Schema graphs were introduced by Sowa [Sow84], who used schemas for plausible reasoning. In this chapter, schemas are only used to suggest to a human annotator information that can be useful to introduce in an annotation.

The main drawback of a labeled graph approach is that it requires (at least partial) manual annotating. As noted in the previous section for RDF annotations, manually annotating a resource has many drawbacks. Nevertheless, in some cases, automatic procedures cannot be used, and manual annotation is required in situations such as:

- the documents are not in electronic form or not directly available through the system (e.g., a traditional book library),
- the documents are images, sounds, videos, etc., so extraction of high level concepts and relations is necessary to take complex queries into account, and there is no (completely) automatic annotating mechanism which allows extracting of high level concepts from documents such as videos (e.g., [Fuh00], [Sme00]),
- the documents are texts but more precise representation than sets of keywords (i.e., those currently obtainable through natural language processing systems) of the content is required because the users are specialists with specific tasks (e.g., editing, cf. Sect. 13.1.2.2)).

Some experiments have shown that manual annotation with labeled graphs is not much more difficult than annotating with other manual annotation systems (cf. [GC97]). The hard part of manual annotating concerns more the analysis of the document than the annotation language.

Finally, such an approach will be directly usable if one day automatic methods for extracting high-level, i.e., conceptual, structures from electronic resources are developed .

Appendix A
Mathematical Background

Overview

The formalization of conceptual graphs relies upon different domains of discrete mathematics, mainly graph theory, ordered set theory and First Order Logic (FOL). This appendix is a short summary of the basic mathematical notions needed to understand the formalization of conceptual graphs presented in the book. More detailed notions are given in the different chapters when necessary.

In Sect. A.1, basic notions concerning the naive set theory (e.g., set operations, mappings, relations) are reviewed. Different sorts of graphs encountered in the book (e.g., directed and undirected graphs, trees, bipartite graphs, hypergraphs) are defined in Sect. A.2. Elementary graph theoretical notions and the fundamental notion of graph homomorphism are also reviewed in Sect. A.2. In conceptual graph models, node label sets are ordered. Section A.3 deals with ordered sets (basic algorithms are given in Chap. 6). Section A.4 reviews basic notions in first order logic. The last section is a very sketchy introduction to algorithm and problem complexity.

For readers who wish a deeper presentation of these mathematical notions, here is a list of background textbooks (with exercises) whose terminology we have generally followed: Introduction to Graph Theory [Wes96], Introduction to Lattices and Order [DP02], Mathematical Logic for Computer Science [BA01] and Introduction to Algorithms [CLRS01].

Our theorization of conceptual graphs is also closely related to several areas of computer science, especially relational database theory and constraint processing. Notions concerning these two domains are given when needed, especially in Chap. 5.

A.1 Sets and Relations

A.1.1 Sets and Elements

We begin with basic notions from (naive) set theory.

A *set* is a collection of objects called *elements* of that set. The statement that an object x is an element of a set S is denoted $x \in S$, and $x \notin S$ means that x is not an element of S. A set S can be described

1. either by listing its elements (e.g., $S = \{a, b, c\}$ is the finite set composed of the three elements a, b and c, the infinite set of positive integers is denoted $N = \{1, 2, \ldots, n, \ldots\}$),
2. or by giving a set U and a property p charaterizing, among the elements of U, those belonging to S, notation $S = \{x \in U \mid p(x)\}$ (e.g., $S = \{x \in N \mid 2 \leq x \leq 5\}$ is the set $\{2, 3, 4, 5\}$).

The *empty set* is the set without any element, it is denoted $\emptyset$. The number of elements of a set S is denoted $|S|$. A set S is a *subset* of a set T, in symbols $S \subseteq T$, if any element of S is also an element of T. The set of subsets of a set S is denoted 2^S.

The *union* of two sets S and T is the set $S \cup T = \{x \mid x \in S \text{ or } x \in T\}$, the *intersection* of S and T is the set $S \cap T = \{x \mid x \in S \text{ and } x \in T\}$, the *difference* of two sets is the set $S \setminus T = \{x \mid x \in S \text{ and } x \notin T\}$. Two sets are *disjoint* if their intersection is the empty set.

A set of non-empty subsets of a set S is a *partition* of S if the union of the subsets is equal to S and if the subsets are pairwise disjoint. More formally,

Definition A.1 (Cover and partition). Let $P = \{S_1, \ldots, S_k\}$ be a set of non-empty subsets of S. P is a *cover* of S if $S = \bigcup_{i=1}^{i=k} S_i$, if, furthermore, $\forall i \neq j$ one has $S_i \cap S_j = \emptyset$, then P is called a *partition* of S.

Example. Let us consider the content of a child toybox. It can be represented as a set over which covers and partitions can be defined, e.g., let us assume that the toys are clustered by colors, if each toy is monocolor, then a partition is obtained, if toys can have several colors then a cover is obtained.

Values of an attribute about objects (e.g., age of persons) can define partitions (e.g., people gathered by age respecting a set of disjoint intervals) or covers (non-disjoint intervals) over a set of these objects. Usually levels in a taxonomy define partitions, e.g., a set of mammals is partitioned into cats, dogs, etc.

A *k-tuple*, k integer ≥ 1, over a set S is a sequence of k not necessarily distinct elements of S, a k-tuple is denoted $(x_1, \ldots, x_k)$. The *cartesian product* of finitely many sets $S_1, \ldots, S_n$, written $S_1 \times \ldots \times S_k$ is the set of all k-tuples $(x_1, \ldots, x_k)$ such that $x_1 \in S_1, \ldots, x_k \in S_k$.

Example.

Let $T = \{car, truck, bicycle\}$ and $U = \{cheap, expensive\}$.

$T \times U = \{(car, cheap), (car, expensive), (truck, cheap), (truck, expensive), (bicycle,$

cheap),(*bicycle,expensive*)}.
Let $I = \{a,b\}$ then $I \times I = \{(a,a),(a,b),(b,b),(b,a)\}$.

A.1.2 Relations and Mappings

The relation notion is the fundamental notion of relational databases and of first order logic models. A *relation* is a subset of a cartesian product $S_1 \times \ldots \times S_k$. The number k of sets in the cartesian product is called the *arity* of the relation. A binary relation over the same set, i.e., a subset of $S \times S$, is simply called a *binary relation*.

Example. Kinship relationships concerning a set of persons can be represented by binary relations, e.g., the relation childOf can be defined by $(x,y) \in$ childOf if x is a child of y, also denoted childOf(x,y).

A ternary relation play3 can be defined over $C \times C \times T$, where C is a set of children and T a set of toys, by $(x,y,z) \in$ play3 if the children x and y are playing with the toy z.

Information about a person can be represented by a 5-tuple (*name,address,age,-profession,income*), in this case the set of 5-tuples of a group of persons is a relation of arity 5.

Given sets S and T a *partial mapping* from S to T is a relation $f \subseteq S \times T$ such that for all $x \in S$ and for all $y,z \in T$, if $(x,y) \in f$ and $(x,z) \in f$, then $y = z$. A partial mapping is also called a *function*. If f is a partial mapping, then $(x,y) \in f$ is also denoted $y = f(x)$. Let f be a partial mapping from S to T.

The *domain* of f is the subset of S defined by: $dom(f) = \{x \in S \mid \exists y \in T, (x,y) \in f\}$, the *range* of f is the subset of T defined by: $range(f) = \{y \in T \mid \exists x \in S, (x,y) \in f\}$.

A partial mapping f such that $dom(f) = S$ is called a *mapping*. A mapping is also called a *total function*. A mapping $f \subseteq S \times T$ is generally denoted $f : S \rightarrow T$.

A mapping f from S to T is *surjective* if $range(f) = T$. f is *injective* if for all x and y in S, $f(x) = f(y)$ implies $x = y$. An injective and surjective mapping is called a *bijective* mapping. An injective (resp.surjective, bijective) mapping is also called an injection (resp. surjection, bijection). Let f be a bijection from S to T then the *inverse* of f, denoted f^{-1}, is the bijection from T to S defined by: $f^{-1}(x) = y$ if $f(y) = x$.

More generally, let f be a partial mapping from S to T and let $x \in T$, then $f^{-1}(x)$ is equal to the set $\{y \in S \mid f(y) = x\}$. f is injective if and only if for all $x \in T$, $\mid f^{-1}(x) \mid \leq 1$.

Example. Let S be a finite set of n elements $\{s_1,\ldots,s_n\}$. The mapping f from 2^S to the cartesian product of n times the set $\{0,1\}$ defined by $\forall S' \subseteq S, f(S') = (x_1,\ldots,x_n)$ with $x_i = 1$ if $s_i \in S'$ and $x_i = 0$, otherwise is a bijection. The image of S' is called the *characteristic vector* of S'.

Let S be a set of companies and phoneNumb be a binary relation, such that (x,y) is in phoneNumb if y is a phone number of the company x. If a company can have only one phone number, then `phoneNumb` is a mapping from the set of companies to the set of phone numbers, otherwise it is a binary relation which is not a mapping. Furthermore, if two companies cannot have the same phone number then `phoneNumb` is injective.

Given a mapping $f : S \rightarrow T$ and $S' \subseteq S$ the *restriction* of f to S' is the mapping $f_{S'} : S' \rightarrow T$ defined as follows, $f_{S'}(x) = f(x)$ for all $x \in S'$. Let $f : S \rightarrow T$ and $g : U \rightarrow T$ be two mappings with $S \subseteq U$. The mapping g is an *extension* of f if for all $x \in S$, $f(x) = g(x)$, i.e., $g_S = f$.

Let R be a k-ary relation $R \subseteq S_1 \times \ldots \times S_k$, and $\{A_1, \ldots, A_k\}$ be a set of k attributes (or names or variables), a tuple $t = (x_1, \ldots, x_k)$ of R can be identified to a mapping, still denoted t, from the attribute set to the union of the S_i defined as follows: for all $i = 1, \ldots, k$, $t(A_i) = x_i$. In this situation, S_i is called the domain of the attribute A_i and a relation R can be identified to a set of mappings from the attribute set to the union of the domain attributes.

Relations are sets, hence set operations (union, intersection,...) apply to relations. Three other basic operations are especially important in relational databases: projection, join, and selection. The cartesian product is not commutative and the same set can occur several times in a cartesian product, hence the named attributes are used to define these operations.

1. *Projection.* Let R be a k-ary relation with attributes $\{A_1, \ldots, A_k\}$ and $\{B_1, \ldots, B_l\}$ a subset of $\{A_1, \ldots, A_k\}$. The projection of R onto $\{B_1, \ldots, B_l\}$ is the set of l-tuples which are restrictions of all tuples t in R to $\{B_1, \ldots, B_l\}$, .

2. *Join.* Let R be a relation with attributes $\mathcal{A} = \{A_1, \ldots, A_k, B_1, \ldots, B_l\}$ and R' a relation with attributes $\mathcal{A}' = \{A'_1, \ldots, A'_{k'}, B_1, \ldots, B_l\}$, with $A_i \neq A'_j$ for all i and j. The join of R and R', denoted $R \bowtie R'$, is the set of tuples t with attributes $\mathcal{A}'' = \{A_1, \ldots, A_k B_1, \ldots, B_l, A'_1, \ldots, A'_{k'}\}$ such that the restriction of t to $\{A_1, \ldots, A_k, B_1, \ldots, B_l\}$ is a tuple of R and the restriction of t to $\{A'_1, \ldots, A'_{k'}, B_1, \ldots, B_l\}$ is a tuple of R'.

3. *Selection.* Let R be a k-ary relation with attributes $\{A_1, \ldots, A_k\}$, $\{B_1, \ldots, B_l\}$ a subset of $\{A_1, \ldots, A_k\}$, and for all $i = 1, \ldots, l$, a_i be an element of the domain of B_i. The selection of R with respect to $(a_1, \ldots, a_l)$ is the set of tuples t of R, such that for all $i = 1, \ldots, l$, $t(B_i) = a_i$.

A binary relation R over a set S is a subset of $S \times S$. Basic properties for a binary relation R over S are:

- reflexivity, i.e., $\forall x \in S\ (x,x) \in R$,
- transitivity, i.e., $\forall x, y, z \in S$ if $(x,y) \in R$ and $(y,z) \in R$, then $(x,z) \in R$,
- symmetry, i.e., $\forall x, y \in S$ if $(x,y) \in R$, then $(y,x) \in R$,
- antisymmetry, i.e., $\forall x, y \in S$ if $(x,y) \in R$ and $(y,x) \in R$, then $x = y$.

A reflexive, transitive and symmetric relation is called an *equivalence* relation. Equivalence relations over a set S and partitions over S are closely related since there is a bijection between them. Indeed, let R be an equivalence relation over S. The set of all elements s' in S equivalent to an element s in S, i.e., such that $(s,s') \in R$, is called the class of s modulo R. The set of classes modulo R is a partition of S. Reciprocally, to any partition of S an equivalence relation on S can be associated as follows: two elements are equivalent iff they belong to the same element of the partition.

A reflexive, transitive and antisymmetric relation is called an *order* or a *partial order*. Section A.3 is devoted to order relations which are important in our approach.

A.2 Graphs

A.2.1 Directed Graphs

Definition A.2 (Directed graph). A *directed graph* is a notion closely related to a binary relation over a set since it is pair (X, U) composed of a set X and a binary relation U over X, i.e., $U \subseteq (X \times X)$. An element of X is called a *vertex* or a *node*, an element of U is called an *arc*.

An arc (x, x) is called a *loop*. If (x, y) is an arc, then y is a *successor* of x and x is a *predecessor* of y.

One very attractive feature of graphs is that they can be represented by drawings: a point is assigned to each vertex and an arc (x, y) is represented by a simple curve from x to y.

Example. Figure A.1 represents two different directed graphs:
$G = (\{a, b, c\}, \{(a, b), (b, a), (a, c), (b, c), (c, c)\})$ and
$H = (\{a, b, c, d\}, \{(a, b), (b, a), (a, c), (b, c), (c, c)\})$ which have the same arc set.

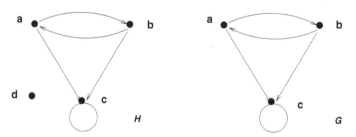

Fig. A.1 Directed graphs

A directed graph is simply called a graph in the sequel of this section. Let $G = (X, U)$ be a graph, $Y \subseteq X$ and $V \subseteq U$. The *subgraph of G induced by Y* is the graph $G(Y) = (Y, U \cap (Y \times Y))$. A *stable set* of a graph $G = (X, U)$ is a subset Y of X such that the subgraph of G induced by Y has no arcs, i.e., $U \cap (Y \times Y) = \emptyset$. A *partial graph* of a graph $G = (X, U)$ is a graph $H = (X, V)$ where $V \subseteq U$.

A *path* of a graph $G = (X, U)$ is a sequence $x_1 \ldots x_k$ of vertices of G such that $k \geq 1$ and for all $i = 1, \ldots, k-1$, (x_i, x_{i+1}) is an arc of G. An *elementary path* is a path having all its vertices distinct. A *circuit* is a sequence $x_1 \ldots x_k x_1$, such that $x_1 \ldots x_k$ is a path and (x_k, x_1) is an arc. If $x_1 \ldots x_k$ is an elementary path, then the circuit is an *elementary circuit*.

A vertex y is a *descendant* of a vertex x if there is a path from x to y, then x is an *ascendant* of y. A *root* r is a vertex which is an ascendant of all vertices, i.e. $\forall x \in X$, there is a path from r to x. An *antiroot* r is a vertex which is a descendant of all vertices, i.e. $\forall x \in X$, there is a path from x to r.

Let $G = (X, U)$ be a graph. The *transitive closure* of G is the graph $G' = (X, U')$, such that $(x, y) \in U'$ if there is a path not reduced to a vertex from x to y in G. The *reflexive closure* of G is $G' = (X, U \cup (X \times X))$. The *reflexo-transitive closure* of G is $G^* = G'^t$.

Example. Figure A.2 presents a graph G and its transitive closure H, r and r' are the roots of G.

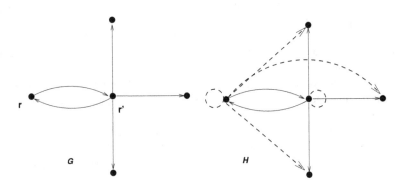

Fig. A.2 A graph G and its transitive closure H

Definition A.3 (Quotient graph). Let $G = (X, U)$ be a graph and $P = \{X_1, \ldots, X_k\}$ be a partition of X. The *quotient graph* G/P is the graph obtained from G as follows. For all $i = 1, \ldots, k$, all vertices in X_i are merged in a new vertex y_i whose successors are the vertices y_j such that a vertex in X_j is a successor of a vertex in X_i. Formally, $G/P = (\{y_1, \ldots, y_k\}, \{(y_i, y_j) \mid i \neq j, \exists x \in X_i, \exists x' \in X_j, (x, x') \in U\})$.

Example. A graph G and its quotient G/P, where $P = \{\{a, b\}, \{c, d\}, \{e, f, g\}\}$, are drawn in Fig. A.3.

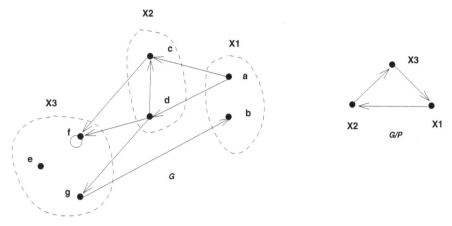

Fig. A.3 A graph G and the quotient graph G/P, $P = \{X_1, X_2, X_3\}$

A.2.2 Homomorphism

The graph homomorphism notion is one of the major mathematical notions used in the book.

Definition A.4 (Graph homomorphism). Let $G = (X, U)$ and $H = (Y, V)$ be two directed graphs. A *homomorphism* from G to H is a mapping f from X to Y such that: $\forall x, y \in X$, if $(x, y) \in U$ then $(f(x), f(y)) \in V$.

An *isomorphism* is a bijective homomorphism f such that its inverse f^{-1} is also a homomorphism.

Two isomorphic directed graphs have the same structure and only differ by the names of their vertices.

Example. In Fig. A.4, f defined as follows, $f(a) = 2$, $f(b) = 4$, $f(c) = 7$, $f(d) = 5$ and $f(e) = 6$, is an injective homomorphism from G to H.

Let H' be the subgraph of H obtained by deleting vertices 1 and 3, then f is a bijective homomorphism from G to H', which is not an isomorphism.

Let H'' be the partial subgraph of H obtained by deleting the vertices 1 and 3 and the arcs $(2, 6)$ and $(6, 7)$, then f is an isomorphism from G to H''.

g defined by: $g(a) = 4$, $g(b) = 4$, $g(c) = 6$, $g(d) = 5$ and $g(e) = 6$ is a non-injective homomorphism from G to H.

Let H''' be the subgraph of H induced by the vertices 4 and 5 (equivalently, obtained by deleting from H the vertices 1,2,3 and 6), then h defined by $h(a) = 4$, $h(b) = 4$, $h(c) = 4$, $h(d) = 5$ and $h(e) = 4$ is a surjective homomorphism from G to H'''.

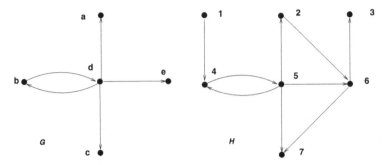

Fig. A.4 Homomorphisms

A.2.3 Different Sorts of Graphs

Definition A.5 (Undirected graph). An *undirected graph* is a pair (X, E), where X is a set, called the vertex set, and any element of E, called the edge set, is a subset of X having either two vertices or one vertex (it is then called a loop).

Paths and cycles are defined for undirected graphs in a similar same way as for directed graphs, e.g., an elementary path of an undirected graph is a sequence of distinct vertices such that two consecutive vertices define an edge.

Definition A.6 (Connected graph). An undirected graph is *connected* if there is a path between any two vertices. A *connected component* of an undirected graph G is a connected subgraph H of G, with H being maximal with this property.

Example. The connected components of the graph G in Fig. A.5 are C, D and E

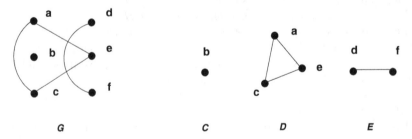

Fig. A.5 Connected components

Definition A.7 (Cut vertex). A *cut vertex (or node)* is a vertex in a graph whose deletion increases the number of connected components.

Example. In Fig. A.6, the vertex x is a cut vertex. The deletion of x yields a graph whose connected components are the graphs C, D, E in Fig. A.5.

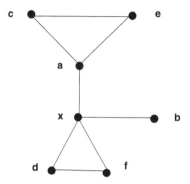

Fig. A.6 Cut vertex

Definition A.8 (Tree and rooted tree). A *tree* is an edge minimal connected graph, i.e., it is a connected graph and the deletion of any edge gives a non-connected graph. There are numerous characterizations of trees, e.g., between any pair of vertices there is one and only one path. A *rooted tree* is an arc minimal directed graph with a root, i.e., it has a root and the deletion of any arc gives a directed graph without a root.

A rooted tree T' can be obtained from a tree T by choosing a vertex r (its root) and by transforming each edge into an arc in such a way that for all x the only (undirected) path from r to x in T is transformed into a (directed) path in T'.

Whenever only directed graphs are considered, then a rooted tree is often called a tree.

Undirected graphs are equivalent to symmetric directed graphs. When dealing with a directed graph (X, U), it can be useful to consider its underlying undirected graph (X, E), which is defined as follows. Loops of E are loops of U and for $x \neq y$, $\{x, y\}$ is an edge of E if $(x, y) \in U$ or $(y, x) \in U$.

For instance, it is convenient to first define the connected component notion on undirected graphs, then to extend this notion to directed graphs: a directed graph is connected if its underlying undirected graph is connected.

Definition A.9 (Bipartite graph). A graph G is *bipartite* if the vertex set is partitioned in two subsets X and Y such that any edge has a vertex in X and the other in Y.

Bipartite graphs are basic structures underlying conceptual graphs.

Example. In Fig. A.7 G is an undirected graph, H is a tree, H' is the rooted tree obtained from H by choosing r for root, and K is a bipartite graph (all these graphs are connected).

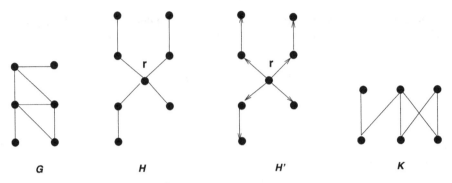

G   H   H'   K

Fig. A.7 Different sorts of graphs

Numerous problems on graphs can be broken down into subproblems on connected components, e.g., a graph is bipartite iff all its connected components are
bipartite. In Fig. A.5, G is not bipartite since D is not bipartite.

Definition A.10 (k-clique). A *k-clique* is a graph $G = (X,E)$ such that $|X| = k \geq 1$
and $E = \{\{x,y\} \mid x,y \in X, x \neq y\}$.

If there can be several arcs (or several edges in the undirected case) from x to y,
i.e. if U is no longer a set of ordered pairs but is a *family* of ordered pairs, then the
structure is called a directed (or undirected) *multigraph.*

Example. In Fig. A.8, G is a directed multigraph and H an undirected multigraph.

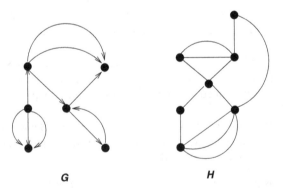

G   H

Fig. A.8 Multigraphs

A.2.4 Hypergraphs

Definition A.11 (Hypergraph).
An *hypergraph* is a pair $(X, \mathcal{E})$ such that X is a set and $\mathcal{E}$ a set of non-empty subsets of X. X is called the vertex set of the hypergraph and $\mathcal{E}$ is called its edge set.

Hypergraphs are a generalization of undirected graphs when an edge can be composed of any non-empty subset of vertices and not necessarily of a pair of vertices or a loop.

Hypergraphs can be drawn as follows: a point is assigned to each vertex and all vertices that constitute an edge are enclosed by a closed curve (Venn diagram). If some vertices belong to several edges, such a diagram can be difficult to draw (and to read) even though it has a small number of vertices.

Example.
In Fig. A.9, H represents the hypergraph $[\{a, b, c, d\}, \{\{a, b\}, \{a, c\}, \{a, b, c, d\}\})$. The hypergraph $K = (\{a, b, c, d\}, \{\{a, b\}, \{a, b, c\}, \{a, c\}, \{a, c, d\}, \{b, c\}, \{b, c, d\}, \{a, b, c, d\}\})$ is more difficult to represent by a Venn diagram.

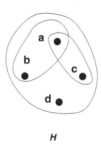

H

Fig. A.9 Hypergraphs

There are different directed hypergraph notions. A directed hypergraph is sometimes defined as an hypergraph whose edges are each split into two subsets, i.e. the positive and the negative parts of the edge. However, in the book, a *directed hypergraph* is an hypergraph whose edges are each totally ordered (see next section which deals with ordered sets).

A.3 Ordered Sets

A.3.1 Basic Notions

Definition A.12 (Order). An *ordered set* is a pair $P = (X, R)$, such that X is a set and R is a binary relation on X which is reflexive, transitive, and antisymmetric. The binary relation R is called an order (on X). A strictly ordered set is a pair (X,R), such that X is a set and R is a binary relation on X which is antireflexive (i.e., $\forall x \in X$ one has $(x,x) \notin R$), transitive, and antisymmetric. The binary relation R is called a strict order (on X).

The classical generic notation $\leq$ is used to denote an order, with $<$ denoting its associated strict order. The notation $\leq_P$ is used whenever it is necessary to specify the order P.

An ordered set is also called a *partially* ordered set by opposition to a *totally or linearly* ordered set which is an ordered set such that any two elements are comparable, i.e., $\forall x, y \in X$ one has $x \leq y$ or $y \leq x$. A totally ordered set is also called a *chain*. An *antichain* is an ordered set (X,R) restricted to the loops, i.e., such that $R = \{(x,x) | x \in X\}$. Whenever strict orders are considered, an antichain has an empty set of arcs.

Definition A.13 (Dual order). Let $P = (X,R)$ be an ordered set then $P^d = (X,R^d)$ is called the *dual* of P, where $(x,y) \in R^d$ iff $(y,x) \in R$.

The dual of $\leq$ is denoted $\geq$.

Let $P = (X,\leq)$ be an ordered set and Y a subset of X. Then Y inherits an order relation from $\leq$, precisely defined as follows. Given $x, y \in Y$, $x \leq y$ in Y iff $x \leq y$ in X. The restriction of the order relation $\leq$ on Y is still noted $\leq$ and the ordered set $Q = (Y,\leq)$ is called the *suborder* of P *induced* by Y. The following sorts of suborders of an order P are especially important for studying P: chains and antichains, and also ideals and filters.

Definition A.14 (Ideal and filter). Let $P = (X,\leq)$ be an ordered set, an *ideal* (resp. a *filter*) of P is a suborder $(Y,\leq)$ of P, such that: $\forall x \in X$ and $y \in Y$ if $x \leq y$ then $x \in Y$ (resp. $\forall x \in X$ and $y \in Y$ if $y \leq x$ then $x \in Y$).

In some situations, an order relation on X is given by a set U of comparable elements, i.e., by a directed graph on X. If this set U is hand-built, then some errors may occur, especially circuits may be constructed. Detecting circuits in a directed graph is a simple problem that can be efficiently solved (cf. e.g., [CLRS01]), and we suppose that U is without circuit.

The order relation $\leq$ induced by U is equal to U^* the reflexo-transitive closure of U, i.e., $t' \leq t$ iff $(t,t') \in U^*$, i.e., t' is a descendant of t, and t is an ascendant of t'.

At the other end, the minimum information needed to compare two elements of an ordered set is given by the *covering relation* of the order.

Definition A.15 (Covering relation). Let $(X,\leq)$ be an ordered set. The *covering relation* of $\leq$ is denoted by $\prec$ and is defined by: $t \prec t'$ iff $t < t'$, and for all t'', $t \leq t'' < t'$ implies $t = t''$. If $t \prec t'$, then t' is called a *successor* of t and t is called a *predecessor* of t'.

The covering relation U^h of an order is also called the *transitive reduction* of that order. If an order is given by a binary relation U whose transitive closure is the order, the transitive reduction can be built directly from U without computing U^*. The transitive reduction of an order on X is the binary relation on X with the minimal cardinality such that its reflexo-transitive closure is that order. One has $t \leq t'$ iff there is a path from t to t' in the transitive reduction of $\leq$. (X, U^h) is also called the Hasse graph of (X, U).

Any $V \subseteq X \times X$ such that $U^h \subseteq V \subseteq U^*$ is a representative of the order since its transitive closure is equal to the order, i.e. U^*, and its transitive reduction is equal to U^h. Thus, the set of representatives of an order is a lattice (cf. the next section) for the set inclusion.

In graph theory terminology, the transitive reduction of the strict order corresponding to a chain is an elementary path, and the strict order corresponding to an antichain is a stable set.

Theorem A.1 (Dilworth's Theorem). *Let P be an ordered set. The maximum cardinality of an antichain of P, called the* width *of P and denoted w(P), is equal to the minimum number of chains of P whose union is equal to P.*

Definition A.16 (Max and min). Let $P = (X, \leq)$ be an ordered set and $Y \subseteq X$. An element x of Y is a *maximal* element of Y if there is no $y \in Y$ such that $x < y$. Dually, $x \in Y$ is a *minimal* element of Y if there is no $y \in Y$ such that $y < x$. The set of maximal (resp. minimal) elements of P is denoted by $max(P)$(resp. $min(P)$).

Definition A.17 (Greatest and least elements). Let $P = (X, \leq)$ be an ordered set. If $x \in X$ is such that $\forall y \in X, y \leq x$, then x is called the *greatest* element (or maximum) of P (the *least* element (or minimum) is dually defined).

An ordered set does not necessarily have a greatest (resp.least) element, but if it has a greatest (resp. least) element then it is unique.

Let $P_1 = (X_1, \leq_1)$ and $P_2 = (X_2, \leq_2)$ be two ordered sets. Then, $(X_1 \times X_2, \leq)$, where $(x, y) \leq (x', y')$ iff $x \leq_1 x'$ and $y \leq_2 y'$) is an ordered set denoted by $P_1 \times P_2$ and is called the cartesian product of the two ordered sets.

Definition A.18 (Order homomorphism). Let $P = (X, \leq_P)$ and $Q = (Y, \leq_Q)$ be two ordered sets. An *order homomorphism* from P to Q is a mapping from X to Y such that $\forall x, y \in X$, if $x \leq_P y$, then $f(x) \leq_Q f(y)$. Such a mapping is also called a *monotone* mapping or an order-preserving mapping.

Note that an order homomorphism is a graph homomorphism, thus the notions concerning graph homomorphism apply to orders, e.g. an order isomorphism is a graph isomorphism.

A.3.2 Lattices

Definition A.19 (Upper and lower bound). Let $P = (X, \leq)$ be an ordered set and $Y \subseteq X$. An element u of P is an *upper bound* of Y if $\forall x \in Y$ one has $x \leq u$. If the set of upper bounds of Y has a least element, then it is called the *least upper bound* of Y and is denoted $lub(Y)$. One can dually define the notions of *lower bound* and *greatest lower bound* denoted $glb(Y)$. If X has a lub, it is the maximum of X, and if X has a glb it is the minimum of X.

Definition A.20 (Lattice). An ordered set $(X, \leq)$ is a *lattice* if any pair $\{x,y\}$ of elements of X has a lub, denoted $x \vee y$ and a glb denoted $x \wedge y$.

A.4 First Order Logic (FOL)

A.4.1 Syntax

A first order language is a set $L = \{a_1, \ldots, a_k, f_1, \ldots, f_m, p_1, \ldots, p_n\}$ in which: a_i are constant symbols, f_i are function symbols and p_i are predicate symbols. Any f_i has an arity, $arity(f_i)$, which is an integer ≥ 1, and any p_i has an arity, $arity(p_i)$, which is an integer ≥ 0

Let L be a first order language and X a set of variable symbols. The set of *terms* over L and X is inductively defined as follows:

1. a constant or a variable is a (elementary) term,
2. if f_i is an n-ary function symbol and if $t_1, \ldots, t_n$ are terms then $f_i(t_1, \ldots, t_n)$ is a term.

The well formed formulas (of FOL) can now be inductively defined.

Definition A.21 (FOL wff). The FOL *well formed formulas (wff)* are defined as follows:

1. An *atom* is either p_i if p_i is a predicate symbol of arity 0 or $p_i(t_1, \ldots, t_n)$ if p_i is a predicate symbol of arity $n \geq 1$ and if $t_1, \ldots, t_n$ are terms; an atom is a wff,
2. if A and B are wffs then $\neg A$, $(A \vee B)$, $(A \wedge B)$, $(A \rightarrow B)$ and more generally $(A * B)$ for any propositional binary connector $*$ are wffs,
3. if A is a wff and x is a variable, then $\exists x A$ and $\forall x A$ are wffs.

Syntactical notions can be inductively defined from the previous inductive definition of wffs, e.g. the syntactic tree of a wff, the depth of a wff, set of sub-wffs of a wff and so on. Here are examples.

The mapping *var* which associates to a term or to a wff the set of their variables is defined as follows.

1. *Term.* Let t be a term. If t is a constant, then $var(t) = \emptyset$; if t is a variable, then $var(t) = \{t\}$; if $t = f(t_1, \ldots, t_n)$, then $var(t) = \bigcup_{i=1}^{i=n} var(t_i)$.
2. *wff.* Let A be a wff.

   a. If A is an atom $p(t_1, \ldots, t_n)$, then $var(A) = \bigcup_{i=1}^{i=n} var(t_i)$ (if $arity(p) = 0$ then $var(p) = \emptyset$),

   b. if $A = \neg B$, then $var(A) = var(B)$,

   c. if $A = (B * C)$, then $var(A) = var(B) \cup var(C)$ for any binary propositional connector $*$,

   d. if $A = \exists xB$ or $A = \forall xB$, then $var(A) = var(B) \cup \{x\}$.

Having defined var, the mapping *free-var* (resp. *bound-var*) which associates to a wff the set of the variables with free occurrences (resp. bound occurrences) can be defined as follows:

1. If A is an atom $p(t_1, \ldots, t_n)$, then $free\text{-}var(A) = var(A)$ and $bound\text{-}var(A) = \emptyset$,
2. if $A = \neg B$, then $free\text{-}var(A) = free\text{-}var(B)$ and $bound\text{-}var(A) = bound\text{-}var(B)$,
3. if $A = (B * C)$, then $free\text{-}var(A) = free\text{-}var(B) \cup free\text{-}var(C)$ and $bound\text{-}var(A) = bound\text{-}var(B) \cup bound\text{-}var(C)$ for any binary propositional connector $*$,
4. if $A = \exists xB$ or $A = \forall xB$, then $free\text{-}var(A) = free\text{-}var(B) \setminus \{x\}$ and $bound\text{-}var(A) = bound\text{-}var(B) \cup \{x\}$

A *closed wff* is a wff A such that $free\text{-}var(A) = \emptyset$.

Let A be a non-closed wff with $free\text{-}var(A) = \{x_1, \ldots, x_k\}$. The existential closure of A is the wff $\exists x_1 \ldots \exists x_k A$. The universal closure is defined in the same way by substituting the universal quantifier for the existential quantifier.

Definition A.22 (Substitution).

A *substitution* is a set of pairs $\{(x_1, t_1), \ldots, (x_n, t_n)\}$, where the x_i are distinct variables and the t_i are terms such that for all $i = 1, \ldots, n$, $t_i \neq x_i$.

Let $\sigma = \{(x_1, t_1), \ldots, (x_n, t_n)\}$ be a substitution, t a term. The term $\sigma(t)$ is defined as follows:

1. if t is a constant, then $\sigma(t) = t$,
2. if t is a variable, then if there is $i \in \{1, \ldots, n\}$ with $t = x_i$, then $\sigma(t) = t_i$ otherwise $\sigma(t) = t$,
3. if $t = f(t_1, \ldots, t_k)$, then $\sigma(t) = f(\sigma(t_1), \ldots, \sigma(t_k))$.

Let A be a wff, the wff $\sigma(A)$ is defined as follows:

1. if $A = p(t_1, \ldots, t_k)$, then $\sigma(A) = p(\sigma(t_1), \ldots, \sigma(t_k))$,
2. if $A = \neg B$, then $\sigma(A) = \neg \sigma(B)$,
3. if $A = (B * C)$, then $\sigma(A) = (\sigma(B) * \sigma(C))$,
4. if $A = \exists xB$ or $A = \forall xB$, then if there is $i \in \{1, \ldots, n\}$ with $x = x_i$ then $\sigma(A) = A$ otherwise $\sigma(A) = \exists x \sigma(B)$ or $\sigma(A) = \forall x \sigma(B)$.

The *substitution product* is defined as follows. Let $\alpha = \{(x_i, u_i)\}_i$ and $\beta = \{(y_j, v_j)\}_j$ be two substitutions, $\beta \circ \alpha$ is the substitution equal to $\{(x, \beta(\alpha(x))) \mid x \in \{x_i\}_i \cup \{y_j\}_j, x \neq \beta(\alpha(x))\}$.

A *unifier* of two terms or of two atoms q and q' is a substitution σ such that $\sigma(q) = \sigma(q')$. A *most general unifier* of q and q', denoted by $mgu(q, q')$, is a unifier of q and q' such that for any unifier σ' of q and q' there is a substitution α such that $\sigma' = \alpha \times \sigma$.

A.4.2 Semantics and Models

Definition A.23 (Model). Let L be a first order language $L = (a_1, a_2, \ldots, a_k, f_1, f_2, \ldots, f_m, p_1, p_2, \ldots, p_n)$. A *model* I, also called an *interpretation*, of L is a relational structure:
$I = (D, I(a_1), I(a_2), \ldots, I(a_k), I(f_1), I(f_2), \ldots, I(f_m), I(p_1), I(p_2), \ldots, I(p_n))$, such that:

- D is a non-empty set called the domain of I,
- for all $i = 1, \ldots, k$, $I(a_i)$ is an element of D,
- for all $i = 1, \ldots, m$, $I(f_i)$ is a mapping from $D^{arity(f_i)}$ to D,
- for all $i = 1, \ldots, n$, $I(p_i)$ is a $arity(p_i)$-ary relation on D.

Let t be a term on a language L, I a model of L and s an *assignation* of the variables of t, i.e., a mapping from $var(t)$ to D the domain of I. The *value of t for I and s*, denoted $v(t, I, s)$, is an element of D inductively defined as follows:

1. if t is a constant, then $v(t, I, s) = I(t)$,
2. if t is a variable, then $v(t, I, s) = s(t)$,
3. if $t = f(t_1, \ldots, t_k)$, then $v(t, I, s) = I(f)(v(t_1, I, s), \ldots, v(t_k, I, s))$.

Let A be a wff on a language L, I a model of L and s an *assignation* of the variables of A, i.e. a mapping from $var(A)$ to D the domain of I. The *value of A for I and s*, denoted $v(A, I, s)$, is inductively defined as follows:

1. let A be an atom $p(t_1, \ldots, t_k)$, $v(A, I, s) = true$ iff $(v(t_1, I, s), \ldots, v(t_k, I, s)) \in I(p)$,
2. $v(\neg A, I, s) = true$ iff $v(A, I, s) = false$,
3. $v((A \wedge B), I, s) = true$ iff $v(A, I, s) = v(B, I, s) = true$,
4. $v(\exists x A, I, s) = true$ iff for some $d \in D$ $v(A, I, s(x, d)) = true)$, where $s(x, d)$ is the assignment s' s.t. for any $y \neq x$ $s'(y) = s(y)$ and $s'(x) = d$.

Values for other propositional connectives or for the universal quantifier are easily derived from the previous definition.

A wff A is *satisfied* by a model I and an assignment s if $v(A, I, s) = true$, which is also denoted $(I, s) \models A$.

A wff is *satisfiable* if it is satisfied by a model and it is *unsatisfiable* otherwise. A wff is *valid* if it is satisfied by any model and assignment.

Two wffs A and B are *equivalent* if they have the same value for all I and all s, i.e. $(I, s) \models A$ iff $(I, s) \models B$.

A wff C is a *logical consequence* of a set of wffs $H_1, \ldots, H_k$ if for all I and s such that for all $i = 1, \ldots, k$ $(I, s) \models H_i$, then $(I, s) \models C$. The fact that C is a logical consequence of $H_1, \ldots, H_k$ is denoted $H_1, \ldots, H_k \models C$.

Theorem A.2. *The following three statements are equivalent:*

1. $H_1, \ldots, H_k \models C$.
2. $(H_1 \wedge \ldots \wedge H_k \rightarrow C)$ *is a valid wff.*
3. $(H_1 \wedge \ldots \wedge H_k \wedge \neg C)$ *is an unsatisfiable wff.*

A.4.3 Clausal Form

A *literal* is either an atom (or positive literal) or the negation of an atom (or negative literal). A *clause* is the universal closure of a disjunction of literals. A clause can be identified to a set of literals.

Any wff is equivalent to a wff in *prenex* conjunctive form which is composed of a sequence of quantifiers (existential and universal) preceding a conjunction of disjunctions of literals, i.e.

$Q_1 x_1 \ldots Q_k x_k (L_{11} \vee \ldots \vee L_{1n_1}) \wedge \ldots \wedge (L_{u1} \vee \ldots \vee L_{un_u})$, where any Q_i is either $\exists$ or $\forall$.

The *skolemization* of a prenex wff A consists of successively, from left to right, deleting an existential quantifier, say $\exists x_i$, and applying the substitution $(x_i, f(x_1, \ldots, x_r))$, where f is a new n-ary function symbol, and $x_1, \ldots, x_r$ are the universally quantified variables preceding x_i. The wff obtained is a *skolem form* of A.

A wff is in *clausal form* if it is the conjunction of a set of clauses, therefore skolemization transforms a wff into a clausal form.

Theorem A.3. *A wff A is unsatisfiable iff any skolem form of A is unsatisfiable.*

Example. Let $A = \exists x \exists y (person(x) \wedge boy(y) \wedge father(x,y))$ be a wff on a language L. Skolemization of A gives the wff: $skol(A) = person(a) \wedge boy(b) \wedge father(a,b)$ where a, b are two constants which do not belong to L.

Let $A = \forall x \exists y (\neg person(x) \vee mother(y,x))$ be a wff on a language L. $skol(A) = \forall x (\neg person(x) \vee mother(f(x),x))$ where f is a unary symbol function $\notin L$.

A.4.4 Resolution and SLD-Resolution

Let C and C' be two clauses with no variables in common. If $p(U_1), \ldots, p(U_k)$ is a set of literals of C and $\neg p(V_1), \ldots, \neg p(V_l)$ is a set of literals of C' (where U_i and V_j are term vectors of length $arity(p)$) and σ is a most general unifier of $p(U_1), \ldots, p(U_k), p(V_1), \ldots, p(V_l)$, then applying the *resolution rule* to C, C' and σ consists of building the clause:

$\{\sigma(C) \setminus \{\sigma(p(U_1)), \ldots, \sigma(p(U_k))\}\} \cup \{\sigma(C') \setminus \{\sigma(\neg p(V_1)), \ldots, \sigma(p(V_l))\}\}$.

Theorem A.4 (Soundness and completeness of the resolution method). *A clausal form is unsatisfiable iff the empty clause can be derived from the initial clauses by a finite number of applications of the resolution rule.*

Different strategies have been proposed to apply the resolution rule. Especially, whenever all the clauses are *Horn clauses*, i.e., clauses with at most one literal that is an atom, and the other are negation of atoms, then the SLD-resolution can be applied. This method is described using the logic programming language reviewed hereafter.

A Horn clause $C = H \vee \neg B_1 \ldots \vee \neg B_k$, where H is the only atom of C is denoted by a rule $H \leftarrow B_1, \ldots, B_k$. H is called the head of the rule, the body of the rule is the sequence of atoms after the left arrow. A Horn clause restricted to an atom A is called a *fact* and is denoted $A \leftarrow$. A set of rules and facts is called a *logic program*. A *goal* is a clause C with no atom, i.e. $C = \neg G_1 \ldots \vee \neg G_k$ and is denoted $\leftarrow G_1, \ldots, G_k$.

Let us consider a logic program P and a goal $G = \leftarrow G_1, \ldots, G_k$. The *SLD-resolution* is defined as a sequence of steps transforming the goal using the facts and rules of P as follows. The first goal is $G_1 = G$. Assume that the goal $G_i = H_1, \ldots, H_l$ has been derived. A SLD-resolution step consists of :

- select a subgoal H_i,
- choose a fact $H \leftarrow$ or a rule $H \leftarrow B_1, \ldots, B_k$ of P,
- compute, if any, a most general unifier σ of H_i and H,
- replace G_i by the new goal:
  $G_{i+1} = \sigma(H_1, \ldots, H_{i-1}, B_1, \ldots, B_k, H_{i+1}, \ldots, H_l)$ (note that if a fact is considered, then the subgoal H_i is deleted).

Theorem A.5 (Soundness and completeness of the SLD-resolution for logic programs). *Let P be a logic program and $G = \leftarrow G_1, \ldots, G_k$ a goal. $G_1 \wedge \ldots \wedge G_k$ is a logical consequence of P iff the empty goal can be derived from P and G by the SLD-resolution method.*

A.5 Algorithm and Problem Complexity

Problem decidability gives results about the solvability of a given decision problem (a problem which can be stated as a question whose answer is YES or NO). For instance, deduction in FOL is undecidable nevertheless semi-decidable. This means that there is no algorithm which stops for any input wff with the answer YES if the wff is valid and with the answer NO if the wff is not valid, but there are algorithms which stop with the answer YES given any valid wff (and if the input wff is not valid they may not stop). Knowing that a problem is decidable does not imply that this problem can be efficiently solved.

Complexity theory provides tools to assess the difficulty of a problem and the efficiency of an algorithm. Let us present the basic notions through an example (and cf. [CLRS01] for a thorough introduction to the domain). Let us consider the homomorphism problem, denoted GRAPH-HOMOMORPHISM between (finite) graphs. It is presented as follows:

GRAPH-HOMOMORPHISM
Instance: two (finite) graphs G and H.
Question: Is there a homomorphism from G to H?

The *size* of a problem is an integer which is a measure of the number of characters of a reasonable (not unary) encoding of a problem instance. The sum of numbers of nodes and vertices of the two graphs is a usual size for the previous example.

The *time complexity* of an algorithm (solving a given problem) is expressed as a function giving the number of (elementary) operations run by the algorithm in terms of the size of the instance. This function is generally an asymptotic function (when the instance size increases) in the worst case instance.

The *O-notation* is classicaly used for expressing this asymptotic worst-case complexity. Let f be a function of a positive integer parameter n. Saying that an algorithm A for solving a problem P has a time (resp. size) complexity in $O(f(n))$ means this: there is a positive constant C and there is a positive integer n_0 such that the number of elementary operations (resp. memory size) of A needed for solving any instance (of P) whose size is $\geq n_0$ is $\leq Cf(n)$.

A problem is in the complexity class **P** if there is a deterministic algorithm, for solving this problem, which has a polynomial time complexity. A problem is in the complexity class **NP** if there is a non-deterministic algorithm, for solving this problem, which has polynomial time complexity. **P$\neq$NP** is one of the most important conjectures in Computer Science. The most difficult problems in the **NP** class are called *NP-complete*. If we can show that a problem is NP-complete, then we know that it is very unlikely that we can solve all instances of this problem in reasonable time.

It is well-known that typical AI problems are computationally difficult. Hence, it might seem worthless to analyze the computational complexity of these problems. Nevertheless, complexity analysis provides insight into the possible sources of complexity and can be used to direct the search for efficient methods to solve the problem (e.g., good heuristics or polynomial algorithms for particular cases).

References

[AB94]   K. R. Apt and R. N. Bol. Logic programming and negation: A survey. *Journal of Logic Programming*, 19/20:9–71, 1994.

[ABC06]   J.-P. Aubert, J.-F. Baget, and M. Chein. Simple Conceptual Graphs and Simple Concept Graphs. In H. Schärfe et al, editor, *Proc. ICCS'06*, volume 4068 of *LNAI*, pages 87–101. Springer, 2006.

[ACP87]   S. Arnborg, D.G. Corneil, and Proskurowski. Complexity of finding embeddings in a k-tree. *SIAM Journal of Algebraic and Discrete Methods*, 8:277–284, 1987.

[AHV95]   S. Abiteboul, R. Hull, and V. Vianu. *Foundations of Databases*. Addison-Wesley, 1995.

[Ann01]   Annotea. Annotea Project. http://www.w3.org/2001/Annotea/, 2001.

[AvBN96]   H. Andréka, J. van Benthem, and I. Németi. Modal languages and bounded fragments of FOL. Research Report ML-96-03, Univ. of Amsterdam, 1996.

[BA01]   M. Ben-Ari. *Mathematical Logic for Computer Science*. Springer, 2001.

[Baa99]   F. Baader. Logic-based knowledge representation. In M.J. Wooldridge and M. Veloso, editors, *Artificial Intelligence Today, Recent Trends and Developments*, number 1600 in Lecture Notes in Computer Science, pages 13–41. Springer Verlag, 1999.

[Bag99]   J.-F. Baget. A Simulation of Co-Identity with Rules in Simple and Nested Graphs. In *Proc. of ICCS'99*, volume 1640 of *LNAI*, pages 442–455. Springer, 1999.

[Bag01]   J.-F. Baget. *Représenter des connaissances et raisonner avec des hypergraphes: de la projection à la dérivation sous contraintes*. PhD thesis, Université Montpellier II, Nov. 2001.

[Bag03]   J.-F. Baget. Simple Conceptual Graphs Revisited: Hypergraphs and Conjunctive Types for Efficient Projection Algorithms. In *Proc. of ICCS'03*, volume 2746 of *LNAI*. Springer, 2003.

[Bag04]   J.-F. Baget. Improving the forward chaining algorithm for conceptual graphs rules. In *KR'04*, pages 407–414. AAAI Press, 2004.

[Bag05]   J. F. Baget. RDF Entailment as a Graph Homomorphism. In *Proc. of ISWC'05*, 2005.

[BBV97]   C. Bos, B. Botella, and P. Vanheeghe. Modeling and Simulating Human Behaviors with Conceptual Graphs. In *Proc. of ICCS'97*, volume 1257 of *LNAI*, pages 275–289. Springer, 1997.

[BC81]   P. A. Bernstein and D.-M. W. Chiu. Using semi-joins to solve relational queries. *J. ACM*, 28(1):25–40, 1981.

[BCM+03]   F. Baader, D. Calvanese, D. L. McGuinness, D. Nardi, and P. F. Patel-Schneider, editors. *The Description Logic Handbook: Theory, Implementation, and Applications*. Cambridge University Press, 2003.

[BCvBW02] F. Bacchus, X. Chen, P. van Beek, and T. Walsh. Binary vs. non-binary constraints. *Artif. Intell.*, 140(1/2):1–37, 2002.

[BFMY83] C. Beeri, R. Fagin, D. Maier, and M. Yannakakis. On the desirability of acyclic database schemes. *J. ACM*, 30(3):479–513, 1983.

[BFR95] C. Bessière, E. C. Freuder, and J.-C. Régin. Using inference to reduce arc consistency computation. In *Proc. IJCAI'95*, pages 592–599, 1995.

[BG81] P. A. Bernstein and N. Goodman. Power of natural semijoins. *SIAM J. Comput.*, 10(4):751–771, 1981.

[BGM99a] J.-F. Baget, D. Genest, and M.-L. Mugnier. Knowledge acquisition with a pure graph-based knowledge representation model—application to the Sisyphus-1 case study. In *Twelfth Workshop on Knowledge Acquisition, Modeling and Management (KAW '99)*, 1999.

[BGM99b] J.-F. Baget, D. Genest, and M.-L. Mugnier. A pure graph-based solution to the SCG-1 initiative. In *Proc. ICCS'99*, volume 1640 of *LNAI*, pages 355–376, 1999.

[BHP+92] C. Beierle, U. Hedtstück, U. Pletat, P. H. Schmitt, and J. H. Siekmann. An order-sorted logic for knowledge representation systems. *Artif. Intell.*, 55(2):149–191, 1992.

[Bid91] N. Bidoit. Negation in rule-based database languages: a survey. *Theoretical Computer Science*, 78(1):3–83, 1991.

[BK99] F. Baader and R. Küsters. Matching in description logics with existential restrictions. In *Proc. of DL'99*, volume 22 of *CEUR Workshop Proceedings*. CEUR-WS.org, 1999.

[BKM99] F. Baader, R. Küsters, and R. Molitor. Computing least common subsumers in description logics with existential restrictions. In *Proc. of IJCAI'99*, pages 96–103. Morgan Kaufmann, 1999.

[BL01] T. Berners-Lee. Conceptual Graphs and the Semantic Web. http://www.w3.org/DesignIssues/CG.html, 2001.

[BL04] R.J. Brachman and H.J. Levesque. *Knowledge, Representation and Reasoning*. Elsevier, 2004.

[BM02] J.-F. Baget and M.-L. Mugnier. The Complexity of Rules and Constraints. *J. of Artif. Intell. Research (JAIR)*, 16:425–465, 2002.

[BMFL02] C. Bessière, P. Meseguer, E. C. Freuder, and J. Larrosa. On forward checking for non-binary constraint satisfaction. *Artif. Intell.*, 141(1/2):205–224, 2002.

[BMS98] F. Baader, R. Molitor, and S.Tobies. The guarded fragment of conceptual graphs. Research Report 98-10, LTCS, 1998.

[BMT99] F. Baader, R. Molitor, and S. Tobies. Tractable and Decidable Fragments of Conceptual Graphs. In *Proc. of ICCS'99*, volume 1640 of *LNAI*, pages 480–493. Springer, 1999.

[Bod] H. L. Bodlaender. TreewidthLIB: A benchmark for algorithms for treewidth and related graph problems. http://people.cs.uu.nl/hansb/treewidthlib/index.php.

[Bod93] H.L. Bodlaender. A tourist guide through treewidth. *Acta Cybernetica*, 11(1–2):1–21, 1993.

[Bod96] H.L. Bodlaender. A linear-time algorithm for finding tree-decompositions of small treewidth. *SIAM Journal of Computing*, 25:1305–1317, 1996.

[Bod98] H. L. Bodlaender. A partial arboretum of graphs with bounded treewidth. *Theoretical Compututer Science*, 209(1-2):1–45, 1998.

[Bor96] A. Borgida. On the relative expressiveness of description logics and predicate logics. *Artif. Intell.*, 82:353–367, 1996.

[Bou71] A. Bouchet. Etude combinatoire des ordonnés finis. Thèse de doctorat d'Etat, Université de Grenoble, 1971.

[BP83] J. Barwise and J. Perry. *Situations and Attitudes*. MIT Press, Cambridge, MA, 1983.

[Bra77] R. Brachman. What's in a concept: Structural foundations for semantic networks. *International Journal Of Man Machine Studies*, 9:127–152, 1977.

[BS85] R.J. Brachman and J.G. Schmolze. An overview of the kl-one knowledge representation system. *Cognitive Science*, 9(2):171–216, 1985.

[BS06] J.-F. Baget and E. Salvat. Rules dependencies in backward chaining of conceptual graphs rules. In H. Schärfe et al, editor, *Proc. ICCS'06*, volume 4068 of *LNAI*, pages 102–116. Springer, 2006.

[BT01]   J.-F. Baget and Y. Tognetti. Backtracking through biconnected components of a constraint graph. In *Proc. IJCAI'01*, volume 1, pages 291–296, 2001.

[BV84]   C. Beeri and M.Y. Vardi. A proof procedure for data dependencies. *Journal of the ACM*, 31(4):718–741, 1984.

[CC06]   M. Croitoru and E. Compatangelo. A tree decomposition algorithm for conceptual graph projection. In *Proc. of KR'06*, pages 271–276. AAAI Press, 2006.

[CCW97]   T. H. Cao, P. N. Creasy, and V. Wuwongse. Fuzzy unification and resolution proof procedure for fuzzy conceptual graph programs. In D. Lukose et al, editor, *Proc. ICCS'97*, volume 1257 of *LNAI*, pages 386–400. Springer, 1997.

[CDH00]   O. Corby, R. Dieng, and C. Hebert. A Conceptual Graph Model for W3C RDF. In *Proceedings of ICCS00*, volume 1867 of *LNAI*, pages 468–482. Springer, 2000.

[CF98]   P. Coupey and C. Faron. Towards correspondence between conceptual graphs and description logics. In *ICCS*, volume 1453 of *LNCS*, pages 165–178. Springer, 1998.

[CG95]   O. Cogis and O. Guinaldo. A linear descriptor for conceptual graphs and a class for polynomial isomorphism test. In *Proc. of ICCS'95*, volume 954 of *LNAI*, pages 263–277. Springer, 1995.

[CGL+05]   D. Calvanese, G. De Giacomo, D. Lembo, M. Lenzerini, and R. Rosati. DL-Lite: Tractable description logics for ontologies. In *AAAI*, pages 602–607, 2005.

[CGL+07]   D. Calvanese, G. De Giacomo, D. Lembo, M. Lenzerini, and R. Rosati. Tractable reasoning and efficient query answering in description logics: The DL-Lite family. *J. Autom. Reasoning*, 39(3):385–429, 2007.

[CH93]   B. Carbonneill and O. Haemmerlé. Implementing a CG Platform for Question/Answer and Database capabilities. In *Proc. 2nd International Workshop on Peirce*, pages 29–32, 1993.

[CH94]   B. Carbonneill and O. Haemmerlé. Rock: Un système question/réponse fondé sur le formalisme des graphes conceptuels. In *Actes du 9-ième congres RFIA*, pages 159–169, 1994.

[CLRS01]   T. A. Cormen, C. E. Leiserson, R. L. Rivest, and C. Stein. *Introduction to Algorithms*. The MIT Press, Cambridge, Ma, USA, 2001.

[CLS07]   Information technology—common logic (cl): a framework for a family of logic-based languages. ISO/IEC 24707:2007, 2007.

[CM92]   M. Chein and M.-L. Mugnier. Conceptual Graphs: Fundamental Notions. *Revue d'Intelligence Artificielle*, 6(4):365–406, 1992.

[CM95]   M. Chein and M.-L. Mugnier. Conceptual Graphs are also Graphs. Research Report RR-LIRMM 95-003, LIRMM, 1995. Available at http://www.lirmm.fr/~mugnier/.

[CM97]   M. Chein and M.-L. Mugnier. Positive Nested Conceptual Graphs. In D. Lukose et al, editor, *Proc. ICCS'97*, volume 1257 of *LNAI*, pages 95–109. Springer, 1997.

[CM04]   M. Chein and M.-L. Mugnier. Concept types and coreference in simple conceptual graphs. In K. E. Wolff et al, editor, *Proc. ICCS'04*, volume 3127 of *LNAI*, pages 303–318. Springer, 2004.

[CMS98]   M. Chein, M.-L. Mugnier, and G. Simonet. Nested Graphs: A Graph-based Knowledge Representation Model with FOL Semantics. In *Proc. of KR'98*, pages 524–534. Morgan Kaufmann, 1998. Revised version available at http://www.lirmm.fr/~mugnier/.

[cog01a]   CoGITaNT, 2001. http://cogitant.sourceforge.net/.

[cog01b]   COGUI, 2001. http://www.lirmm.fr/cogui/.

[CR97]   Ch. Chekuri and A. Rajaraman. Conjunctive Query Containment Revisited. In *Proc. ICDT'97, LNCS, vol. 1186*, pages 56–70. Springer, 1997.

[CS98]   S. Coulondre and E. Salvat. Piece Resolution: Towards Larger Perspectives. In *Proc. of ICCS'98*, volume 1453 of *LNAI*, pages 179–193. Springer, 1998.

[CW87]   D. Coppersmith and S. Winograd. Matrix multiplication via arithmetic progression. In *Proceedings of* 19-th Ann. Symp. on the Theory of Computation, pages 1–6, 1987.

[Dau02]   F. Dau. *The Logic System of Concept Graphs with Negation And Its Relationship to Predicate Logic*. PhD thesis, Tech. Univer. Darmstadt, 2002.

[Dau03a]   F. Dau. Concept graphs without negations: Standardmodels and standardgraphs. In *Proc. ICCS'03*, volume 2746 of *LNAI*. Springer, 2003.

[Dau03b]   F. Dau. *The Logic System of Concept Graphs with Negation And Its Relationship to Predicate Logic*, volume 2892 of *LNCS*. Springer, 2003.

[Dec03]   R. Dechter. *Constraint Processing*. Morgan Kaufmann Publishers Inc., San Francisco, CA, USA, 2003.

[DHL98]   J. Dibie, O. Haemmerlé, and S. Loiseau. A Semantic Validation of Conceptual Graphs. In *Proc. of ICCS'98*, volume 1453 of *LNAI*, pages 80–93. Springer, 1998.

[Dic94]   J.P. Dick. Using Contexts to Represent Text. In *Proc. ICCS'94*, volume 835 of *LNAI*, pages 196–213, 1994.

[DP02]   B.A. Davey and H.A. Priestley. *Introduction to Lattices and Order*. Cambridge University Press, Cambridge, UK, 2002.

[ELR94]   G. Ellis, R. Levinson, and P. J. Robinson. Managing Complex Objects in Peirce. *Int. J. Hum.-Comput. Stud.*, 41(1-2):109–148, 1994.

[FFG01]   J. Flum, M. Frick, and M. Grohe. Query evaluation via tree decomposition. *ICDT'01*, 1973:22–38, 2001.

[FFG02]   J. Flum, M. Frick, and M. Grohe. Query evaluation via tree-decompositions. *J. ACM*, 49(6):716–752, 2002.

[Fit69]   M. C. Fitting. *Intuitionistic Logic, Model Theory and Forcing*. North Holland, Amsterdam, 1969.

[FLDC86]   J. Fargues, M.-C. Landau, A. Dugourd, and L. Catach. Conceptual graphs for semantic and information processing. *IBM Journal of Research and Development*, 30(1):70–79, 1986.

[FLS96]   H. Fargier, J. Lang, and T. Schiex. Mixed constraint satisfaction: a framework for decision problems under incomplete knowledge. In *Proc. of AAAI'96*, pages 175–180, 1996.

[FNTU07]   C. Farré, W. Nutt, E. Teniente, and T. Urpí. Containment of conjunctive queries over databases with null values. In *ICDT 2007*, volume 4353 of *LNCS*, pages 389–403. Springer, 2007.

[Fuh00]   N. Fuhr. Models in information retrieval. In G. Pasi M. Agosti, F. Crestani, editor, *Lecture on Information Retrieval*, volume 1980 of *Lecture Notes in Computer Science*, pages 21–50. Springer, 2000.

[FV93]   T. Feder and M.Y. Vardi. Monotone Monadic SNP and Constraint Satisfaction. In *Proc. the 25th ACM STOC*, pages 612–622, 1993.

[Gal85]   J.H. Gallier. *Logic for Computer Science*. Wiley, 1985.

[GC97]   D. Genest and M. Chein. An experiment in document retrieval using conceptual graphs. In D. Lukose et al, editor, *Proc. ICCS'97*, volume 1257 of *LNAI*, pages 489–504. Springer, 1997.

[GC05]   D. Genest and M. Chein. A content-search information retrieval process based on conceptual graphs. *Knowledge and Information Systems (KAIS)*, 8:292–309, 2005.

[Gen00]   D. Genest. *Extension du modèle des graphes conceptuels pour la recherche d'informations*. PhD thesis, Université Montpellier II, Dec. 2000.

[GHP95]   P. Galinier, M. Habib, and C. Paul. Chordal graphs and their clique graphs. In *Proc. Workshop on Graph-Theoretic Concepts in Computer Science*, pages 358–371, 1995.

[GJ79]   M. R. Garey and D. S. Johnson. *Computers and Intractability: a Guide to the Theory of NP-Completeness*. W.H. Freeman, 1979.

[GLS99]   G. Gottlob, N. Leone, and F. Scarcello. Hypertree decompositions and tractable queries. In *Proc. PODS'99*, pages 21–32, 1999.

[GLS01]   G. Gottlob, N. Leone, and F. Scarcello. Robbers, marshals and guards: Game-theoretic and logical characterizations of hypertree width. *PODS'01*, 2001.

[Grä99]   E. Grädel. On the restraining power of guards. *Journal of Symbolic Logic*, 64(4):1719–1742, 1999.

[Guh91]   R.V. Guha. Contexts: a formalization and some applications. PhD Thesis, Stanford University, 1991.

[GW95]   B. C. Ghosh and V. Wuwongse. A Direct Proof Procedure for Definite Concep-
           tual Graphs Programs. In *Proc. of ICCS'95*, volume 954 of *LNAI*, pages 158–172.
           Springer, 1995.
[GW99]   B. Ganter and R. Wille. *Formal Concept Analysis*. Springer-Verlag, 1999.
[Hay04]   P. Hayes. Rdf semantics. w3c recommendation. http://www.w3.org/TR/2004/REC-
           rdf-mt-20040210/, 2004.
[HE80]   R. M. Haralick and G. L. Elliott. Increasing tree search efficiency for constraint
           satisfaction problems. *Artif. Intell.*, 14(3):263–313, 1980.
[HHN95]   M. Habib, M. Huchard, and L. Nourine. Embedding partially ordered sets into chain-
           products. In *Proc. of the International KRUSE Symposium (KRUSE'95*, pages 147–
           161. University of Santa-Cruz, 1995.
[HM01]   G. Hahn and G. MacGillivray. Graph homomorphisms: Computational aspects and
           infinite graphs. Draft, 2001.
[HMR93]   M. Habib, M. Morvan, and X. Rampon. On the calculation of transitive reduction-
           closure of orders. *Discrete mathematics*, 111:289–303, 1993.
[HN90]   P. Hell and J. Nešetřil. On the complexity of H-coloring. *J. Combin. Th.*, (B)(48):33–
           42, 1990.
[HN04]   P. Hell and J. Nesetril. *Graphs and Homomorphisms*, volume 121. Oxford University
           Press, 2004.
[HNR97]   M. Habib, L. Nourine, and O. Raynaud. A new lattice-based heuristic for taxonomy
           encoding. In *Proc. of the International KRUSE Symposium (KRUSE'97)*, pages 60–
           71. University of Vancouver, 1997.
[HNZ94]   P. Hell, J. Nešetřil, and X. Zhu. Duality and Polynomial Testing of Tree Homomor-
           phisms. Manuscript, 1994.
[HS03a]   S. Handschuh and S. Staab, editors. *Annotation for the Semantic Web*, volume 96
           of *Frontiers in Artificial Intelligence and Applications*. IOS Press, Amsterdam, The
           Netherlands, 2003.
[HS03b]   S. Handschuh and S. Staab. Cream—creating metadata for the semantic web. *Com-
           puter Networks*, 42(1):579–598, 2003.
[HT97]   G. Hahn and C. Tardif. Graph homomorphism: structure and symmetry. In G. Hahn
           and G.Sabidussi, editors, *Graph Symmetry*, pages 107–166. Kluwer, 1997.
[Jac88]   M. K. Jackman. Inference and the Conceptual Graph Knowledge Representation
           Language. In *Research and Development in Expert Systems IV*, 1988.
[Jea98]   P. G. Jeavons. On the algebraic structure of combinatorial problems. *Theoretical
           Computer Science (TCS)*, 200:185–204, 1998.
[JN98]   F. Lepage J.Y. Nie. Toward a broader logical model for information retrieval, 1998.
[KBvH01]   A.M.C.A. Koster, H.L. Bodlaender, and S.P.M. van Hoesel. Treewidth: Computa-
           tional experiments. Draft, 2001.
[Ker01]   G. Kerdiles. *Saying it with Pictures: a logical landscape of conceptual graphs*. PhD
           thesis, Univ. Amsterdam, Nov. 2001.
[Kli05]   Julia Klinger. Local negation in concept graphs. In *Proc. ICCS'05*, volume 3596 of
           LNAI, pages 209–222, 2005.
[KV98]   P. G. Kolaitis and M. Y. Vardi. Conjunctive-Query Containment and Constraint Sat-
           isfaction. In *Proc.PODS'98*, pages 205–213, 1998.
[KV00]   P. G. Kolaitis and M. Y. Vardi. Conjunctive-Query Containment and Constraint Sat-
           isfaction. *Journal of Computer and System Sciences*, 61:302–332, 2000.
[LB94]   M. Liquière and O. Brissac. A class of conceptual graphs with polynomial isoprojec-
           tion. In *Suppl. Proc. ICCS'94*, 1994. Revised version.
[LE92]   R. Levinson and G. Ellis. Multilevel hierarchical retrieval. *Knowl.-Based Syst.*,
           5(3):233–244, 1992.
[Lec95]   M. Leclère. *Les connaissances du niveau terminologique du modele des graphes
           conceptuels: construction et exploitation*. PhD thesis, Université Montpellier II, Dec.
           1995.
[Lec97]   M. Leclère. Reasoning with type definitions. In *Proc. ICCS'97*, volume 1257 of
           LNAI, pages 401–415. Springer, 1997.

[Lec98]   M. Leclère. Raisonner avec des définitions de types dans le modèle des graphes conceptuels. *Revue d'Intelligence Artificielle*, 12:243–278, 1998.

[Leh92]   F. Lehman. *Semantics Networks in Artificial Intelligence*. Pergamon Press, 1992.

[LM06]   M. Leclère and M.-L. Mugnier. Simple conceptual graphs with atomic negation and difference. In H. Schärfe et al, editor, *Proc. ICCS'06*, volume 4068 of *LNAI*, pages 331–345, 2006.

[LM07]   M. Leclère and M.-L. Mugnier. Some algorithmic improvements for the containment problem of conjunctive queries with negation. In *Proc. ICDT'07*, volume 4353 of *LNCS*, pages 404–418. Springer, 2007.

[LR96]   A. Y. Levy and M.-C. Rousset. Verification of Knowledge Bases Based on Containment Checking. In *Proc. of AAAI'96*, pages 585–591, 1996.

[LR98]   A. Y. Levy and M.-C. Rousset. Verification of Knowledge Bases Based on Containment Checking. *Artificial Intelligence*, 101(1-2):227–250, 1998.

[LS98]   M. Liquière and J. Sallantin. Structural machine learning with galois lattice and graphs. In *Proc. ICML'98*, pages 305–313, 1998.

[Mac77]   A. K. Mackworth. Consistency in networks of relations. *Artif. Intell.*, 8(1):99–118, 1977.

[Mai83]   D. Maier. *The Theory of Relational Databases*. Computer Science Press, 1983.

[Mar97]   P. Martin. CGKAT: a knowledge acquisition tool and an information retrieval tool which exploits structured documents, conceptual graphs and ontologies. In *Proc. CGTOOLS'97*. Univ. of Washington, 1997.

[MC92]   M.L. Mugnier and M. Chein. Polynomial algorithms for projection and matching. In H. D. Pfeiffer and T. E. Nagle, editors, *Proc. the 7th Workshop on Conceptual Graphs*, volume 754 of *LNAI*, pages 49–58, New Mexico State University, Las Cruces, New Mexico, 1992. Springer.

[MC93]   M.L. Mugnier and M. Chein. Characterization and Algorithmic Recognition of Canonical Conceptual Graphs. In *Proc. ICCS'93*, volume 699 of *LNAI*, pages 294–311. Springer Verlag, 1993.

[MC96]   M.-L. Mugnier and M. Chein. Représenter des connaissances et raisonner avec des graphes. *Revue d'Intelligence Artificielle*, 10(1):7–56, 1996.

[McC93]   J. McCarthy. Notes on Formalizing Contexts. In *Proc. IJCAI'93*, pages 555–560, 1993.

[ME99]   P. Martin and P. Eklund. Embedding Knowledge in Web Documents. In *Proc. of the 8th Int. World Wide Web Conference (WWW8)*, pages 1403–1419, 1999.

[Min75]   M. Minsky. A framework for representing knowledge. In P. Winston, editor, *The Psychology of Computer Vision*. McGraw-Hill, New York, 1975.

[ML07]   M.-L. Mugnier and M. Leclère. On querying simple conceptual graphs with negation. *Data Knowl. Eng.*, 60(3):468–493, 2007.

[MLCG07] N. Moreau, M. Leclère, M. Chein, and A. Gutierrez. Formal and Graphical Annotations for Digital Objects. In *Proc. of International Workshop on Semantically Aware Document Processing and Indexing (SADPI'07)*, volume 259 of *ACM Digital Library, ACM International Conference Proceeding Series*, pages 69–78. ACM, 2007.

[MM97]   G. W. Mineau and R. Missaoui. The Representation of Semantic Constraints in Conceptual Graphs Systems. In *Proc. of ICCS'97*, volume 1257 of *LNAI*, pages 138–152. Springer, 1997.

[Moh89]   R.H. Mohring. Computationally tractable classes of ordered sets. *Algorithms and Orders*, 255:105–193, 1989.

[Mug92]   M.L. Mugnier. *Contributions algorithmiques pour les graphes d'héritage et les graphes conceptuels*. PhD thesis, Univ. Montpellier II, Oct. 1992.

[Mug95]   M.-L. Mugnier. On Generalization / Specialization for Conceptual Graphs. *Journal of Experimental and Theoretical Artificial Intelligence*, 7:325–344, 1995.

[Mug00]   M.-L. Mugnier. Knowledge Representation and Reasoning based on Graph Homomorphism. In *Proc. ICCS'00*, volume 1867 of *LNAI*, pages 172–192. Springer, 2000.

[Mug07]   M.-L. Mugnier. On the π_p^2-completeness of the containment problem of conjunctive queries with negation and other problems. Research Report 07004, LIRMM, 2007.

[Naz93]   A. Nazarenko. Representing Natural Language Causality in Conceptual Graphs . In *Proc. ICCS'93*, volume 699 of *LNAI*, pages 205–222, 1993.

[Pap94]   C. H. Papadimitriou. *Computational Complexity*. Addison-Wesley, 1994.

[Pau98]   C. Paul. *Parcours en largeur lexicographique : un algorithme de partitionnement.* PhD thesis, Université Montpellier II, 1998.

[PMC98]   A. Preller, M.-L. Mugnier, and M. Chein. Logic for nested graphs. *Computational Intelligence: An International Journal*, 14-3:335–357, 1998.

[Pos47]   E. L. Post. Recursive unsolvability of a problem of Thue. *Journal of Symbolic Logic*, 12(1):1–11, 1947. Reprinted in: M. Davis (Ed.), The Collected Works of Emil L. Post, Birkhauser, Boston 1994, pp. 503-513.

[Pre98a]   S. Prediger. *Kontextuelle Urteilslogik mit Begriffsgraphen. Ein Beitrag zur Restrukturierung der Mathematischen Logik.* PhD thesis, Technische Universitat Darmstadt, 1998.

[Pre98b]   S. Prediger. Simple concept graphs: A logic approach. In M.-L. Mugnier and M. Chein, editors, *Proc. ICCS'98*, volume 1453 of *LNAI*, pages 225–239. Springer, 1998.

[Pre00]   S. Prediger. Nested concept graphs and triadic power context families. In *Proc. ICCS'00*, volume 1867 of *LNAI*, pages 249–262, 2000.

[QV07]   V. Quint and I. Vatton. Structured Templates for Authoring Semantically Rich Documents. In *Proc. of International Workshop on Semantically Aware Document Processing and Indexing (SADPI'07)*, volume 259 of *ACM Digital Library, ACM International Conference Proceeding Series*, pages 41–48. ACM, 2007.

[RM00]   O. Roussel and P. Mathieu. The achievement of knowledge bases by cycle search. *Information and Computation(IC)*, 162:43–58, 2000.

[RN03]   S. J. Russell and P. Norvig. *Artificial Intelligence, A Modern Approach*. Prentice Hall, 2003.

[Rob92]   D.D. Roberts. The Existential Graphs. In F. Lehman, editor, *Semantics Networks in Artificial Intelligence*, pages 639–663. Pergamon Press, 1992.

[Rob03]   D.D. Roberts. *The Existential Graphs of Charles S. Peirce*. Mouton, La Hague, 2003.

[Roz97]   G. Rozenberg, editor. *Handbook of Graph Grammars and Computing by Graph Transformations, Volume 1: Foundations*. World Scientific, 1997.

[RS86]   N. Robertson and P. D. Seymour. Graph minors ii. algorithmic aspects of tree-width. *Journal of Algorithms*, 7:309–322, 1986.

[RS04]   N. Robertson and P. D. Seymour. Graph minors. xx. wagner's conjecture. *J. Comb. Theory, Ser. B*, 92(2):325–357, 2004.

[RvW06]   F. Rossi, P. van Beek, and T. Walsh, editors. *Handbook of Constraint Programming*. Foundations of Artificial Intelligence. Elsevier Science Publishers, Amsterdam, The Netherlands, 2006.

[Sal98]   E. Salvat. Theorem proving using graph operations in the conceptual graphs formalism. In *Proc. of ECAI'98*, pages 356–360, 1998.

[Sar95]   Y. P. Saraiya. On the efficiency of transforming database logic programs. *J. Comput. Syst. Sci.*, 51(1):87–109, 1995.

[SCM98]   G. Simonet, M. Chein, and M.-L. Mugnier. Projection in Conceptual Graphs and Query Containment in nr-Datalog. R.R. LIRMM, 1998.

[Sim98]   G. Simonet. Two FOL Semantics for Simple and Nested Conceptual Graphs. In *Proc. of ICCS'98*, volume 1453 of *LNAI*, pages 240–254. Springer, 1998.

[SM96]   E. Salvat and M.-L. Mugnier. Sound and Complete Forward and Backward Chainings of Graph Rules. In *Proc. of ICCS'96*, volume 1115 of *LNAI*, pages 248–262. Springer, 1996.

[Sme00]   A. F. Smeaton. Indexing, browsing and searching of digital video and digital audio information. In G. Pasi M. Agosti, F. Crestani, editor, *Lecture on Information Retrieval*, volume 1980 of *Lecture Notes in Computer Science*, pages 93–110. Springer, 2000.

[Sow76]   J. F. Sowa. Conceptual Graphs. *IBM Journal of Research and Development*, 20(4):336–357, 1976.

[Sow84]   J. F. Sowa. *Conceptual Structures: Information Processing in Mind and Machine.* Addison-Wesley, 1984.

[Sow92]   J. Sowa. Conceptual graphs as a universal knowledge representation. In F. Lehmann, editor, *Semantic Networks*, pages 75–94. Pergamon Press, 1992.

[Sow99]   J. F. Sowa. *Knowledge Representation: Logical, Philosophical and Computational Foundations.* Brooks/Cole, 1999.

[SPT83]   L.K. Schubert, M.A. Papalaskaris, and J. Taugher. Determining type, part, color and time relationships. *Computer*, 16:53–60, 1983.

[Sto77]   L. J. Stockmeyer. The polynomial-time hierarchy. *Theoretical Computer Science*, 3:1–22, 1977.

[SU02]   M. Schaefer and C. Umans. Completeness in the polynomial-time hierarchy: A compendium. Sigact News. Available on M. Schaefer's homepage, 2002.

[SW86]   J. Sowa and E. C. Way. Inplementing a semantic interpreter using conceptual graphs. *IBM Journal of Research and Development*, 30(1):57–69, 1986.

[SY80]   Y. Sagiv and M. Yannakakis. Equivalences among relational expressions with the union and difference operators. *J. ACM*, 27(4):633–655, 1980.

[Tar72]   R.E. Tarjan. Depth-first search and linear graph algorithms. *SIAM Journal of Computing*, 1(2):146–160, 1972.

[TBH03a]   R. Thomopoulos, P. Buche, and O. Haemmerlé. Different kinds of comparisons between fuzzy conceptual graphs. In *Proc. ICCS'03*, volume 2746 of *LNAI*, pages 54–68, 2003.

[TBH03b]   R. Thomopoulos, P. Buche, and O. Haemmerlé. Representation of weakly structured imprecise data for fuzzy querying. *Fuzzy Sets and Systems*, 140(1):111–128, 2003.

[tH04]   H.J. ter Horst. Extending the RDFS Entailment Lemma. In *Proc. of the Third International Semantic Web Conference (ISWC'04)*, volume 3298 of *LNCS*, page 7791. Springer, 2004.

[Thi01]   E. Thierry. *Sur quelques interactions entre structures de données et algorithmes efficaces.* PhD thesis, Université Montpellier II, Oct. 2001.

[Tho07]   M. Thomazo. Complexité et propriétés algorithmiques de la déduction dans les graphes polarisés. First year research report, ENS Cachan, 2007.

[Thu14]   A. Thue. Probleme über Veränderungen von Zeichenreihen nach gegebenen Regeln. *Skr. Viedensk. Selsk. I*, 10, 1914.

[TML06]   R. Thomopoulos, M.-L. Mugnier, and M. Leclère. Mapping contexts to vocabularies to represent intentions. In *Proceedings of ECAI2006 Workshop on Contexts and Ontologies: Theory, Practice and Applications*, pages 44–46, 2006.

[TV97]   M. Talamo and P. Vocca. A data structure for lattice representation. *Theoretical Computer Science*, 175:373–392, 1997.

[TY84]   R.E. Tarjan and M. Yannakakis. Simple linear-time algorithms to test chordality of graphs, test acyclicity of hypergraphs, and selectively reduce acyclic hypergraphs. *SIAM Journal of Computing*, 13(3):566–579, 1984.

[Var84]   M.Y. Vardi. The implication and finite implication problems for typed template dependencies. *Journal of Computer and System Science*, 28(1):3–28, 1984.

[vB97]   J. van Benthem. Dynamic bits and pieces. ILLC Research Report LP-1997-01, Univ. of Amsterdam, 1997.

[vR79]   C.J. van Rijsbergen. *Information retrieval.* Butterworths, 1979.

[vR86]   C. J. van Rijsbergen. A non-classical logic for information retrieval. *The Computer Journal*, 29(6):481–485, 1986.

[Wal93]   R. J. Wallace. Why ac-3 is almost always better than ac4 for establishing arc consistency in csps. In *Proc. IJCAI'93*, pages 239–247, 1993.

[Wer95a]   M. Wermelinger. Conceptual Graphs and First-Order Logic. In *Proc. of ICCS'95*, volume 954 of *LNAI*, pages 323–337, 1995.

[Wer95b]   M. Wermelinger. Teorica básica das estruturas conceptuais. Master's thesis, Universidade Nova de Lisboa, 1995.

[Wes96]   D.B. West, editor. *Introduction to Graph Theory.* Prentice Hall, 1996.

[Wil95]   M. Willems. Projection and unification for conceptual graphs. In *Proc. of ICCS'95*, volume 954 of *LNCS*, pages 278–292. Springer, 1995.

[Wil97]   R. Wille. Conceptual graphs and formal context analysis. In D. Lukose et al, editor, *Proc. ICCS'97*, volume 1257 of *LNAI*, pages 290–303. Springer, 1997.

[WL94]   M. Wermelinger and J. G. Lopes. Basic conceptual structures theory. In M.-L. Mugnier and M. Chein, editors, *Proc. ICCS'98*, volume 835 of *LNAI*, pages 144–159. Springer, 1994.

[Woo75]   W.A. Woods. What's in a link? In D. G. Bobrow and A. M. Collins, editors, *Representation and Understanding: Studies in Cognitive Science*, pages 3–50. Academic press, 1975.

Index